T0292466

CAMBRIDGE LIBRARY COLLECTION

Books of enduring scholarly value

Physical Sciences

From ancient times, humans have tried to understand the workings of the world around them. The roots of modern physical science go back to the very earliest mechanical devices such as levers and rollers, the mixing of paints and dyes, and the importance of the heavenly bodies in early religious observance and navigation. The physical sciences as we know them today began to emerge as independent academic subjects during the early modern period, in the work of Newton and other 'natural philosophers', and numerous sub-disciplines developed during the centuries that followed. This part of the Cambridge Library Collection is devoted to landmark publications in this area which will be of interest to historians of science concerned with individual scientists, particular discoveries, and advances in scientific method, or with the establishment and development of scientific institutions around the world.

The World of Comets

Written in 1877 by the French journalist Amédée Guillemin, this work appeared on British bookshelves at a time of intense interest in space, the solar system and stars. In the same year, Schiaparelli made his infamous 'discovery' of Martian canals, whetting the public's appetite for all things astronomical. Guillemin's account of comets was equally ambitious and, ultimately, more valuable. His subjects range from comet superstitions in Renaissance Italy to an accessible explanation of their orbits, constitution and brilliance. As James Glaisher notes in his Preface, 'there is no work that at all occupies the ground covered' by Guillemin. The author's imaginative prose, exemplified by his description of comets as 'long disowned stars', was translated sympathetically by Glaisher. Accompanied by eighty-five striking illustrations, including Halley's Comet as depicted in the Bayeux Tapestry, *The World of Comets* provides a fascinating insight into both astronomy and nineteenth-century scientific enquiry.

Cambridge University Press has long been a pioneer in the reissuing of out-of-print titles from its own backlist, producing digital reprints of books that are still sought after by scholars and students but could not be reprinted economically using traditional technology. The Cambridge Library Collection extends this activity to a wider range of books which are still of importance to researchers and professionals, either for the source material they contain, or as landmarks in the history of their academic discipline.

Drawing from the world-renowned collections in the Cambridge University Library, and guided by the advice of experts in each subject area, Cambridge University Press is using state-of-the-art scanning machines in its own Printing House to capture the content of each book selected for inclusion. The files are processed to give a consistently clear, crisp image, and the books finished to the high quality standard for which the Press is recognised around the world. The latest print-on-demand technology ensures that the books will remain available indefinitely, and that orders for single or multiple copies can quickly be supplied.

The Cambridge Library Collection will bring back to life books of enduring scholarly value (including out-of-copyright works originally issued by other publishers) across a wide range of disciplines in the humanities and social sciences and in science and technology.

The World of Comets

Amédée Guillemin

CAMBRIDGE UNIVERSITY PRESS

Cambridge, New York, Melbourne, Madrid, Cape Town, Singapore,
São Paolo, Delhi, Dubai, Tokyo

Published in the United States of America by Cambridge University Press, New York

www.cambridge.org
Information on this title: www.cambridge.org/9781108014151

This edition first published 1877
This digitally printed version 2010

ISBN 978-1-108-01415-1 Paperback

COMETS

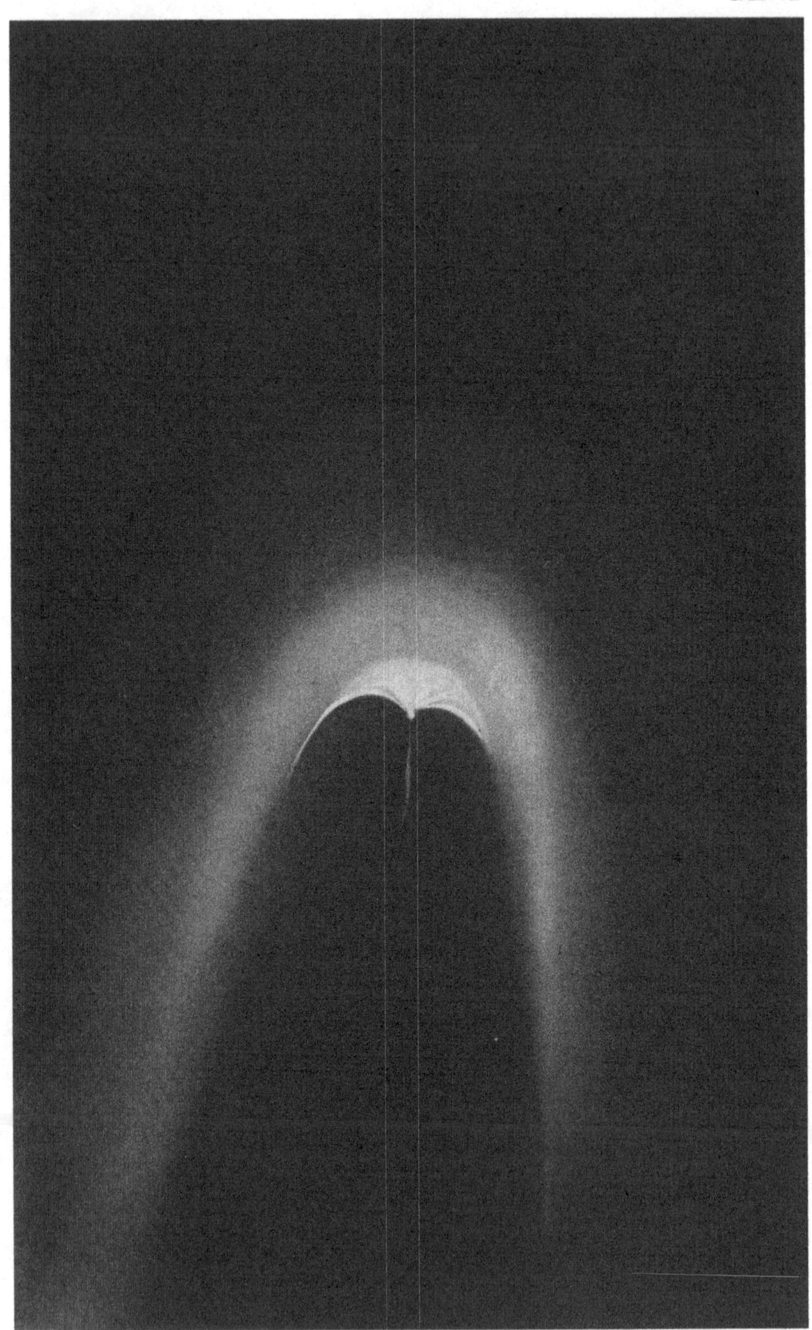

Warren De La Rue del. J. Basire, s.c.

July 2 10.h 10.m G.M.T

THE GREAT COMET OF 1861

AS SEEN BY WARREN DE LA RUE D.C.L., F.R.S.

WITH HIS NEWTONIAN EQUATOREAL

OF 13 INCHES APERTURE

THE

WORLD OF COMETS

BY

AMÉDÉE GUILLEMIN

AUTHOR OF 'THE HEAVENS'

TRANSLATED AND EDITED BY

JAMES GLAISHER, F.R.S.

NUMEROUS WOODCUT ILLUSTRATIONS and CHROMOLITHOGRAPHS

LONDON

SAMPSON LOW, MARSTON, SEARLE, & RIVINGTON

CROWN BUILDINGS, 188 FLEET STREET

1877

EDITOR'S PREFACE.

I HAVE great pleasure in introducing to English readers M. GUILLEMIN'S valuable and interesting work on comets. When rapid progress has been made in any branch of science, it is generally very difficult for anyone, who has not been actually concerned in the investigations in question, to obtain accurate information of the state of our knowledge; and for this reason a book, such as the present, which gives an account of the new results that we owe to very recent researches, really confers a benefit upon many persons who, though taking a strong interest in the subject, have necessarily been quite unable to follow its development in the periodical publications of English and foreign scientific societies. There is no work that at all occupies the ground covered by that of M. GUILLEMIN; and as the subject is one which, always of high interest, has in the last few years acquired great importance in consequence of Schiaparelli's discovery of a connexion between comets and shooting-stars, I was anxious that it should appear in our language.

Whenever I have thought that additional explanation was desirable, or that the researches of the two years that have elapsed since the publication of the original work threw further light upon the subject, I have added a note of my own, such notes being always enclosed in square parentheses [ ]; and although I must not be understood to endorse M. GUILLEMIN'S conclusions in every case where I have not added a note, still I may say that there are very few of his views from which I should feel at all inclined to dissent. Of course I have corrected in the text all errors I have met with, which were evidently purely accidental, and such as always will occur in the first edition of any work. In two cases I have ventured to make more lengthy additions of my own—these relate to Coggia's comet, which had only just left us when M. GUILLEMIN'S work was published, and the connexion of comets and shooting-stars. Some remarks will also be found in the note which follows the catalogue of comets at the end of the book.

In conclusion, I must express my thanks to Dr. WARREN DE LA RUE, F.R.S., for having very kindly placed at my disposal copies of his beautiful drawings of the great comet of 1861, which add greatly to the value of the work.

JAMES GLAISHER.

BLACKHEATH, S.E.: *Nov.* 15, 1876.

PREFACE.

THE UNIVERSE is formed of an infinity of worlds similar to our own. The thousands of stars which meet our gaze in the azure vault of the heavens when we contemplate it with the naked eye, and which may be reckoned by hundreds of millions when we explore its depths by the aid of the telescope, are suns. . These foci of light, these sources of heat, and incontestably of life, are not isolated; they are distributed into groups or clusters; sometimes by twos or threes, sometimes by hundreds, sometimes by myriads; the clouds of vaporous light called nebulæ are for the most part thus constituted.

Isolated or in groups, the stars seem to us immovable, so prodigious is the distance by which they are separated from the earth and from our sun. They move nevertheless; and amongst those whose velocities have as yet been measured may be reckoned some which are moving ten times and even fifty times quicker than a cannon-ball when it leaves the cannon. Movement is, therefore, the most universal law of the stars.

In like manner our sun moves through space and compels the earth to follow. He bears along with him, in this voyage through the boundless ether, the globes which form his *cortége* and gravitate about his enormous mass. During the thousands of years that man has been a witness—an unconscious witness, it is true—of this circumnavigation of the universe, he has seen no change in the aspect of the surrounding worlds; the sidereal shores of the ocean in which this fleet of more than a hundred celestial bodies pursues its way preserves to all appearance its unchanging front. The immensity of the sidereal distances, it is well known, is the sole cause of this apparent immobility.

The solar world is, therefore, separated from all other worlds by unfathomable abysses; the sun is as it were isolated, lost in a corner of space, far from the millions of stars with which nevertheless it forms a system. Member of an immense association, integral molecule of the most vast, to all appearance, of the nebulæ, the Milky Way, the central star of our group seems to have no other mode of communicating with its co-associates than by the reciprocal exchange of undulations, that is to say, by the exchange of light and heat. Like disciplined and devoted soldiers, the earth and the planets march in company with the sun, effecting with marvellous regularity their nearly circular revolutions around their common focus, and never deviating from the limits imposed upon them by the law of gravitation.

They remain, therefore, isolated like the sun, separated from other sidereal systems by distances so enormous that the mind is powerless to conceive of them.

A relation, however, exists between our system and these systems, as we have just mentioned: the sun is a star of the Milky Way. But, we may ask, has the solar world no closer and more direct connexion with the rest of the visible universe?

The movement of translation by which it is animated proves at least that in some quarter of the heavens there is either an unknown celestial body, or a system of celestial bodies, around which gravitation causes the group to describe an orbit of undetermined period. And this movement of the whole results from the concurrent action of all the stars in the universe. The force of gravitation is, therefore, a common bond of union between our world and all others.

Is it, therefore, steadily advancing to some celestial archipelago which it will finally attain in a few millions of years? Are, then, future generations destined to see other suns, from other points of view? These are questions whose solution may be considered inaccessible to us.

But, amongst the stars of which the solar system is composed, are there not some less immutably attached than the planets and the earth to the focus of their movement? Are there not some which depart to a greater distance from their focus, and which, like messengers detached from the group, carry to neighbouring world news of our own?

Such a hypothesis is not without foundation.

Astronomers, in fact, have for the last two centuries studied the movement of certain celestial bodies, which come to us and gravitate about the sun, but which, after having, so to speak, saluted on their way the ruler of the planets, return and plunge again to immeasurable distances in the depths of ether. A small number of these stars, retained by the solar power, diverted from their path by the influence of some of the larger planets, have remained tributaries of the group of which they now form an integral part.

These singular stars, long disowned, are Comets.

I have said long disowned. Comets, indeed, have only been considered during the last two centuries as properly belonging to the family of the stars: before Newton's time they were regarded even by astronomers as transient meteors, whose appearance, disappearance, and movements were subject to no law. For the ancients, and the world in general during the Middle Ages, and even during the Renaissance, they were objects of fear, miraculous apparitions, signs the precursors of terrible calamities, flaming symbols of the Divine anger. For the savants of former times comets were the monsters of the sky.

Two centuries of scientific progress exhibit these calumniated stars in a very different light. Thanks to Newton's discovery of gravitation, and to the united efforts of mathematicians and observers, their courses have been reduced to the same laws as those which govern the movements of the planets. These vagabonds of the sky have testified, some by their regular motion, others by their return at the predicted dates, their submission to the laws of celestial mechanics. Their very deviations have been recognised as the legitimate consequence of foreign influences.

For some years more has been done: an endeavour has been made, and with some success, to penetrate the mystery of their organisation, and to discover the physical and chemical nature of their light; but more especially the part which they perform in the solar system and in the universe itself is beginning to be viewed in its proper light.

Comets, as Laplace had foreseen, are of different origin to the planets. The eccentricity of their orbits, the inclinations of the planes in which they move, their course, sometimes direct, sometimes retrograde, mark a profound distinction between them and the planets. Their interior structure, the nebulous appearance of nearly all, the rapid changes observed both in their nuclei and atmospheres, remove them no less from the permanent and globular figure, either solid or liquid, of the majority of the planets. Some comets move in infinite orbits. They are, therefore, strangers to our world, which they visit on their journey. Those which are periodical have most frequently such lengthened orbits that, after voyages the durations of which are measured by thousands of centuries, they are cast adrift far from the sun, far from the directing focus of movement. Is it certain that they will rejoin him on their return, and that these wanderers will not be, in the end, stars lost to our world?

In any case, they come to us out of the depths of infinite space. And if the views of M. Hoek, a Dutch savant, are well founded, it is not singly but in groups that these nebulosities—let us say nebulæ, since the structure of comets appears analogous to that of the nebulæ properly so called—quit the sidereal depths and penetrate to the heart of our system. Here, then, is the material bond, the real connexion establishing a direct and uninterrupted communication between the solar world and the millions, the thousands of millions of stars which constitute the splendour of the heavens.

But the physical nature of these frail messengers of space is such, that they cannot without injury pass through the regions traversed by the planets and the sun. So great is the 'swell' engendered by the motions of these massive stars, that comets when navigating in these agitated deeps of the ethereal sea are there subjected to considerable damage; sometimes they are shattered and broken into fragments; frequently they leave behind them *débris* which follows in their wake. The interplanetary spaces are in this way strewn with the particles which the planets meet with in their periodical course, and which cause our nights to be illuminated momentarily with brilliant trains of light.

The SHOOTING-STARS are due to these rencontres.

Ten years have hardly elapsed since Schiaparelli, a learned Italian astronomer, by a happy idea connected meteoric with cometary astronomy. If this bold theory, involving, if not the identity, at least the community of origin between meteors and comets, be true, how important do the latter suddenly become in the economy of the universe! Travelling from world to world, scattering upon their route in the neighbourhood of the permanent stars of each system the dust of the elements of which they are composed, may it not be that they modify in the course of time the structure of these stars themselves? If spectral analysis does not err, the matter of comets is chiefly composed of carbon combined with some other element, such as hydrogen. Here, then, are comets, shedding these substances so important to vegetable and animal life, first in the interplanetary spaces, and then, by the fall and combustion of meteors, in the atmospheres of the planets ; thereby, perhaps, maintaining life upon them.

The unformed portions of matter distributed in immense masses in certain unresolvable nebulæ, and successively detached in separate globules, would continue to describe hyperbolæ or other curves of endless branches, on their passage from star to star, and from world to world: these masses of vapour would be presented to us under the form of comets or long trains of vapour.

Comets, therefore, which during the reign of ignorance and superstition were looked upon as scourges, are, more probably, not only inoffensive stars but, perhaps, even beneficent regenerators of life in more advanced worlds.

These views, it is true, are only hypotheses: we know so little of that department of astronomical science which a great writer of our time has called by anticipation Celestial Organism, in opposition to Celestial Mechanics, of which our knowledge is now so far advanced. But they suffice to show what interest, scientific and philosophic, attaches to the subject of this work. Comets up to the present time have furnished only one chapter, and that the briefest, to the science of the heavens. The preceding considerations will have sufficed to show that they perform a part of great importance in the universe, and that their history merits more ample development than has been accorded to it in most treatises.

Besides, apart from the scientific interest of the subject, a historic interest attaches to it. Considered from this point of view, comets would furnish matter for an interesting work. In the volume before the reader will be found a few chapters devoted to a brief history of what may be called Cometary Astrology. It forms a necessary introduction to the purely astronomical portion; and, were it omitted, it would be difficult to understand how the world has passed from the most extravagant prejudices to the calm and reassuring conceptions of contemporary science.

In ancient Greece, in heroic times, comets, as indeed all celestial phenomena, excited only graceful ideas. Take, for example Homer: it is Minerva and Apollo, the two brilliant deities of Olympus, who thus manifest themselves to mortals. Later on, they became fatal presages. The Romans, more austere, had already interpreted them as signs of fatal augury, forerunners of calamity. In the Middle Ages the ideas connected with them continued to increase in gloom: comets were then stars only of misfortune, ruin, and death. The terrible and grandiose idea of the end of the world, so universal at that period of darkness, predominated over all and set its seal on all. At last, with the revival of learning, scientific observation slowly dissipated these prejudices. In the eighteenth century the light of a free interpretation of nature resumed its empire: comets were spoken of without awe, and these stars, but lately so formidable, became even a theme for satire.

Contemporary science, more profound, restores to comets their majesty and importance, but it also despoils these apparitions for ever of all significance derived from idle superstition and terror.

CONTENTS.

CHAPTER I.

BELIEFS AND SUPERSTITIONS RELATIVE TO COMETS.

SECTION I.

COMETS CONSIDERED AS PRESAGES.

PAGE

SECTION II.

COMETS IN GREEK AND ROMAN ANTIQUITY.

SECTION III.

THE COMETS OF THE MIDDLE AGES.

SECTION IV.

COMETS FROM THE RENAISSANCE TO THE PRESENT DAY.

CHAPTER II.

COMETARY ASTRONOMY UP TO THE TIME OF NEWTON.

SECTION I.

COMETS AND THE ASTRONOMERS OF EGYPT AND CHALDEA.

SECTION II.

COMETARY ASTRONOMY IN THE TIME OF SENECA.

SECTION III.

COMETS DURING THE RENAISSANCE AND UP TO THE TIME OF NEWTON AND HALLEY.

SECTION IV.

NEWTON DISCOVERS THE TRUE NATURE OF COMETARY ORBITS.

CHAPTER III.

THE MOTIONS AND ORBITS OF COMETS.

SECTION I.

SECTION II.

MOTIONS OF COMETS.

SECTION III.

IRREGULARITIES IN THE MOTIONS OF COMETS.

SECTION IV.

THE ORBITS OF COMETS.

SECTION V.

THE ORBITS OF COMETS COMPARED WITH THE ORBITS OF THE PLANETS.

SECTION VI.

DETERMINATION OF THE PARABOLIC ORBIT OF A COMET.

CHAPTER IV.

PERIODICAL COMETS.

SECTION I.

COMETS WHOSE RETURN HAS BEEN OBSERVED.

SECTION II.

HALLEY'S COMET.

SECTION III.

ENCKE'S COMET; OR, THE SHORT PERIOD COMET.

SECTION IV.

BIELA'S OR GAMBART'S COMET.

SECTION V.

FAYE'S COMET.

SECTION VI.

BRORSEN'S COMET.

SECTION VII.

D'ARREST'S COMET.

SECTION VIII.

TUTTLE'S COMET.

SECTION IX.

WINNECKE'S PERIODICAL COMET.

SECTION X.

TEMPEL'S SHORT PERIOD COMET.

CHAPTER V.

PERIODICAL COMETS.

SECTION I.

COMETS WHOSE RETURN HAS NOT BEEN VERIFIED BY OBSERVATION.

SECTION II.

INTERIOR COMETS, OR COMETS OF SHORT PERIOD, THAT HAVE NOT YET RETURNED.

SECTION III.

COMETS OF MEAN PERIOD.

SECTION IV.

COMETS OF LONG PERIOD.

CONTENTS.

CHAPTER VI.

THE WORLD OF COMETS AND COMETARY SYSTEMS.

SECTION I.

THE NUMBER OF COMETS.

SECTION II.

COMETS WITH HYPERBOLIC ORBITS.

SECTION III.

REMARKS ON THE ORIGIN OF COMETS.

SECTION IV.

SYSTEMS OF COMETS.

SECTION V.

COMETARY STATISTICS.

CHAPTER VII.

PHYSICAL AND CHEMICAL CONSTITUTION OF COMETS.

SECTION I.

COMETS PHYSICALLY CONSIDERED.

SECTION II.

COMETARY NUCLEI, TAILS, AND COMÆ.

SECTION III.

COMETS DEVOID OF NUCLEUS AND TAIL.

SECTION IV.

DIRECTION OF THE TAILS OF COMETS.

SECTION V.

NUMBER OF TAILS.

SECTION VI.

DIFFERENT FORMS OF TAILS.

SECTION VII.

LENGTH OF TAILS.

SECTION VIII.

FORMATION AND DEVELOPMENT OF TAILS.

SECTION IX.

BRILLIANCY OF COMETS.

SECTION X.

DIMENSIONS OF NUCLEI AND TAILS.

CHAPTER VIII.

PHYSICAL TRANSFORMATIONS OF COMETS.

SECTION I.

AIGRETTES—LUMINOUS SECTORS—NUCLEAL EMISSIONS.

SECTION II.

OSCILLATIONS OF LUMINOUS SECTORS: COMET OF 1862.

SECTION III.

DUPLICATION OF BIELA'S COMET.

SECTION IV.

DOUBLE COMETS MENTIONED IN HISTORY.

CHAPTER IX.

MASS AND DENSITY OF COMETS.

SECTION I.

FIRST DETERMINATION OF THE MASSES OF COMETS.

SECTION II.

METHOD OF ESTIMATING THE MASSES OF COMETS BY OPTICAL CONSIDERATIONS.

SECTION III.

THIRD METHOD OF DETERMINING THE MASSES OF COMETS.

CHAPTER X.

THE LIGHT OF COMETS.

SECTION I.

SECTION II.

TRANSPARENCY OF NUCLEI, ATMOSPHERES, AND TAILS.

SECTION III.

COLOUR OF COMETARY LIGHT.

SECTION IV.

SUDDEN CHANGES OF BRILLIANCY IN THE LIGHT OF COMETARY TAILS.

SECTION V.

DO COMETS SHINE BY THEIR OWN OR BY REFLECTED LIGHT?

SECTION VI.

SPECTRAL ANALYSIS OF THE LIGHT OF COMETS.

SECTION VII.

THE COMET OF 1874, OR COGGIA'S COMET.

CHAPTER XI.

THEORY OF COMETARY PHENOMENA.

SECTION I.

WHAT IS A COMET?

SECTION II.

CARDAN'S HYPOTHESIS.

SECTION III.

THEORY OF THE IMPULSION OF THE SOLAR RAYS.

SECTION IV.

HYPOTHESIS OF AN APPARENT REPULSION.

SECTION V.

THEORY OF OLBERS AND BESSEL.

SECTION VI.

THEORY OF COMETARY PHENOMENA.

SECTION VII.

THE REPULSIVE FORCE A REAL PHYSICAL FORCE.

SECTION VIII.

THEORY OF THE ACTINIC ACTION OF THE SOLAR RAYS.

SECTION IX.

COMETS AND THE RESISTANCE OF THE ETHER.

CHAPTER XII.

COMETS AND SHOOTING STARS.

SECTION I.

WHAT IS A COMET?

SECTION II.

SECTION III.

COMETS AND SWARMS OF SHOOTING STARS.

SECTION IV.

COMMON ORIGIN OF SHOOTING STARS AND COMETS.

CHAPTER XIII.

COMETS AND THE EARTH.

SECTION I.

COMETS WHICH HAVE APPROACHED NEAREST TO THE EARTH.

SECTION II.

COMETS AND THE END OF THE WORLD.

SECTION III.

MECHANICAL AND PHYSICAL EFFECTS OF A COLLISION WITH A COMET.

SECTION IV.

SECTION V.

THE COMET OF 1680, THE DELUGE, AND THE END OF THE WORLD.

SECTION VI.

PASSAGE OF THE EARTH THROUGH THE TAIL OF A COMET IN 1861.

CHAPTER XIV.

PHYSICAL INFLUENCES OF COMETS.

SECTION I.

SUPPOSED PHYSICAL INFLUENCES OF COMETS.

SECTION II.

DO COMETS EXERCISE ANY INFLUENCE UPON THE SEASONS?

SECTION III.

PENETRATION OF COMETARY MATTER INTO THE TERRESTRIAL ATMOSPHERE.

SECTION IV.

CHEMICAL INFLUENCES OF COMETS.

CHAPTER XV.

SOME QUESTIONS ABOUT COMETS.

SECTION I.

ARE COMETS HABITABLE?

SECTION II.

WHAT WOULD BECOME OF THE EARTH IF A COMET WERE TO MAKE IT ITS SATELLITE?

SECTION III.

IS THE MOON AN ANCIENT COMET?

TABLE I.

TABLE II.

LIST OF ILLUSTRATIONS.

PLATES.

WHOLE-PAGE WOODCUTS.

WOODCUTS IN TEXT.

LIST OF ILLUSTRATIONS.

LIST OF ILLUSTRATIONS.

LIST OF ILLUSTRATIONS.

CHAPTER I.

BELIEFS AND SUPERSTITIONS RELATIVE TO COMETS.

SECTION I.

COMETS CONSIDERED AS PRESAGES.

Comets have been considered in all times and in all countries as signs, precursors of fatal events—Antiquity and universality of this belief; its probable origin—Opinion of Seneca; habitual and regular phenomena fail to attract the attention of the multitude; meteors and comets, on the contrary, make a profound impression—The moderns in this respect resemble the ancients contemporary with Seneca—The incorruptible heavens of the ancients, in contradistinction to the sublunary or atmospheric regions; stars and meteors—Inevitable confusion of certain celestial or cosmical phenomena with atmospheric meteors.

In all countries and in all times the apparition of a comet has been considered as a presage: a presage fortunate or unfortunate according to the circumstances, the popular state of mind, the prevailing degree of superstition, the imbecility of princes or the calculation of courtiers. Science itself has helped to confirm the formidable and terrible signification most frequently accorded by common belief to the sudden and unexpected arrival of one of these remarkable stars. Not two centuries ago, as we shall shortly see, learned men and astronomers of undoubted merit continued to believe in the influence of comets over human events. What wonder, then, if we should find existing in our own time, in the midst of the nineteenth century, numerous vestiges of a superstition as old as the world?

How has this superstition originated? This is a question we shall not undertake to resolve: we leave it to others more learned and competent than ourselves in similar matters to reply. Let us confine ourselves to a simple and by no means new remark. The things which we see every day, the phenomena which are constantly or regularly reproduced under our eyes do not strike us, and fail to excite either our attention or curiosity. D'Alembert has said: 'It is not without reason that philosophers are astonished to see a stone fall to the ground, and people who laugh at their astonishment will upon the smallest reflection share it themselves.' Yes, it is necessary to be a philosopher, or man of science, as we should say at the present day; it is necessary to reflect in order to discover the why and the how of facts, of those at least whose production is frequent and regular. The most admirable phenomena remain unperceived. Habit blunts the impression we derive from them and renders us indifferent.

As applied to comets, this idea has been perfectly expressed by Seneca, at the commencement of Book vii. of his *Quæstiones Naturales*: 'There is,' he observes, 'no mortal so apathetic, so obtuse, so bowed down towards the earth, that he does not erect himself and tend with all the powers of his mind towards divine things, particularly when some new phenomenon makes its appearance in the heavens. Whilst all above follows its daily course, the recurrence of the spectacle robs it of its grandeur. For man is thus constituted: that which he sees every day, however admirable it may be, he passes with indifference, whilst the least important things as soon as they depart from the accustomed order captivate and interest him. The whole choir of heavenly constellations under this immense vault, whose beauty they diversify, fails to attract the attention of the multitude; but should anything extraordinary appear, all faces are turned towards the heavens. The sun has spectators only when he is eclipsed. The moon is observed only when she undergoes a similar crisis. Cities then raise a cry of alarm, and everyone in panic fear trembles for himself. . . . So much is it in our nature to admire the new rather than the great. The same thing takes place in respect to comets. If one of these flaming bodies should appear of rare and unusual form, everyone is anxious to see what it is; all the rest are forgotten whilst everyone enquires concerning the new arrival: no one knows whether to admire or to tremble; for there are not wanting people who draw from thence grave prognostics and disseminate fears.'

Is it not with us to-day as with the contemporaries of Seneca? Doubtless thoughtful and reflective minds yield themselves readily to a sentiment of contemplative admiration before the majestic spectacle of the heavens. The solemn march of the heavenly bodies, the well-ordered harmony of worlds, are for them the symbol of eternal laws governing the universe; from the unalterability of these laws they derive confidence. But the mass of the people ordinarily remains indifferent before impassable and immutable nature. It is reserved for one unusual apparition to rouse all from this indifference, to awaken curiosity in some, fear in others, and, if the phenomenon should be of unwonted proportions, admiration in every one.

Moreover, whether it be a comet, or any other remarkable meteor, bolide, aurora borealis, or stone fallen from heaven, the sentiments of fear inspired by these phenomena are always the same, the superstitious interpretation similar, but closely proportioned in degree to the brilliancy and the more or less whimsical or extraordinary form of the apparition

Amongst the Greeks and Romans, as we all know, a number of the most ordinary and familiar actions, singular rencontres, the cries of animals, the flight and the song of birds, were looked upon as omens, as so many means made use of by the gods to communicate with man, to warn him of their decrees, to signify to him their thoughts and will. But they regarded the importance of the warning as proportional to the grandeur of the sign and the brilliancy of the phenomenon, and it is not difficult to understand that comets amongst these manifestations of the divine will appeared the most significant and formidable.

A comet, moreover, not being a simple local phenomenon, seen only by some, but exhibiting itself to all, brilliant as a star, and of unusual dimensions, varying from day to day in form, position and size, had all the appearance of a sign fraught with significance to the entire people: this portent addressed itself to those who played an important part in public affairs, and concerned kings, or at least great personages. It had a certain resemblance to the stars, which it sometimes surpassed in the brilliancy of its light, but it differed from them in its erratic course; it had a certain resemblance likewise to atmospheric meteors, by its sudden appearance, and oftentimes as sudden disappearance, and by the rapid changes to which it was ordinarily subjected. The heavens, with their countless hosts, the sun, the moon, the fixed stars and planets, were for the ancients the domain of the incorruptible—*cœli incorrupti.* Under this name it was the dwelling-place of the gods, the habitation of the immortals. On the contrary, the air, the atmosphere, the sublunary space—for the ancients it was all one—was the region of meteors and of things corruptible and fleeting; and in the same manner as the thunder-bolt of Jupiter was the chosen instrument of his vengeance, comets were the selected messengers of fate, sent to announce to mortals, on the part of the gods, events that were inevitable.

In this confusion of certain celestial phenomena with atmospheric meteors lies the chief source of the difficulty experienced by the astronomers of antiquity and the Middle Ages, and even of modern times, in solving the complicated problem of cometary movements. Down to the sixteenth century, we shall see men of undoubted science refusing to comets the quality of stars. They were confirmed in this error alike by the prejudices we have just mentioned and by the superstitious beliefs which are found to be so persistent amongst all people and, as we have already said, in all times: doubtless because these beliefs have the same foundation or common origin, faith in the supernatural intervention of the gods in human affairs.

But let us see what signification was given to the apparition of comets by the ancient Greeks and Romans. It is a curious and instructive side of cometary science; it will aid us further on perhaps in comprehending the ideas entertained by their astronomers concerning the physical nature of these stars.

SECTION II.

COMETS IN GREEK AND ROMAN ANTIQUITY.

The apparition of a comet or a bolide is a warning from the gods: the Iliad and the Æneid—Supposed physical influences of comets; Earthquakes in Achaia; submersion of Helicè and Bura; comet of the year 371—Comets, presages of happy augury; Cæsar transported to the heavens under the form of a comet; popular credulity turned to account; opinion of Bayle—Pliny, Virgil, Tacitus, Seneca—The comet of the year 79 and the Emperor Vespasian—Comet of the year 400 and the siege of Constantinople.

A COMET is thought to have appeared in the last year of the siege of Troy. By Pingré and Lalande it is considered an apparition of the famous comet of 1680, and whilst the former cites in support of his opinion a passage from Homer, the latter draws attention to certain lines in the Æneid probably referring to the same comet.* The following is the passage in the fourth book of the Iliad to which Lalande refers:—

'Thus having spoken, he urged on Athenè already inclined; she hastening descended the heights of Olympus. As the star which the son of wily Saturn sends, a sign either to mariners or to a wide host of nations, and from which many sparks are emitted: so Pallas Athenè hastened to the earth and leaped into the midst [of the army]; and astonishment seized the horse-breaking Trojans and the well-greaved Greeks

* In the time of Pingré and Lalande the period of revolution of the comet of 1680 was believed to be 575 years. Encke has since assigned it the much longer period of 8814 years.

looking on. And thus said one to another: "Doubtless evil war and dreadful battle din will take place again, or Zeus, the arbiter of war amongst men, is establishing friendship between both sides."' *

In our opinion the star to which Homer compares Athenè was no comet, but a bolide, the explosion of which is frequently attended with an emission of sparks, and can itself be seen in broad daylight. This remark applies in all respects to the verses of Virgil, who, moreover, mentions the noise of the detonation, oftentimes similar, in explosions of bolides, to the rumbling of thunder. Anchises, in the passage we now quote, has just ceased to invoke Jupiter:—

'Scarcely had my aged sire thus said, when, with a sudden peal, it thundered on the left, and a star, that fell from the skies, drawing a fiery train, shot through the shade with a profusion of light. We could see it, gliding over the high tops of the palace, lose itself in the woods of Mount Ida, full in our view, and marking out the way; then all along its course an indented path shines, and all the place a great way round smokes with sulphureous steam. And now my father, overcome, raises himself to heaven, addresses the gods, and pays adoration to the holy star.' †

The confusion above indicated is of frequent occurrence amongst the ancient writers, with whom the Aurora Borealis, bolides, and comets are evidently phenomena of the same nature, and, so far as their supernatural interpretation is concerned, this is not difficult to understand. Even in our day the public is not more exact. Who can have forgotten the impression caused by the splendid Aurora Borealis of November 24, 1870, during the Siege of Paris? That reddened glare in the heavens, those shifting paths of light, were they not,

* Iliad, iv. 52–84.

† Æneid, ii. 674–709.

for weak and credulous minds, under the stimulus besides of passing events, certain presages of the blood about to be shed? A comet would have produced a similar effect.

Be this as it may, the meteor, be it either bolide or comet, is in the eyes of all beholders, both in Virgil and Homer, a presage, or warning from Jupiter, a sign, the precursor of events auspicious or inauspicious, according to the interpretation or circumstances.

Three hundred and seventy-one years before our era a very brilliant comet appeared, which Aristotle has described, and which Diodorus Siculus refers to in the following terms: 'In the first year of the hundred and second Olympiad, Alcisthenes being then Archon of Athens, several prodigies announced to the Lacedæmonians their approaching humiliation: a burning torch of extraordinary size, which was spoken of as a fiery beam, appeared for several nights.' This comet, of which we shall speak again, had more than one claim to notice. According to Ephorus it divided into two; and about the time of its apparition the earthquakes took place which caused Helicè and Bura, two towns of Achaia, to be swallowed up by the sea. Comets, therefore, for the ancients were not only the precursors of fatal events, but they had direct power to occasion them. Such is certainly the opinion of Seneca when he remarks: 'This comet, so anxiously observed by everyone, *because of the great catastrophe which it produced as soon as it appeared*, the submersion of Bura and Helicè.'

Comets not only announced fatal events, disasters, and wars; portents of evil for some, they were naturally presages of happy augury for others. Thus, according to Diodorus Siculus and Plutarch, the comet of the year B.C. 344 was for Timoleon of Corinth a token of the success of the expedition which he directed that year against Sicily. 'The gods by an extraordinary prodigy announced his success and future greatness:

a burning torch appeared in the heavens throughout the night and preceded the fleet of Timoleon until it arrived off the coast of Sicily.'

The births and deaths of princes, especially of those remembered in history for the evil they have done, could not fail to be distinguished by the apparition of prodigies, and by comets, the most striking of all prodigies. Thus the comets of B.C. 134 or 137 and B.C. 118 were referred, the former to the birth and the latter to the accession of Mithridates, and the comet of the year B.C. 43 was supposed to be nothing less than the soul of Cæsar transported to the heavens. Bodin (*Universæ Naturæ Theatrum*) attributes to Democritus the opinion that such is, in fact, the part performed by some of these stars, and confesses that he is not far from sharing the same opinion. 'I reflect,' he says, 'upon the idea of Democritus, and I am led to believe with him that comets are the souls of illustrious persons, which, after having lived upon the earth for a long succession of ages, ready at last to perish, are borne along in a sort of triumph, or are called to the starry heavens, as brilliant lights. This is why famine, epidemic maladies, and civil wars follow the apparition of comets; cities and nations then find themselves deprived of the help of those excellent leaders who strove to allay their intestine troubles.' We willingly place amongst the beliefs and superstitions of the ancients this triumphant explanation of the supposed disasters which, according to the universal opinion, were certain to follow the apparition of a comet. Nor is it in all probability of earlier date than the sixteenth century, for it is very likely that Bodin calumniated Democritus and was himself its true author. Pingré, remarking on the above passage and the apotheosis of Cæsar, observes with justice: 'The preceding should be added to the number of base and indecent flatteries, rather than be classed among philosophical opinions.'

Less than a century after Bodin we find Bayle protesting against this superstition, which appeared singular indeed to a man so enlightened as he who has been compared to Montesquieu. In his *Pensées sur la Comète*, in which so much good sense is blended with so much irony, Bayle ingeniously shows with what skill popular credulity was turned to account, and how the same comet was made to subserve several ends. 'Augustus,' he says, 'from policy, was well pleased that the people should believe it to be the soul of Cæsar; because it was a great advantage for his party to have it believed that they were pursuing the murderers of a man who was then amongst the gods. For this reason he caused a temple to be built to this comet, and publicly declared that he looked upon it as a very auspicious omen. . . . Those who were still republican at heart said, on the contrary, that the gods testified by it their displeasure that the liberators of their country were not supported.'

In the Natural History of Pliny we find several passages attesting the terrible significance attached to comets by the ancients. 'A comet,' he observes, 'is ordinarily a very fearful star; it announces no small effusion of blood. We have seen an example of this during the civil commotion in the consulate of Octavius.' This refers to the comet which appeared B.C. 86. The following alludes to the comet of B.C. 48, and perhaps no less to the apparition of remarkable bolides and Auroræ Boreales: 'We have, in the war between Cæsar and Pompey, an example of the terrible effects which follow the apparition of a comet. Towards the commencement of this war the darkest nights were made light, according to Lucan, by unknown stars; the heavens appeared on fire, burning torches traversed in all directions the depths of space; the comet, that fearful star, which overthrows the powers of the earth, showed its terrible locks.'

Virgil, at the end of the first Georgic, expresses in his harmonious language all the horror caused in the superstitious minds of the vulgar by the prodigies so skilfully turned to account by politicians and sceptics. After speaking of the prognostics which may be drawn from the aspect of the setting sun in reference to the weather, he adds:—

'Who dares to call the sun deceiver? He even forewarns often that hidden tumults are at hand, and that treachery and secret wars are swelling to a head. He also pitied Rome at Cæsar's death, when he covered his bright head with murky iron hue, and the impious age feared eternal night; though at that time the earth too, and ocean's plains, ill-omened dogs, and presaging birds, gave ominous signs. How often have we seen Ætna from its burst furnaces boil over in waves on the lands of the Cyclops and shoot up globes of flame and molten rocks! Germany heard a clashing of arms over all the sky; the Alps trembled with unwonted earthquakes. A mighty voice, too, was commonly heard through all the silent groves, and spectres strangely pale were seen under the cloud of night; and the very cattle (Oh horrible!) spoke; rivers stopped their courses, the earth yawned wide; the mourning ivory weeps in the temples, and the brazen statues sweat. Eridanus, king of rivers, overflowed, whirling in mad eddy whole woods along, and bore away the herds with their stalls over all the plains. Nor at the same time did either the fibres fail to appear threatening in the baleful entrails, or blood to flow from the wells, and cities to resound aloud with wolves howling by night. Never did more lightnings fall from a serene sky nor direful comets so often blaze.' *

All these prodigies, this mixture of facts natural and true, and the whimsical beliefs of popular credulity, are for the poet

* Georgic, i. 463–488.

testimonies of the anger and vengeance of the gods, forerunners of fresh disasters, the precursors of that battle of Philippi in which Roman armies inflamed by civil discord are about to encounter and shed each other's blood. Nature acts in unison with man, and her manifestations are a reflex of his fury. Everything, moreover, concurs to render the divine intervention striking; earthquakes, volcanic eruptions, and inundations. The comets and bolides with which Virgil concludes his enumeration appear to be the supreme signs of this menacing intervention:

Non alias cœlo ceciderunt plura sereno
Fulgura; nec diri toties arsere cometæ.

Later on, comets were not only presages: they became pretexts for the persecutions of imperial tyranny. Thus, Tacitus says, in regard to the comet of the year 64: 'At the close of this year people discoursed only of prodigies, the forerunners of approaching calamities; of thunderbolts more frequent than at any other epoch, and of the apparition of a comet, a kind of presage that Nero always expiated with illustrious blood.' Several comets, in fact, appeared during the reign of this monster, and it is concerning one of them that Seneca had the audacity to say, 'that having appeared in the reign of Nero, it has removed infamy from comets.' It does not seem, however—and we shall find other proofs of it later on—that the author of the *Quæstiones Naturales* shared the prevailing prejudices on the subject of comets. He does not deny that they cause disasters, but he manifestly inclines towards a physical explanation of these phenomena. Speaking of the comet of the year 62, he observes: 'The comet which appeared under the consulate of Paterculus and Vopiscus has been attended with the consequences that Aristotle and Theophrastus have attributed to this kind of star. Everywhere there have been violent and continual storms: in Achaia and Macedonia several towns have been overthrown by earthquakes.'

Let us conclude what more we have to say of the superstitious beliefs of the ancients concerning comets with the mention of two or three famous apparitions; they will suffice to show that from the most remote antiquity down to the Middle Ages, from the erroneous ideas of the pagans to those of Christian nations, during this long dark night of history we pass without interruption or sensible modification.

In the year 69, according to Josephus, several prodigies announced the destruction of Jerusalem. 'Amongst other warnings, a comet, one of the kind called Xiphias, because their tails appear to represent the blade of a sword, was seen above the city for the space of a whole year.'

Pingré quotes, in reference to the comet of the year 79, this curious passage from Dion Cassius, which proves that there were *esprits forts* even amongst the Roman emperors: 'Several prodigies preceded the death of Vespasian; a comet was for a long time visible; the tomb of Augustus opened of itself. When the physicians reproved the Emperor Vespasian for continuing to live as usual and attend to the business of the state, although attacked by a serious malady, he replied, "It is fitting that an emperor should die standing." Perceiving some courtiers conversing together in a low tone of voice about the comet, "This hairy star," he remarked, "does not concern me; it menaces rather the King of the Parthians, for he is hairy and I am bald." Feeling his end approach, "I think," said he, "that I am becoming a god."'

The death of the Emperor Constantine was announced by the comet of the year 336.

In the year 400 the misfortunes with which Gainas menaced Constantinople were so great, say the historians Socrates and

Sozomenes, that they were announced by the most terrible comet mentioned in history; it shone above the city, and reached from the highest heavens to the earth. The same comet was also regarded as the presage of a plague which broke out about the same time.

Lastly, the invasions of barbarians, at a time when moral disorder and anarchy of ideas were in unison with the disorganisation of the Empire, could not fail to be signalised by various prodigies, birds of evil augury, frequent thunderbolts, monstrous hailstones, fires, and likewise apparitions of comets, 'that spectacle which the earth has never seen with impunity.' In the Middle Ages, therefore, we shall find that beliefs in the supernatural and the intervention of the gods in human affairs are further strengthened and increased by the mysticism which the ascendency of religious ideas tended to foster in the minds of the people.

SECTION III.

THE COMETS OF THE MIDDLE AGES.

Prevalence of popular superstitions—Comets announce wars, plagues, the deaths of sovereigns—Terrors of the year 1000; comets and the end of the world—Gian Galeazzo Visconti and the comet of 1402—Ambrose Paré; celestial monsters—Halley's comet and the Turks; origin of the *Angelus de Midi*—The comet of 1066 and the conquest of England by the Normans; apostrophe to the comet by a monk of Malmesbury.

If a complete history were desired of all the superstitions which, during the Middle Ages and in modern times, have obtained with respect to comets, it would be necessary to pass in review every apparition of these stars, together with such incidental phenomena as the Aurora Borealis, new and temporary stars, bolides, &c., all of which have been converted by popular credulity into as many prodigies. Interesting in a scientific point of view, this long enumeration derived from the naïve chronicles of the time, the only documents available in the absence of a more complete and intelligent record, would be but a tedious study of human errors; a constant and monotonous repetition of the same absurd beliefs. To this state of things savants have themselves contributed, as at the epoch when these voluminous records were compiled cometary influences were still believed in, and the erudite of the day shared the universal prejudice.

I will here limit myself to a few characteristic traits of this tenacious superstition, in order to exhibit the progress, I was about to say the revolution, of ideas which has taken place under the increasing influence of science, and more especially of astronomy and physics. Wherever the light of science has been able to make its way the phantoms of the supernatural have vanished, and the most extraordinary apparitions, even if they continue unexplained, are no longer regarded as prodigies, presages, or Divine manifestations, but natural phenomena concerning which all men of science, without exception, are at one in their endeavour to trace the laws that govern them.

Let us come now to the facts we have to mention.

In ancient times, especially amongst the Greeks, comets, as it has been seen, were not invariably regarded as of evil omen. The darker and more gloomy spirit of the Middle Ages only saw in these apparitions the announcement of terrible events, wars, pestilence, and especially the deaths of sovereigns. The comet of 451 or 453 announced the death of Attila, and the comet of 455 that of the Emperor Valentinian; comets appeared successively to announce the death of Meroveus in 577, of Chilperic in 584, of the Emperor Maurice in 602, of Mahomet in 632, of Louis le Débonnaire in 837, of the Emperor Louis II. in 875. That the apparition of comets was connected with the death of the great is an idea so widely spread that many chroniclers appear to have recorded, perhaps in good faith, comets which were never seen; such, according to Pingré, was the comet of 814, which presaged the death of Charlemagne.

In the year 1024 a comet appeared, an augury, it was supposed, of the death of the King of Poland, Boleslas I.; an eclipse of the sun and a comet marked in 1033 that of Robert, King of France; comets appeared in 1058, the year of the death of Casimir, King of Poland; in 1060, the year in which died Henry I., King of France, and in the years 1181, 1198, 1223, 1250, 1254, 1264, 1337, 1402, 1476, 1505, 1516, and 1560.

Under these respective dates we find the deaths of the following sovereigns: Pope Alexander III.; Richard I., King of England; Philip Augustus, King of France; the Emperor Frederick, deposed and excommunicated; Pope Urban IV.; Gian Galeazzo Visconti, Duke of Milan; Charles the Bold; Philip I. of Spain; Ferdinand the Catholic; and Francis II., King of France. This list might be considerably extended. Amongst the chroniclers or historians who relate these coincidences we find no shadow of a doubt as to the certainty or signification of the presage. The mention of these signs, forerunners of the deaths of sovereigns, very frequently occurs with a curious naïveté, of which we will give two or three examples.

'At the commencement of July,' says an old French chronicle, 'a little before the half, appeared a sign in the heavens called a comet denoting a convulsion of the kingdom; for Philip, the king, who for a long time had lain ill of a quartan ague, at Mantua, closed his last day on the 14th of July, 1223.' Gian Galeazzo Visconti was sick when the comet of 1402 appeared. As soon as he perceived the fatal star he despaired of life: 'For,' said he, 'our father revealed to us on his deathbed that, according to the testimony of the astrologers, a similar star would appear for eight days at the time of our death.' 'This prince was not deceived,' adds the historian, from whom we borrow this account; 'surprised by an unexpected malady, he died a few days after.' Another chronicler gives us to understand that the comet only appeared when Galeas was already attacked by the malady of which he died. But the faith of the duke in the celestial warning was not less complete. 'At this time a great comet was seen. Galeas was told of it. His friends helped him to leave his bed; he saw the comet, and exclaimed, "I render thanks to God for having decreed that my death should be announced to men by this celestial sign."

His malady increasing, he died shortly afterwards, at Marignan, on the 3rd of September.'

Pingré, in quoting the first of these accounts, observes that the unexpected malady of Galeas might well have been *occasioned* by the chimerical fear of this prince; he might have added, *or aggravated*. This simple remark of the Canon of St. Geneviève sufficiently marks the difference of the times. Till the eighteenth century the writers who record the coincidences of comets and great events have implicit faith in them, and naïvely describe as a self-evident fact the connexion between the comet and the event itself. Pingré, writing in the eighteenth century, less than a century after the labours of Newton, and in quest of dates to enable him to calculate various cometary orbits, esteems it fortunate that in these times of ignorance such absurd beliefs should have existed, as without them history perhaps would never have recorded one of these apparitions so valuable to science.

There have been degrees nevertheless, according to the times, in the superstitious terror created by the apparition of a comet; this terror was also proportioned to the degree of brilliancy of the star, the magnitude of its tail, and the more or less singular form of the coma and luminous appendage. In the year 1000, at that melancholy epoch when the end of the world was so confidently looked for, the most simple phenomena, if unexpected, must have assumed terrible proportions. About this time we are told of earthquakes, and a comet was visible for the space of nine days. 'The heavens having opened, a kind of burning torch fell upon the earth, leaving behind it a long train of light similar to a flash of lightning. Such was its light that it frightened not only those who were in the open country but those who were within doors. As this opening in the heavens closed imperceptibly there became visible the figure of a dragon, whose feet were blue, and whose head seemed continually to increase.' This evidently relates to the apparition of a bolide, and also perhaps to an Aurora Borealis, but not to the comet, whose apparition lasted nine days.

The drawing of these 'frightful' meteors, which we here reproduce from the *Theatrum Cometicum* of Lubienietzki, is interesting in various respects. It shows the height to which imagination can attain under the stimulus of terror; it proves also the little value to be attached, scientifically speaking, to the descriptions of the time, whether written or portrayed. This drawing is comparatively modern, probably of much later date than the epoch at which the apparition represented by it took place; but the next which we give (fig. 2) is taken from

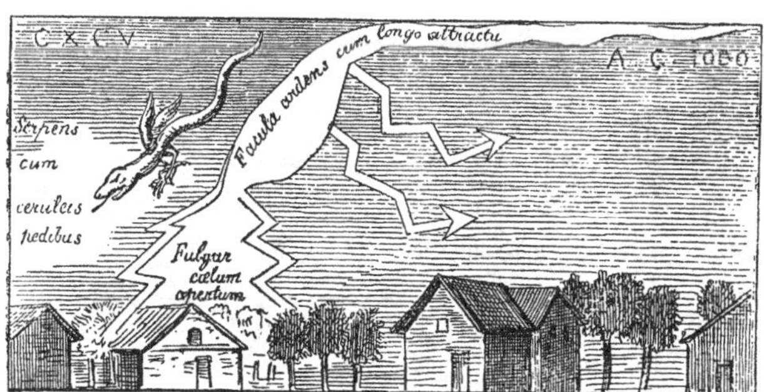

Fig. 1.—Phenomena of the Year 1000. Fac-simile of a drawing in the *Theatrum Cometicum* of Lubienietzki.

a work by Ambrose Paré, a contemporary of the apparition. The decapitated heads, the sabres, the arms which accompany the drawing of the hairy star, are only the translation of what the over-excited popular imagination believed itself to have seen in comets or other meteors, signs from heaven.*

* In his admirable work *The Universe* our late learned naturalist M. F. A. Pouchet with much justice remarks, in a note: 'In Ambrose Paré may be seen to what extent the mightiest minds of these latter centuries allowed themselves to be led astray on the subject of comets. The illustrious surgeon, who was by no means superstitious, gives in his important work the most fan-

Observe in what terms the historian Nicetas describes the comet (or meteor) of the year 1182: 'After the Romans were driven from Constantinople a prognostic was seen of the excesses and crimes to which Andronicus was to abandon himself. A comet appeared in the heavens similar to a writhing

Fig. 2.—Comet of 1528. Fac-simile of a drawing of Celestial Monsters from the work of Ambrose Paré.

tastic drawings of some of these bodies. In his chapter entitled *Celestial Monsters* Ambrose Paré speaks of comets as hairy, bearded, buckler-shaped, lance-shaped, dragon-like, or resembling a battle of the clouds. And he in particular describes and represents in all its details a blood-red comet which appeared in 1528 (the figure above represented (fig. 2). 'This comet,' said he, 'was so horrible, so frightful, and it produced such great terror in the vulgar, that some died of fear, serpent; sometimes it extended itself, sometimes it drew itself in; sometimes, to the great terror of the spectators, it opened a huge mouth; it seemed that as if, thirsting for human blood, it was upon the point of satiating itself.'

'Comiers,' says Pingré, 'makes a *horrible comet* appear in the month of October, 1508, very red, representing human heads, dissevered members, instruments of war, and in the midst a sword.' May it not be, with an error of date on the part of one or other of the chroniclers, the comet of which we have spoken, and a fac-simile of which we have reproduced in fig. 2?

Under the heading of periodical comets we shall see that one of the most famous in history is that which is now called Halley's comet, from the name of the astronomer who calculated and first predicted its return. This comet has, in fact, made its appearance twenty-four times within sight of the earth since the year 12 before our era, the most remote date on record of its apparition. Let us here transcribe, according to Babinet, the most remarkable particulars of the events which have been connected with it by popular belief.

'The Mussulmans, with Mahomet at their head, were besieging Belgrade, which was defended by Huniades, surnamed the *Exterminator of the Turks.* The comet of Halley appeared, and the two armies were alike seized with fear. Pope Calixtus III., himself struck with the general terror, ordered public prayers to be offered up, and launched a timid anathema against the comet and the enemies of Christianity. He instituted the prayer called the *Angelus de Midi*, the use of which still con-

and others fell sick. It appeared to be of excessive length, and was of the colour of blood. At the summit of it was seen the figure of a bent arm, holding in its hand a great sword, as if about to strike. At the end of the point there were three stars. On both sides of the rays of this comet were seen a great number of axes, knives, blood-coloured swords, among which were a great number of hideous human faces, with beards and bristling hair.'

tinues in all Catholic churches. The Franciscans brought 40,000 defenders to Belgrade, besieged by the conqueror of Constantinople, the destroyer of the Empire of the East. The battle took place, and lasted two days without intermission. This conflict of two days caused the loss of more than 40,000 combatants. The Franciscans, without arms, crucifix in hand, appeared in the foremost ranks of the defenders, invoking the exorcism of the Pope against the comet, and turned against the enemy the Divine anger of which no man at this time doubted. What primitive astronomers!'

Fig. 3.—Halley's Comet on its apparition in 1066. From the Bayeux Tapestry.

But let us go back to an earlier date in the history of this comet. It appeared in the month of April 1066. 'The Normans had at their head their Duke William, since surnamed the Conqueror, and were ready to invade England, the throne of which was at that time usurped by Harold in spite of the faith sworn to William.' That the comet was the precursor of the Conquest no one doubted. A new star, a new sovereign. *Nova stella, novus rex*! Such was the proverb of the time. The chroniclers say unanimously, 'The Normans, guided by a comet, invaded England.' Fig. 3 reproduces from the celebrated Bayeux tapestry, attributed to Queen Matilda, wife of William the Conqueror, the episode in which the apparition of the comet appears.

Halley's comet, by its apparition in 1066, gave rise to the objurgations of the monk of Malmesbury, which have been quoted by Pingré from an old English chronicle: 'Seeing his country on the point of being attacked on the one side by Harold, King of Norway, on the other side by William, and judging that bloodshed would ensue, "Here art thou, then," said he, apostrophising the comet, "here art thou, source of the tears of many mothers. Long have I seen thee; but now thou appearest to me more terrible, for thou menacest my country with complete ruin."'

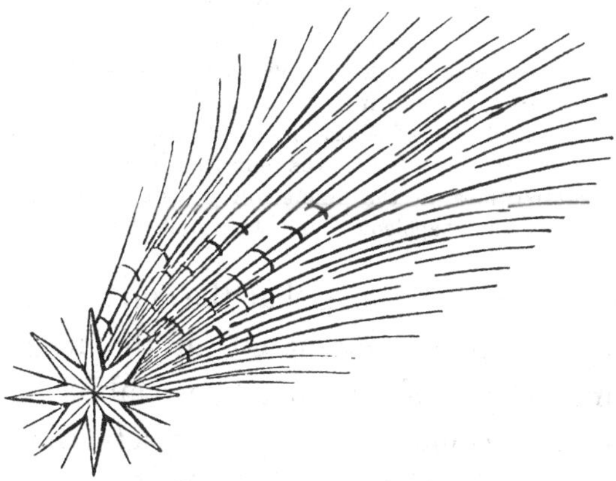

Fig. 4.—Halley's Comet in 684. Fac-simile of a drawing in the *Chronicle of Nuremberg.*

Going back further still, we find that Halley's comet is that which announced the death of Louis le Débonnaire, which came to pass three years later. Lastly, the comet of 684 (fig. 4) is also one of its apparitions.

We will say nothing of the famous comet of 1556, to the influence of which was long attributed the abdication of the Emperor Charles V., because it happens that the celebrated emperor had already descended from the throne when the comet made its appearance. We shall have occasion to speak further on of the announcement of its return between 1848 and 1860, and of its non-reappearance.

SECTION IV.

COMETS FROM THE RENAISSANCE TO THE PRESENT DAY.

Slow improvement in the beliefs relative to comets—Bayle's remarks upon the comet of 1680—Passage from Madame de Sevigné's letter referring to this comet and the last hours of Mazarin—In the eighteenth century belief in the supernatural exchanged for belief in the physical influence of comets—Remains of cometary superstitions in the nineteenth century—The comet of 1812 and the Russian campaign; Napoleon I. and the comet of 1769; the great comet of 1861 in Italy.

We have just seen that the superstitious ideas of the Middle Ages were yet dominant in the height of the Renaissance, since a man of learning like Ambrose Paré—no astronomer, it is true—could attribute to comets the same malign influences as those ascribed to them in the year 1000, when the end of the world was confidently expected.* Nor could it be otherwise, science not having then assigned to comets, in common with other extraordinary meteors, their true place in the order of nature.

Little by little, however, healthier ideas make their way, and to the supernatural influence of comets we shall now see gradually succeed in the minds of men of science and the

[* Milton has finely expressed the popular superstition with regard to comets in the well-known lines—

'On the other side,
Incensed with indignation, Satan stood
Unterrified; and like a comet burned,
That fires the length of Ophiuchus huge
In the arctic sky, and from his horrid hair
Shakes pestilence and war.'—*Paradise Lost*, book ii.—Ed.]

more enlightened of the people the idea of an influence purely physical, at first under the form of simple hypotheses, and afterwards as a probability deduced from observations and facts. This progress was slow, like that of cometary astronomy, and owed much of its advance to the assistance of men of original thought, who, without being astronomers, were yet conversant with the scientific knowledge of their time.* Such was Bayle. We have already quoted several passages from his *Pensées sur la Comète*, and we will now complete what still remains to be said in reference to this essay.

The *Pensées diverses écrites à un professeur de Sorbonne* were composed during the public excitement caused in France and Europe by the apparition of the famous comet of December 1680. From the beginning Bayle adopts the opinion of Seneca, and thus renews the train of rational and sound ideas. 'Comets,' he remarks, 'are bodies subject to the ordinary laws of nature, and not prodigies amenable to no law.' Supposing his correspondent to share the current prejudices of the time, he is astonished that so great a doctor should nevertheless suffer himself to be carried along with the stream, and imagine like the rest of the world, in spite of the arguments of the chosen few, that comets are heralds-at-arms sent by God to declare war against the human race.

He then examines the value of the historical testimonies which different writers have applied to the support of the current prejudice on comets.

'The testimony of historians,' he remarks, 'proves only

* The following anecdote which we borrow from Bayle proves that the wits of the seventeenth century began to treat with ridicule this long-cherished superstition. 'It seems to me,' says M. de Bassompierre, writing to M. de Luynes, in 1621, shortly after the death of Philip III., 'that the comet we laughed at at St. Germain is no laughing matter, as it has buried in two months a pope, a grand duke, and a king of Spain.' A belief which is expressed in these terms may be considered as drawing to its end.

that comets have appeared, and that afterwards there have been many disorders in the world, which is very far from proving that the former are to be looked upon as the cause or the prognostic of the latter, unless we are willing to admit that a woman who never looks out of window in the Rue St. Honoré without seeing carriages pass along the street is to imagine that she is the cause of their passing, or that when she shows herself at the window it is a sign to the whole quarter that carriages will soon pass.'

Bayle next attacks astrology and its pretended principles, as the source of all the extravagant beliefs relative to heavenly phenomena; and, indeed, prejudices in respect to comets form but a portion of the whole, and are contained in a separate chapter, which might well be entitled *Cometary Astrology*.

'The details of cometary warnings, resting only upon the principles of astrology, cannot fail to be ridiculous, because there never has been anything more impertinent, more chimerical or more ignominious to human nature, to the eternal shame of which it must be related, that there have been men base enough to deceive others under the pretext of knowing the affairs of heaven, and men foolish enough to believe in them even to the extent of instituting the office of Astrologer, and of not daring to wear a new coat, or plant a tree, without the approbation of that functionary.

'The astrologer will tell you to what people, or to what animals, the comet has reference, and the kind of evil that may be expected. In Aries it signifies great wars and mortality, the fall of the great and the exaltation of the little, together with fearful droughts in places under the dominion of that sign. In Virgo it signifies dangerous childbirth, imprisonments, sterility and death amongst women. In Scorpio, in addition to the preceding evils, reptiles and innumerable locusts.

In Pisces disputes concerning points of faith, frightful apparitions in the air, wars and pestilence among the great, etc. . . .

'It is not in our own time only that astrologers have reasoned upon such extravagances. The same thing prevailed in the time of Pliny. "It is," says he, "thought to be a matter not unimportant whether comets dart their beams towards certain quarters, or derive their power from certain stars, or represent certain things, or shine in particular parts of the heavens. If they resemble a flute, the omen relates to music; when they appear in certain parts of a sign, the omen has reference to the immodest; if they are so situated as to form an equilateral triangle or a square with some of the fixed stars, they are addressed to learning and wit. They distribute poison when they appear in the head of either the Northern or the Southern Serpent.' (Pliny, book ii. chap. xxv.)

Bayle cites a remark attributed to Henry IV. which might be applied, at the present day, to many so-called predictions. Speaking of the astrologers who had forewarned him of his death, Henry IV. is said to have exclaimed, 'They will be right some day, and the public will remember the one prediction that has come true, better than all the rest that have proved false.'

The letter of the celebrated writer is long; it touches upon very many considerations which, though of interest as regards the history of ideas at the end of the sixteenth century, would appear in the present day far removed from our subject; but the philosophic thought which has inspired him is always true. It may be summed up in these eloquent lines, the last that we shall quote:—

'The more we study man the more does it appear that pride is his ruling passion, and that he affects grandeur even in his saddest misery. Mean and perishable creature that he is, he has been able to persuade himself that he cannot die without disturbing the whole of nature and obliging the heavens to put themselves to fresh expense to light his funeral pomp. Foolish and ridiculous vanity! If we had a just idea of the universe we should soon comprehend that the death or birth of a prince is so insignificant a matter, compared to the whole of nature, that it is not an event to stir the heavens.'

Madame de Sevigné, writing on January 2, 1681, to the Comte de Bussy, mentions the same comet, then in sight, and concludes with a remark which in reality is the same as Bayle's. The following is the passage:—

'We have here a comet—it has the most beautiful tail that could possibly be seen. All the greatest personages are alarmed, and firmly believe that heaven, occupied with their loss, is giving intelligence of it by this comet. It is said that Cardinal Mazarin being despaired of by his physicians, his courtiers considered it necessary to honour his last hours by a prodigy, and to tell him that a great comet had appeared which filled them with alarm for him. He had strength enough to laugh at them, and jestingly replied that the comet did him too much honour. In truth, everyone should say the same, and human pride does itself too much honour in believing that when perforce we die it is a great event amongst the stars.'

At the present day what man of education, what enlightened mind would fail to subscribe to the views of the celebrated author and the *spirituelle* marquise? Nevertheless false beliefs relative to comets, celestial and even atmospheric meteors, are not entirely destroyed. We might have found traces of them in the last century, but in an epoch so favourable to science, we must seek under another form for the errors of which we have given a rapid sketch from the most ancient down to comparatively modern times; and in the chapter which we shall devote to the possible influences of comets upon the earth it will be seen that if the popular fears were then of a different kind they were none the less vivid. In our nineteenth century these fears have been openly revived; the idea that the end of the world could be brought about by the meeting of the earth and a comet has found minds disposed to receive it with blind acceptance. Further, the old superstition of the supernatural influence or signification of comets is always rife amongst the ignorant masses of the people, whose minds remain unaffected by the advance of science, because to them science is a dead letter. The following is a fact which occurred in Russia, and hardly more than sixty years ago:—

'It was not by the exchange of diplomatic notes that the inhabitants of Moscow derived a presentiment of some approaching calamity. The famous comet of 1812 first gave them warning of it. Let us see what reflections it inspired in the minds of the Abbess of the *Dievitchi Monastir*, and the nun Antonina, formerly the slave of the Apraxines. 'One evening, as we were on our way to a commemorative service at the Church of the *Décollation de Saint-Jean*, I suddenly perceived on the other side of the church what appeared to be a resplendent sheaf of flame. I uttered a cry and nearly let fall the lantern. The Lady Abbess came to me and said, "What art thou doing? What ails thee?" Then she stepped three paces forward, perceived the meteor likewise, and paused a long time to contemplate it. "*Matouchka*," I asked, "what star is that?" She replied, "It is not a star, it is a comet." I then asked again, "But what is a comet? I have never heard that word." The mother then said, "They are signs in the heavens which God sends before misfortunes." Every night the comet blazed in the heavens, and we all asked ourselves, what misfortunes does it bring?'—'*La Grande Armée à Moscou d'après les témoignages moscovites.*'—*Revue des Deux Mondes*, July 1, 1873.

Can anyone deny that such credulity exists at the present day and elsewhere than in Russia? Are there not persons still who believe that the great comet of 1769, which appeared in the year that Napoleon was born, presaged the era of war which drenched in blood the end of the eighteenth century and the beginning of the nineteenth, and all the disasters which that too famous despot let loose on Europe and at last upon France herself? Have we not seen quite recently, in 1861, when the great comet of that year appeared, how it was currently reported in Italy, and doubtless elsewhere, that the new star was a sign of the speedy return of Francis II. and his restoration to the throne of the Two Sicilies; and also that it presaged the fall of the temporal power and the death of Pope Pius IX.?

We ought not to be astonished at the persistence of these superstitions, which only the spread of science can annihilate for ever. After seeing, in the following chapter, with what great difficulty true ideas on the subject of comets, suspected centuries ago, have achieved their final victory, we shall not be surprised to find that errors still remain in our own nineteenth century, in the midst of what we regard as enlightened populations, but which will never be truly enlightened until primary instruction shall have given to them more definite notions of physics, natural history, and astronomy.

CHAPTER II.

COMETARY ASTRONOMY UP TO THE TIME OF NEWTON.

SECTION I.

COMETS AND THE ASTRONOMERS OF EGYPT AND CHALDEA.

Had the Egyptians and Chaldeans any positive knowledge concerning comets?—Apollonius of Myndus; the Pythagoreans considered comets to be true stars—According to Aristotle they are transient meteors; fatal influence of the authority of this great philosopher upon the development of Cometary Astronomy.

Such is a very brief history of the errors into which the human mind—we should rather say the human imagination—has fallen with respect to comets. We have now to show how little by little, and by very slow degrees, truth disengaged itself from error, and to supplement the history of superstitions and prejudices by that of science. Both are instructive and throw light upon each other at all stages of their mutual development. Thus, for example, we may readily conceive that the irregular movements of comets, their sudden and unforeseen apparition, to say nothing of the singularity of their aspect, for a long time precluded the idea of their being true stars, subjected to fixed laws, like the planets. Centuries of work, observation, and research were required for the discovery of the true system of the world as far as the sun, the planets, and the earth were concerned; but a difficulty of another kind stood in the way of the discovery of the true movements and nature of comets, since no trouble was taken to make exact and continuous observations of them. These difficulties, which were so great an impediment to science, gave, on the contrary, singular encouragement to the prejudices, the superstitions, and the hypotheses which appear so ridiculous in our day. And, in addition, the predominance of mystic ideas contributed to deter astronomers from a study which fell rather within the province of the diviner than the savant.

It is on this account all the more interesting to see a few just ideas, a few true conceptions, break through the dark night of ignorance and superstition. This happened, it is true, at a time and in countries where philosophy, not yet obscured by scholastic subtleties, was employed in explaining facts according to natural hypotheses; and where, by a bold and happy intuition, the Pythagorean school guessed without proving the true system of the world.

Are we to attribute to the Chaldeans and to the ancient Egyptians the first true conceptions concerning the nature of comets? That they regarded comets as stars subjected to regular movements, and not as simple meteors, we may believe, if it be true that they were in possession of means for predicting their return. Passages in Diodorus Siculus prove that the Chaldean and Egyptian astronomers hazarded such predictions; but, so far as our means enable us to judge, there is reason to suppose that these predictions were based upon particular beliefs, more astrological than astronomical. The passage which occurs in Diodorus Siculus relative to the Chaldeans is as follows:—

'The Chaldeans,' says he, 'by a long series of observations have acquired a superior knowledge of the celestial bodies and their movements: a knowledge that enables them to announce future events in the lives of men; but according to them, five stars, which they call interpreters, and which others call planets, deserve particular consideration; their movement is of singular efficacy. They announce likewise the apparition of comets, eclipses of the sun and moon, and earthquakes; all changes that take place in the atmosphere, whether salutary or pernicious, both to whole nations and kings and simple individuals.' Diodorus, also, speaking of the astronomical observations of the Egyptians and their knowledge of the movements of the celestial bodies, assures us 'that they often predicted to men what would happen to them in the course of their lives, the event following the prediction.' 'It is not unusual,' he adds, 'to hear of them announcing the maladies which are about to attack men or animals. In short, by means of accumulated observations, they predict earthquakes, inundations, the births of comets, and, indeed, all that seems to transcend the limits of the human mind.'

It is clear that, in the opinion of the historian, the predictions relative to comets which he attributes to the Egyptians and Chaldeans have no connexion with astronomy. Comets are confounded with other atmospheric meteors, whose return, according to them, was connected with the course of the stars by rare and mysterious coincidences, with which astrologers had far more to do than astronomers.

Nevertheless, we may suppose that the Chaldeans possessed some just ideas on the subject of comets. From them, indeed, and from the Egyptians * the Greeks derived their first knowledge of astronomy; from them, if Seneca is to be trusted, Apollonius of Myndus obtained his ideas concerning these stars. According to Apollonius 'comets are placed by the

* 'Eudoxus first brought with him from Egypt into Greece a knowledge of their movements [the planets]. Nevertheless, he makes no mention of comets. Hence it follows that even the Egyptians, a people more curious than any other in all matters of astronomy, had occupied themselves but little with the study of these bodies. At a later period, Conon, a most accurate observer, drew up a catalogue of the various eclipses of the sun recorded by the Egyptians, but he makes no mention of comets, which he would hardly have omitted if he had found any facts respecting them.'—Seneca, *Quæstiones Naturales*, vii. 3.

Chaldeans amongst the number of wandering stars, and they know their course.' Seneca then explains in detail the opinion of this ancient astronomer. 'A comet is not an assemblage of planets, but many comets are planets. They are not false appearances, nor fires burning on the confines of two stars; they are proper stars, like the sun and moon. Their form is not exactly round, but slender and extended lengthwise. Moreover, their orbits are not visible; they traverse the highest regions of the heavens, and only become apparent at the lowest part of their course. We are not to suppose that the comet which appeared under Nero, and removed infamy from comets, bore resemblance to the comet which, after the murder of Julius Cæsar, during the games of Venus Genetrix, rose above the horizon about the eleventh hour of the day. Comets are in great number and of more than one kind; their dimensions are unequal, their colours are different; some are red, without lustre; others are white, and shine with a pure liquid light; others again present a flame neither pure nor fine, but enveloped in much smoky fire. Some are blood-red, sinister presage of the blood soon to be shed. Their light augments and decreases like that of other stars, which throw out more light and appear larger and more luminous in proportion as they descend and come nearer to us, and are smaller and less luminous when they are returning and increasing their distance from us.'

Seneca, as we shall soon see, adopts this system, in which observations and conjectures nearly approaching the truth are mixed with various errors and traces of the reigning superstitions. The assimilation of comets to the planets as far as concerns their movements is a luminous idea, which is all the more truthful because Apollonius points out at the same time a characteristic difference between the two kinds of celestial bodies—viz., that comets are only visible in a small portion of their orbits.

Amongst the ancient philosophers who believed comets to be stars—stars wandering like the planets—must be mentioned Diogenes, chief of the Ionic school after Anaxagoras (Plutarch), Hippocrates of Chios, and several Pythagoreans. A passage in Stobæus, 5th century A.D., proves, as also book vii. of the *Quæstiones Naturales* of Seneca, that this opinion of the ancients concerning the true nature of comets remained uselessly chronicled in the books which have come down to us through the Middle Ages. Astronomers derived from it no benefit, so general was the superstition and so profoundly was it rooted in all minds. The passage in Stobæus runs: 'The Chaldeans believed comets to be other planets, stars which are hidden for a period, because they are too far distant, and which sometimes appear when they descend towards us, according to the law prescribed for them; they consider that they are called comets by persons ignorant of their being true stars, which only seem to be annihilated when they return to their own region and plunge into the profound abyss of ether, as fishes plunge to the bottom of the sea.'

What was required to render fruitful these remarkable views? Simply to the observation of comets to apply the rules long known and followed by astronomers for noting with precision all the circumstances of the movements of the planets. How precious would such observations now be to us for cometary theories! We must admit, however, that to have extracted from them all that they could yield, it would have been also requisite to have risen at one bound to the conception of the true system of the world, dimly seen by the Pythagorean school, and allowed to repose in the shade till the days of Copernicus and Galileo.

What were the obstacles which opposed so natural a progress in science? First, and most powerful of all, the enslavement of minds to the belief in the supernatural, and the prevailing misconceptions on the subject of comets; prejudices which increased in strength from the time of the Greek philosophers to the Middle Ages, when astrological folly attained its maximum intensity. There was at work also the influence of a powerful genius, who adopted—not very decidedly, it is true—the erroneous theory of the comet-meteors. In those ages, when everyone was always ready to swear *per verba magistri*, the word of Aristotle sufficed to ensure conviction, and the ideas of Apollonius of Myndus and of Seneca were regarded as tainted with heresy.

Pingré divides the opinions of the ancients about comets into three principal systems: that which we have just noticed, and which is, as it were, a rough sketch of the true system; that of Panætius, who regarded comets as destitute of all reality—a simple optical appearance only; and, lastly, the system according to which comets are simple atmospheric meteors, transient and sublunary. Amongst the authors of these different systems some, like Heraclides of Pontus and Xenophanes, regarded comets as very elevated clouds illuminated by the sun, the moon, or stars, or even as burning clouds. Transport these clouds from the atmosphere into the heavens themselves, into the regions where the planets perform their revolutions, and we have nearly the opinion of contemporary astronomers. The same might be said of the notion of Strato of Lampsacus, who regarded comets as lights sunk deep in the midst of clouds of great density, thus comparing them in some sort to lanterns. Does not the luminous nucleus in the centre of the nebulosity, which the telescope of modern times has revealed, correspond, in fact, to the hypothesis of the peripatetic philosopher?

We now come to the views of Aristotle concerning comets, views absolutely false, though maintained but two centuries ago, but yet important, on account of the great influence they exercised over the astronomers of the Middle Ages, and even over those of the Renaissance. In the opinion of this great philosopher comets are exhalations rising from the earth, which, having reached the upper regions of the air, adjoining the region of fire,* are drawn along by the movement of the surrounding medium. They at last unite with it, condense, and catch fire ; so long as the combustible matter lasts the fire burns ; when there is no more aliment for its supply the fire becomes extinct and the comet disappears.

It is useless to refute this hypothesis, which is entirely without foundation, or to record the objections which have been made to it by writers even of the time of Seneca. But it is well to devote a few words to this last philosopher. The book of the *Quæstiones Naturales* in which he relates all that was known in his time of comets, their movements and influence, is of great historic value, and the views of the author himself are certainly worthy of attention on their own account.

* According to Aristotle the air is divided into three regions: that in which animals and plants exist; this is the lower region, which is immovable, like the earth upon which it rests; the intermediate region, intensely cold, participates in the immobility of the first; but the upper region, contiguous to the region of fire or the heavens themselves, is carried along by the diurnal movement of the latter. The exhalations arising from the earth ascend to this higher region, and there, heated by the medium they have entered and by their own movement, they engender igneous meteors to which class comets belong.

SECTION II.

COMETARY ASTRONOMY IN THE TIME OF SENECA.

Book vii. of Seneca's *Quæstiones Naturales* relates to comets—Seneca defends in it the system of Apollonius of Myndus; he puts forth just views concerning the nature of comets and their movements—His predictions respecting future discoveries in regard to comets—The astronomers of the future.

From the beginning of his book Seneca fully appreciates the importance of the question, and the connexion that must necessarily exist between the nature of the comets and the system of the universe itself. He is led to ask 'if comets are of the same nature as bodies placed higher than themselves. They have points of resemblance with them, ascension and declination, and also outward form, if we except the diffusion and the luminous prolongation; they have likewise the same fire, the same light.' Here, then, we have comets assimilated to the planetary bodies as regards their movements, the only points of difference being the nebulosities and tails of the former. Seneca is sensible how important it would be 'to discover, if possible, whether the world revolves about the motionless earth, or if the world is fixed and the earth revolves; whether it is not the heavens but our globe which rises and sets.' 'It would be necessary,' he adds, 'to possess a table of all the comets which have appeared; for their rarity up to the present time has been a hindrance to our understanding the laws which regulate their course, and assuring ourselves if their course is periodical, and if a constant order brings them back to an appointed day. Now, the observation of these celestial bodies is of recent date, and has only been introduced very lately into Greece.' It does not appear that Seneca himself assisted at all the realisation of this reasonable and intelligent desire. In his time several comets appeared, but he hardly mentions them in his book, and relates no circumstance of the apparitions capable of informing us with any certainty of their apparent course.

After these preliminary considerations, which indicate so just a presentiment of the truth in the mind of the Roman philosopher, he proceeds to the explanation of the principal systems of his time, conceived for the explanation of comets. He applies himself to refute the system of Epigenes, who, like Apollonius of Myndus, had consulted the astronomers of Chaldea, but with a very different result, the theory of Epigenes being very nearly the same as that of Aristotle, with the exception of a few details equally false. Seneca, in combating these views, opposes to them objections that are sometimes very just, as, for example, when speaking of cometary movements: 'There is nothing confused,' he says, 'nor tumultuous in their behaviour; nothing by which it might be inferred that they obey elements of disturbance or inconstant principles. And then, even if whirlwinds should be strong enough to seize upon the humid exhalations of the earth and bear them upwards to such heights, they would not rise above the moon; at the level of the clouds the action would cease. Now we see that comets move in the highest heavens amongst the stars.'

Seneca has carefully noted one of the characteristic differences between comets and the planets. 'Let us bear in mind,' he observes, 'that comets do not show themselves in one region of the heavens alone, nor exclusively in the circle of the zodiac. They appear in the east and also in the west, but most frequently towards the north. The comet has its own region; it completes its course; it is not extinguished; it withdraws from our range of sight. If it were a planet, its path, it will be said, would be in the zodiac. But who can assign an exclusive limit to the stars, and confine and restrict these divine beings? The planets themselves, which alone seem to us to move, describe orbits different from each other. Why should there not be stars following courses of their own far removed from the planets? Why should any region of the heavens be inaccessible?'

Further on he explains with sufficient clearness the cause of the retrogressions observed in the movement of the stars and comets, and also of their occasional stationary positions.

'Why,' he says, 'do certain stars seem to turn back upon their journey? It is their meeting with the sun which gives an appearance of slowness to their movements; it is the nature of their orbits and of circles disposed in such manner that at certain moments there is an optical illusion. Thus, vessels even when in full sail appear to be immovable.' This is in effect the true explanation, and equally applies to the movements of the comets.

Seneca enumerates and describes the varied forms presented by their aspect, and then affirms that all comets have the same origin, an opinion altogether arbitrary, and relating to a matter still undetermined at the present day. Upon many points he has caught glimpses of the truth, sometimes supporting his views by reasons dictated by good sense, sometimes maintaining his opinion by explanations which in our day create a smile, borrowed as they are from the ideas of meteorology, astronomy, or physics received at that time, ideas quite without value, and which can only be looked upon as the crude utterances of an infant science.

He quotes the passage of the historian Ephorus concerning the comet of B.C. 371, a passage of extreme value, as it testifies to a phenomenon we have seen repeated in our own day, viz., the division of a comet into two parts. But it is only to treat the narrator as a dupe or an impostor. Let us, however, be just: thirty years ago our astronomers held the same opinion as Seneca, and Pingré does not fail in this case to applaud his discernment. The doubling of Biela's comet under our own eyes was requisite in order to obtain for the testimony of Ephorus the authority which Seneca and, after him, so many modern astronomers had refused to it.

The analysis given by our philosopher of the opinion of Apollonius of Myndus affords him an opportunity of pronouncing in favour of a system of which the cometary theories of modern times are the infinitely extended development. But he is not contented with telling us what seems to him most probable; he boldly prophesies in the name of the science of the future. These passages from the *Quæstiones Naturales* do great honour to Seneca, and deserve to be quoted as testimonies of the power and penetration of his intellect.

'Why,' he observes, 'should we be surprised that comets, phenomena so seldom presented to the world, are for us not yet submitted to fixed laws, and that it is still unknown from whence come and where remain these bodies whose return takes place only at immense intervals? Fifteen centuries have not elapsed since

Greece counted the stars by their names.

How many people, at the present day, know nothing of the heavens except their aspect, and cannot tell why the moon is eclipsed and coverèd with darkness! We ourselves in this matter have but lately attained to certainty. An age will come when that which is mysterious for us will have been made clear by time and by the accumulated studies of centuries. For such researches the life of one man would not suffice were it wholly devoted to the examination of the heavens. How then should it be, when we so unequally divide these few years between study and vile pleasures? The time will come when our descendants will wonder that we were ignorant of things so simple. Some day there will arise a man who will demonstrate in what region of the heavens the comets take their way; why they journey so far apart from other planets, what their size, their nature. Let us, then, be content with what is already known; let posterity also have its share of truth to discover.' *

* [Gibbon makes the following excellent remark (Decline and Fall, ch. xliii.) 'Seneca's seventh book of Natural Questions displays, in the theory of comets, a philosophic mind. Yet should we not too candidly confound a vague prediction, a *veniet tempus*, *&c.*, with the merit of real discoveries.'—Ed.]

SECTION III.

COMETS DURING THE RENAISSANCE AND UP TO THE TIME OF NEWTON AND HALLEY.

Apian observes that the tails of comets are invariably directed from the sun—Observations of Tycho Brahé; his views and hypotheses concerning the nature of comets—Kepler regards them as transient meteors, moving in straight lines through space—Galileo shares the opinion of Kepler—Systems of Cassini and Hevelius.

SIXTEEN CENTURIES passed away between the prediction of Seneca and its full realisation through the accumulated researches of many astronomers and the publication of the *Principia*, in which Newton demonstrated the law of cometary movements. There is nothing to tell of the history of comets and of systems during this long and dreary period in which the doctrine of Aristotle prevailed, except that it is entirely filled with astrological predictions. Our first chapter contains a *résumé* of all that the learned have found of interest concerning the apparition of comets and their formidable signification.

Towards the middle of the sixteenth century the movement of the Renaissance, so favourable to letters and the arts, extended its beneficent influence to the science of observation. At the end of the fifteenth century, we find Regiomontanus describing with care the movements of comets, Apian observing that cometary tails are always turned in a direction from the sun; Cardan remarking that comets are situated in a region far beyond the moon, founding his opinion upon the smallness or absence of parallax. The time had arrived when, instead of proceeding by way of conjecture and hypothesis, astronomers began to multiply observations and to give them that character of exactness and precision which they had hitherto so much needed. Many erroneous hypotheses were yet to be made, but they were subjected to discussion, and the geometrical conclusions to which they led were compared with the facts of observation. Astronomers of high repute like Tycho Brahé, Kepler, Galileo, Hevelius and Cassini were to err as to the true nature of cometary orbits; philosophers like Descartes were to seek to connect them with their bold but false conceptions of the system of the world. But the great principle that was destined to bind in one majestic whole the entire edifice of accumulated astronomical knowledge, the principle of gravitation, was ere long to give Newton a right to regard these bodies as members of the solar system, or at least as bodies subject to the same laws as the planets. From this moment cometary astronomy begins, and rises rapidly to a degree of development comparable to that of other branches of astronomy.

We will first give a rapid sketch of the principal phases of this history up to the time of Newton, and then proceed to the study of comets in connexion with their movements, their physical and chemical constitution, &c.

The apparition of the comet of 1577 may be regarded as the starting-point of the new period. Tycho, who had carefully observed the temporary star of 1572, which had suddenly appeared in Cassiopeia, now applied himself to make numerous observations of the new comet; he determined its parallax, and thus proved beyond a doubt that comets move in regions more remote than the moon, as Cardan had already remarked. Tycho endeavoured to represent the movement of the comet by making it describe around the sun an orbit external to Venus. With respect to its physical nature he regarded it as a meteor, but not an atmospheric meteor, since he supposed it to have been engendered in the depths of space. This was a first blow to the ideas of Aristotle, which other contemporary astronomers, such as Mæstlinus and Rothmann, continued to profess.

The comets of 1607 and 1618 furnished Kepler with an opportunity of explaining their apparent movements, and inventing an hypothesis which, although false, was ingenious. According to the immortal author of the three great laws of the planetary motions, comets traverse the solar system in rectilinear orbits, and Pingré justly remarks that the apparent movement of the comets of 1607 and 1618 is more naturally explained by this hypothesis than by that of Tycho, which is equivalent to saying that the paths of the two comets were more nearly straight lines than circles. As to the physical nature of comets, believed by Kepler to be as numerous in the heavens as fishes in the sea, his remarks on the subject taken from the second book of his work upon comets are as follows: 'They are not eternal, as Seneca imagined; they are formed of celestial matter. This matter is not always equally pure; it often collects like a kind of filth, tarnishing the brightness of the sun and stars. It is necessary that the air should be purified and discharge itself of this species of filth, and this is effected by means of an animal or vital faculty inherent in the substance of the ether itself. This gross matter collects under a spherical form; it receives and reflects the light of the sun, and is set in motion like a star. The direct rays of the sun strike upon it, penetrate its substance, draw away with them a portion of this matter, and issue thence to form the track of light which we call the tail of the comet. This action of the solar rays attenuates the particles which compose the body of the comet. It drives them away; it dissipates them. In this manner the comet is consumed by breathing out, so to speak, its own tail.' We see that although, in the opinion of Tycho and Kepler, comets are raised to the rank of heavenly bodies, they continue to regard them as stars of temporary origin, destined to disappear.

Some of the views of Kepler are affected by the singular and mystic conceptions of the great astronomer concerning the heavenly bodies; yet those relating to the formation of cometary tails, as we shall see further on, have been perfected and adopted by contemporary astronomers, and form the starting-point of one of the most accredited modern theories of cometary phenomena.

Galileo also believed that comets move in straight lines, but he was unable to rise above the common opinion, according to which they were mere transient meteors, exhalations of the earth.

The remarkable comets which appeared about the middle of the sixteenth century—namely, those of 1664, 1665, and 1680—attracted the attention of all men of science; the idea that they were veritable stars more and more gained ground, and, after the lapse of fifteen centuries, a definitive return was made to the system of Apollonius of Myndus; but modern astronomy was more exacting than the science of the ancient Greek philosophers. It was necessary to satisfy numerous and precise observations and to pass beyond vague ideas and conjectures. Henceforth the whole question reduced itself to the investigation of the geometrical form of the orbit described by comets and to the determination of the laws governing their movement.

Cassini attacked this great problem, but he did not arrive at its solution, which is not surprising, when we bear in mind that this illustrious astronomer did not yet dare to abjure the beliefs that Copernicus and Galileo had overthrown concerning the system of the world. By regarding the earth always as a fixed observatory he could not but confound the apparent motions of comets with their real motions. Cassini rightly supposed them to be stars, old as the world, but he made them describe circular orbits very eccentric to the earth, in order to account for the slight portion of the orbit that is visible during the brief durations of their apparitions.

Hevelius, a laborious observer, came back very nearly to Kepler's system, that is to say, to rectilinear orbits, or orbits sensibly rectilinear. Comets, in his opinion also, are the products of exhalations rising from the earth, the planets, or the sun. Drawn away at first by an ascensional movement, combined with the rotatory movement of the planet that has given it birth, the mass, after having described a spiral, finally attains the limit of the vortex which surrounds the planet; there it dies or escapes along the tangent to the limiting surface. The resistance opposed to it by the ether modifies the form of its orbit, which would otherwise be rectilinear, and causes it to take the form of a parabola. The whole of this system is purely imaginary, and must have made great demands upon the imagination of its author; it rests upon no solid basis of astronomical mechanics. The ideas of Hevelius found but few partisans amongst men of science; the work in which they are developed, valuable for the historic details it contains, and for various observations of comets, more especially those of 1652, 1664, and 1665, is little more than an object of curiosity in the history of science.

Newton, moreover, was about to put an end to all these hypotheses, by connecting the movements of comets with the laws that govern the motions of all the heavenly bodies which move within the sphere of the sun's attraction.

SECTION IV.

NEWTON DISCOVERS THE TRUE NATURE OF COMETARY ORBITS.

Newton's *Principia* and the theory of universal gravitation—Why Kepler did not apply to comets the laws of the planetary movements—Newton discovers the true system of cometary orbits—Halley and the comet of 1682; prediction of its return.

KEPLER, in 1618, had already discovered the three laws upon which his fame rests, and which will render his name immortal. These laws govern the movements of bodies which, like the planets and the earth, revolve about the sun in regular periods. In virtue of the first law the orbit described about the sun is an ellipse, of which the sun itself occupies one of the foci; the second relates to the velocity of the planet, a velocity which is greater the nearer the planet is to the sun, and less in proportion as it is further removed; or more accurately the velocity is such that the areas of the sectors swept out by the radius vector of the planet are equal in equal times; hence it follows that the maximum of speed takes place at the perihelion, and the minimum at the aphelion. The third law expresses the constant relation which connects the duration of each periodic revolution with the longest diameter, or major axis of the orbit.

Why did not Kepler apply the planetary laws to the movements of comets? Why did he leave to Newton the merit of an extension which now appears so natural? Because those portions of the cometary orbits visible from the earth are nearly always small fragments only of the immense and elongated curve described by comets in their total revolution; because in Kepler's time no instance was known of a comet having effected its return; and, lastly, because the powerful mind of Kepler himself was, doubtless, enslaved by the general belief that comets were passing, transitory meteors.

Newton, aided by the recent progress of mathematical and physical science, attained to a higher conception of the movements of the celestial bodies; he discovered the *reason* of those laws which the genius of Kepler had extracted from Tycho Brahé's observations and from his own; he gave them a mechanical interpretation; in short, he deduced from them the celestial movements as so many necessary consequences of a single principle—the mutual gravitation of the masses of these bodies and the earth.

From that time comets no longer eluded the investigations of science. Obeying the law of gravitation, describing orbits like the planets, owning the sun for their common focus, their movements are distinguished from those of the planetary bodies chiefly by two important differences, the first of which arises from the inclination of their orbits to the plane of the earth's motion: instead of being confined within narrow limits this inclination may assume any value whatever. From the earth comets can be seen, and indeed are seen, in all regions of the heavens, whilst the apparent paths of the planets are confined to the narrow zone called the zodiac. The second difference arises from the fact that a comet generally performs its revolution in a very elongated ellipse; for this reason we see only a very restricted portion of its orbit; beyond this arc of visibility, on either side, the comet is plunged into depths of space so remote from the earth that it is lost to view. And then, again, the duration of a comet's revolution is generally so great as to render impossible the recognition of the same comet on two successive apparitions; at any rate, this had been the case up to the time of Newton. Ellipses so elongated if we confine ourselves only to the arc described in the neighbourhood of the perihelion, are undistinguishable from parabolas having the same focus and the same vertex. Newton, taking advantage of this approximate assimilation, gave the means of determining, by the employment of a small number of observations, the elements of a comet's orbit regarded as a parabola, a problem much more simple than that which has for its object the investigation of the complete ellipse.

It still remains to point out another difference between the motions of comets and the planets. The movements of the latter are always direct, and invariably take place, for an observer situated upon the northern side of the plane of the ecliptic, from left to right, or from west to east. The movement of some comets is direct, and that of others retrograde. This circumstance had great weight in securing the adoption of Newton's *Principia* in preference to the vortices of Descartes. If the planetary heavens were filled with vortices of matter circulating in the same direction around the sun and around each body belonging to the system, how could we explain the fact that comets are able to traverse these media in a direction opposite to that in which the latter are moving?

All these views, so simple, and at the same time so grand in their entirety, were not, as we know, readily admitted by the philosophers and astronomers of the time of Newton. Still imbued with the spirit of systems and sects, some inclined to the old doctrines derived from Aristotle, and others to the bold novelties of Cartesianism.

But the actual truth was very shortly to be made clear.

Halley, an illustrious contemporary of Newton, contributed to its triumph in the matter of cometary theories. He undertook the calculation—at that time a very laborious task—of the orbits of twenty-four comets of which the observations appeared to be sufficiently numerous and accurate. He compared them with one another, and thought he recognised the identity of several amongst them. A comet lately observed—that of 1682—appeared to him similar to the comets of 1607 and 1531. He satisfied himself of this agreement; he affirmed it to be the same comet, observed on several successive apparitions, and finally predicted its return. Neither Halley nor Newton were able to see the prediction verified by the event. But the year 1759, when the return of the comet of 1682 did actually take place, marks an important date in the history of cometary astronomy, and, from this memorable epoch, there was no longer room for hypotheses—at all events, so far as the motions of comets are concerned.

The time has now come for us to enter upon the scientific portion of our subject.

CHAPTER III.

THE MOTIONS AND ORBITS OF COMETS.

SECTION I.

COMETS PARTICIPATE IN THE DIURNAL MOTION.

Comets participate in the diurnal motion of the heavens. During the time of their apparition they rise and set like the sun, the moon, the stars, and the planets. In this respect, therefore, they do not differ from other celestial bodies.

Let the observer, when a comet is in sight, note the point in the heavens which it occupies when his attention is first directed to it. This is easily done by referring the *nucleus*, the brilliant point from which the tail proceeds, to two adjacent stars. Let a certain time elapse—an hour, for example; at the end of that time the three luminous points, the two stars and the comet, will be found to have changed their position with respect to the horizon, each having described an arc of a circle in the heavens. The common centre of these arcs is the celestial pole, a point situated within a very small distance of the pole-star; the lengths of these arcs depend upon the interval of time between the observations, and the angular distance of each body from the pole. The direction is that of the general movement of the heavens and the stars; that is to say, from east to west.

We have here, then, a fact which clearly teaches us that a comet moves in regions beyond the atmosphere of the earth; for the diurnal motion is an apparent motion, foreign to the comet, and belongs in reality to the observer, or, as we may say, to the observatory. It is caused by the rotation of the earth upon its axis. The entire atmosphere of the earth participates in this movement, and a body immersed in it—although it might, of course, have a separate motion of its own—would not participate in the diurnal motion. This is so elementary a fact that there is no need to insist upon it further.

The ancients, and even those amongst the moderns who have regarded comets as meteors of atmospheric origin, have been compelled either to consider the earth as immovable or to admit that comets, after being formed within the atmosphere, withdraw from our globe, and, becoming independent, move in the heavens—a theory, as we have already seen, adopted by Hevelius.

SECTION II.

MOTIONS OF COMETS.

Distinction between comets, nebulæ, and temporary stars—Comets, in their motions, are subject to stationary periods and retrogressions—The apparent complications arise, as in the case of the planets, from the simultaneous movement of these bodies and the earth.

There is nothing in the foregoing section to distinguish comets from the multitude of brilliant stars which nightly illuminate the azure vault of heaven. Comets, it is true, appear in regions where before they had not been visible, and after a time they disappear; but in this respect they resemble those remarkable stars which have been seen to shine out suddenly in the midst of a constellation, to increase in brilliancy for a time, and afterwards to become faint and disappear; such as the famous temporary stars of 1572 (the Pilgrim), 1604, 1670, and 1866, which appeared and became extinct in the constellations of Cassiopeia, Serpens, Vulpecula, and Corona Borealis respectively. These stars, however, have, without exception, been distinguished by this peculiarity, that from the first to the last day of their apparition they continued immovable in the spot where they first appeared; or, more correctly, that their only motion was that due to the diurnal revolution of the heavens. Situated, like the fixed stars, at immense distances from our system, they had no appreciable movement of their own during the whole time of their visibility—in some instances of considerable duration. The same is true of the nebulæ, which are distinguished from comets by the fact of their immobility. Hence comet-seekers have only to pursue a method analogous to that which astronomers follow for the discovery of small planets.

Comets, on the contrary, have a motion of their own, a motion oftentimes of great rapidity; we can see that they perceptibly change their places from day to day, and sometimes hour by hour, amongst the constellations. This movement they have in common with the planets, and it is due, as we are about to see, to the same causes.

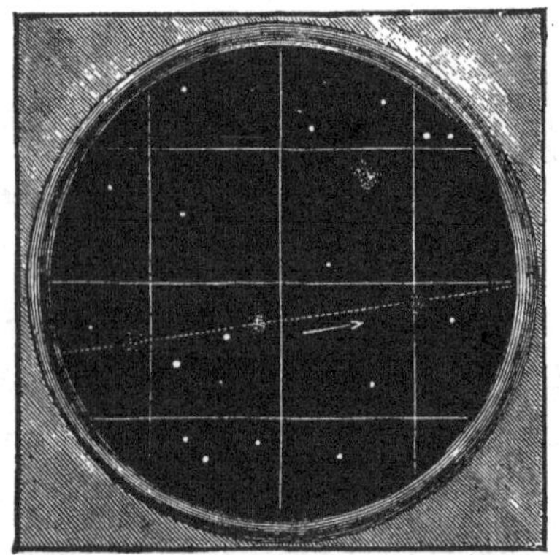

Fig. 5.—Proper motion of a Comet; distinction between a Comet and a Nebula.

In the first place, to confine ourselves to the real movement of a celestial body and its gradual change of place in space. Let us for a moment suppose the earth at rest. The observer situated on its surface would in that case see the body in motion gradually overtake and pass the different stars in its course, and describe upon the concave sphere of the heavens a curve whose form, position, and apparent dimensions would depend upon the actual path of the body, and its velocity of motion. For example, the moon, which describes an oval-shaped curve or ellipse around the earth, in about a month would appear to describe a great circle in the heavens from west to east. The planets Mercury and Venus, which revolve about the sun, and describe closed orbits differing more or less from a circle, but enclosed by the earth's orbit, would appear to move from one side to the other of the central luminary of our system, oscillating periodically to the east and west of it. The superior planets, Mars, Jupiter, and Saturn, as seen from the earth, would make the tour of the heavens in unequal periods of time, because these planets describe orbits exterior to that of the earth, and the actual time of their revolution depends upon the dimensions of their orbits.

But this simplicity of motion does not exist for an observer situated upon the earth, and for the following reasons.

The real and regular motion of the planets becomes combined with the motion of the earth; in the interval of a year our globe itself moves likewise round the sun in a closed curve or orbit differing but slightly from a circle; in fact, our earth moves in an ellipse whose focus is the sun. This displacement of the earth, it will be readily understood, has the effect of complicating the apparent motion of the planets; that is, their change of position upon the starry vault. Sometimes this motion appears accelerated, as will naturally happen when the planet and the earth are describing arcs in opposite directions; the two velocities are then added together, just as to a traveller in a railway train a second train, moving in the contrary direction, appears to pass with a speed equal to the sum of the velocities. But should the two trains be moving in the same direction, they then separate with a speed equal to the difference only of their velocities; and if the velocities are equal, each appears to the other motionless. This is what occurs in the case of the planets as seen from the earth; for we observe that their velocities sometimes decrease and become *nil*, in which case the planet is to all appearance stationary among the stars; and at other times it appears to retrograde.

Thus these effects admit of a very simple explanation. They are merely the result of the combination of the respective movements of the planet and of the earth in their orbits. Whatever may be the true orbit of a comet in the heavens, its apparent path will always be modified by the continual change of position of our earth.

In order, then, to determine the orbit of a comet we must take into account the motion of the earth in its orbit during the time of the comet's apparition. The stationary periods and retrogressions—although, as we have seen, admitting of a most simple explanation—long embarrassed astronomers; but when the true system of the universe was discovered by Copernicus, and more fully developed by Kepler, these apparent complications of the celestial movements, which had always been stumbling-blocks in the way of the erroneous systems, became so many striking confirmations of the true theory.

Difficulties analogous in kind, but much more numerous and grave, long prevented astronomers from discovering the true nature of comets and the laws which regulate their movements. We shall now see why.

SECTION III.

IRREGULARITIES IN THE MOTIONS OF COMETS.

Comets appear in all regions of the heavens—Effects of parallax—Apparent motion of a comet, in opposition and in perihelion, moving in a direction opposite to the earth—Hypothetical comet of Lacaille; calculations of Lacaille and Olbers concerning the maximum relative movement of this hypothetical comet and the earth.

THE orbits which the planets describe about the sun are not circles, but oval curves, termed *ellipses*; these ellipses differ but little from circles; that is to say, their *eccentricities* are small. Moreover, the planes of the orbits in which they move are inclined at small angles to the plane of the ecliptic. Hence it follows that their apparent paths are confined to a comparatively narrow zone of the heavens, which zone is called the zodiac. If we imagine these curves pressed down, as it were, upon the ecliptic they will appear as nearly concentric circles described about the sun, and so disposed as not to intersect each other. The distances of the earth and of each of the planets vary according to the position occupied by these bodies in their respective orbits; but these variations are confined within very narrow limits, and hence it follows that the velocities of the planets change so slightly that the difference is all but imperceptible. The mean diurnal motion of Mercury, which of all the planets moves the most rapidly, amounts to only 4°5′.

With comets the case is very different. These bodies, as we

have seen, are restricted to no region of the starry vault, and traverse the heavens in all directions, and with very different velocities. The third comet of 1739, and the comet of 1472, mentioned by Pingré, described in a single day, the first an arc of 120 degrees—that is to say, the third part of the whole celestial circumference—the second, an arc of 41 degrees and a half in longitude and nearly 4 degrees in latitude. Their real movement was, it is true, in a direction contrary to that of the earth, so that their apparent velocities were in both cases made up of the sum of their own and the earth's velocity combined. Here, then, we have an instance of what is called parallax; that is to say, the apparent movement of the object is affected by the observer's own displacement. We might multiply examples of a similar kind, but the following will suffice. 'The comet of 1729,' says Lalande, 'observed by Cassini during several months, after advancing more than 15 degrees towards the west from the head of Equuleus to the constellation Aquila, suddenly curved round to retrace its path towards the east, thus showing in a very striking manner the effect of the annual parallax.'

These rapid movements are produced by very simple causes, the most important of which are the near proximity of the comet to our globe, and the direction of its movement in relation to that of the earth. The following is a supposititious case, imagined by Lacaille, in which the apparent angular velocity of a comet would be enormous.

This astronomer supposes a comet to be moving in a direction contrary to that of our globe, and in the plane of the ecliptic; it is in perihelion, or at its least distance from the sun, and consequently at that point of its orbit in which its velocity is at its maximum. At the same time the earth is supposed to be in perihelion, and is also moving in its orbit with its greatest velocity. Lastly, the comet is to be not more distant from the earth than the moon, and it is to be in opposition. It is, of course, extremely improbable that all these hypotheses should be realised in the same comet, but there is nothing impossible in them. Under these exceptional conditions the comet, seen from the earth, would describe in the heavens an arc of nearly 39 degrees in longitude during the first hour, and of 32 degrees in the hour following. In three hours the total arc described would amount to 92° 58′, and this independently

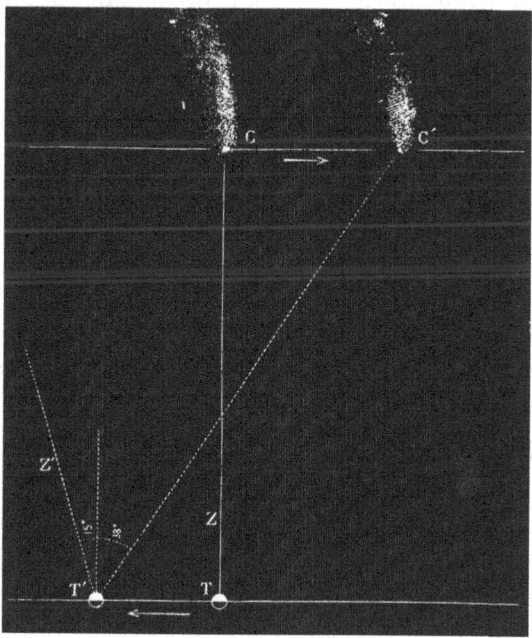

Fig. 6.—Maximum apparent movement of a Comet and the Earth.

of the diurnal movement, which would further increase the velocity by 15 degrees per hour. To an observer situated near the tropics the comet would ascend from the horizon to the zenith in less than two hours; it would, however, take a somewhat longer time to perform the second half of its journey and pass from the zenith to the horizon.

The calculation of Lacaille (modified by Olbers, on account of an error) is by no means difficult to verify; and there is nothing surprising in the result, if we reflect that the velocity of each of the two bodies, the comet and the earth, is then at its maximum; that our globe in one hour at its perihelion passes over in space a distance nearly equal to nine times its own diameter (or 67,000 miles); that the comet has a velocity greater than that of the earth, and passes over 94,000 miles in an hour; so that, in the direction of their motion, these two bodies are receding from one another at the rate of 161,000 miles per hour. At the end of a day the comet and the earth would be more than 3,800,000 miles apart.

It is, therefore, easy to comprehend to what irregularities of movement comets may be subject. Traversing the heavens in all directions, in orbits the planes of which cut the orbit of the earth at every possible inclination, approaching to and receding from the earth in very short spaces of time, influenced by the diurnal motion and their own proper motion, in combination with the earth's displacement, they sometimes suddenly appear, pursuing a rapid course amongst the stars; then, to all appearance they relax their speed, and after coming to a momentary stop reverse their motion, and continue their journey in an opposite direction, sometimes disappearing at a distance from the sun, sometimes being drowned in his rays.

It was these movements, these singular appearances, which so long baffled astronomers, and which the genius of Newton, guided by a higher conception, finally explained. We will now proceed to define geometrically the movements and orbits of comets.

SECTION IV.

THE ORBITS OF COMETS.

Kepler's Laws: ellipses described around the sun; the law of areas—Gravitation, or weight, the force that maintains the planets in their orbits—The law of universal gravitation confirmed by the planetary perturbations—Circular, elliptic, and parabolic velocity explained; the nature of an orbit depends upon this velocity—Parabolic elements of a cometary orbit.

WHAT is the nature of a true cometary orbit? In other terms, what is the geometrical form of curve which a comet describes in space—what is its velocity—how does this velocity vary—and what, in short, are the laws governing the movement of a comet?

In order to reply to these questions, and to enable them to be clearly understood, we must first call to mind a few notions of simple geometry, and also the principal laws which govern the motions of the planets.

Kepler, as we have already said, discovered the form of the planetary orbits, hitherto supposed to be circles more or less eccentric to the sun. This great man demonstrated that the form of a planetary orbit is actually an ellipse, that the sun is at one of the foci of the curve, and that the planet makes its complete revolutions in equal periods of time, but with variable velocity; in fact, that in equal intervals the elliptic sectors described by the radius vector * directed from the sun to the planet are of the same area.

* [The straight line joining the sun to a planet or other body moving under its action is called a *radius vector*.—ED.]

Let us take an example. S being the sun, the closed curve APB ... is the ellipse described by a planet. The distance of the planet from the sun is variable, as we see: it attains its minimum value at A, and its maximum value at B, that is to say, at one or other extremity of the greatest diameter of the orbit.

For this reason A is called the perihelion (from περί, near, and ἥλιος, the sun); B is called the aphelion (from ἀπό, from, and ἥλιος, the sun). For brevity the radius vector AS is called the perihelion distance; the radius vector SB the aphelion distance; and the two united, or the sum of these two distances, forms the major axis AB of the orbit. Lastly, the

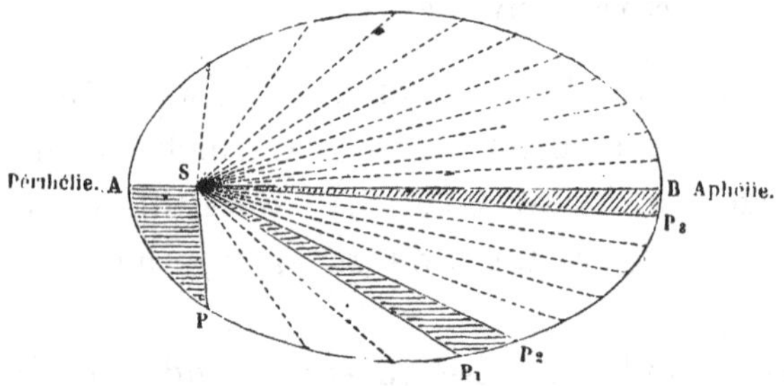

Fig. 7.—Second Law of Kepler. The areas swept out by the radius vector are proportional to the time.

mean distance of the planet from the sun is exactly equal to half the major axis.

Let us suppose that the arcs AP, P_1P_2, and P_3B have been described by the planet in equal spaces of time. According to the second law of Kepler, mentioned above, the three sectorial areas ASP, P_1SP_2, and P_3SB are equal. If the curve were a circle, of which the sun occupied the centre, it is clear that the equality of these areas would involve the equality of the corresponding arcs; and as the arcs are described by the planet in equal times, it would necessarily follow that the velocity would be the same throughout the entire orbit. In other words, a circular orbit presupposes an uniform movement.

But, as a matter of fact, the planets, without exception, describe around the sun ellipses more or less elongated, that is to say, orbits differing more or less from a circle. In every case their velocity is variable ; it is the greatest possible in perihelion; it then decreases by imperceptible degrees until the aphelion is attained, when the minimum of speed takes place. This is a direct consequence of the second law of Kepler.

A third law, discovered after years of research by this powerful genius, connects the duration of the planetary revolutions with the length of the major axis of their orbits. This law we have given elsewhere,* as well as some numerical examples for making it more intelligible to the non-scientific reader. We shall not return to it here, but confine ourselves to the remark that, the time of the revolution of a planet being known, the dimensions of the major axis of the orbit—that is to say, of twice the planet's mean distance from the sun—can be deduced by a simple calculation.† These laws are not rigorously obeyed by the planets in their movements. The strictly elliptic motion

* *Le Ciel*, 4th edition, p. 602.

† [Kepler's third law is that the squares of the periodic times are as the cubes of the mean distances, that is to say, that if r and R be the mean distances of two planets from the sun, and t and T be the durations of their revolutions round the sun, then—

$$t \times t : T \times T :: r \times r \times r : R \times R \times R.$$

For example, taking the mean distance of the earth from the sun as unity, the mean distance of Venus is 0·7233; and the earth performs its revolution round the sun in 365·26 days, Venus in 224·70 days; so that, according to Kepler's law,

$$224{\cdot}7 \times 224{\cdot}7 : 365{\cdot}26 \times 365{\cdot}26 :: 0{\cdot}7233 \times 0{\cdot}7233 \times 0{\cdot}7233 : 1;$$

or, working out the multiplications indicated,

$$50{,}490 : 133{,}415 :: 0{\cdot}37845 : 1,$$

and by division it will be found that each ratio of this proportion is equal to 2·642.

As another example, suppose there were two planets whose periods of revolution were found to be to one another as 27 to 8, then we should know that their mean distances were as 9 to 4 ; for

$$27 \times 27 : 8 \times 8 :: 9 \times 9 \times 9 : 4 \times 4 \times 4.\text{—Ed.]}$$

supposes ideal conditions that are not present in nature. But, by advancing them at an epoch when observations were so far from accurate, Kepler left it for astronomers and mathematicians coming after him to discover the cause of that mechanism of which he had only been able to detect the general laws. Huygens, Newton, and later many illustrious mathematicians (foremost among them Euler, D'Alembert, Clairaut, Lagrange, and Laplace), have explained the reasons not only for the general movements of the celestial bodies, but also for all the irregularities and inequalities which their movements undergo in the course of time.

Ultimately the whole matter resolves itself into a question of two causes, or of two forces. One of these forces is none other than weight or gravitation—the tendency that two bodies or two stars have to become united, a tendency which is proportional to the product of their masses, and which varies inversely as the square of the distance that they are apart. By their weight bodies fall to the surface of the earth when left to themselves in the atmosphere. If the force of gravitation alone existed, the moon would fall upon the earth, and both would together fall with ever increasing speed into the sun, and so likewise would the planets and all the bodies of the solar system.

But, in addition to the central force of gravitation, each planet is animated by another force,* which of itself would cause

* [It is perhaps well to explain that this so-called centrifugal force is not a force in the sense in which gravitation is, *i.e.*, it is not an external force acting upon the body. If a body were projected in space and were not interfered with by any external force, it would continue to move in a straight line. In order, therefore, that it may deviate from a straight line it must be acted upon by some external influence or force, and the resistance this force would have to overcome for the body to change its direction of motion is called 'centrifugal force.' Thus the 'centrifugal force' measures the tendency the body has to continue to move in the direction in which it is moving at the instant. If then a body describes a curve, some external force must be continually acting upon it, as it is

the planet to escape in a straight line in the direction of a tangent to the point of its orbit occupied by the planet. By combining these two forces, and seeking by geometry and analysis to determine the actual motion resulting from their simultaneous and constant action, Newton demonstrated that the laws of this movement were in conformity with those which Kepler had discovered. If one planet alone existed and circulated around the sun, and if its mass were inappreciable in comparison with the enormous mass of that luminary, the elliptic movement would conform rigorously to Kepler's laws. But the planets are more than one in number; they act and react upon each other; their dimensions and masses are more or less unequal; they recede from and approach one another in the course of their revolutions, and their mutual action upon one another is an incessant cause of disturbances and perturbations. It is important to notice that these perturbations are not exceptions in the true meaning of the word; far from invalidating the theory, they afford the most striking confirmation of it, since each of these deviations may be calculated beforehand by the theory of universal gravitation.

But let us here terminate this necessary digression and return to the comets.

Newton, as we have seen, by a bold but logical generalisation, supposed comets to be subject to the same influences as the planets, to be borne along by a primitive force of impulsion, and continually drawn by gravitation towards the sun, the focus of all the movements of our system. Let us endeavour to ex-

continually changing its direction of motion. In the case of a planet or other body describing an ellipse round the sun, the sun is continually pulling it towards itself; and this continued action is necessary to overcome the centrifugal force, *i.e.*, to balance its tendency to move at every instant in the tangent to its path; in fact, if the action of the sun suddenly ceased, the planet would immediately move off along the tangent to the ellipse at the point where it was, and with the velocity it had, at the instant.—ED.]

plain by some simple examples what must be the orbit of a body acted on by such influences; to *explain*, let it be understood, not to demonstrate.

Consider, then, a heavy mass, a planet *M*, gravitating towards the sun, and at the same time moving with a certain velocity due to an impulsion foreign to gravitation; and suppose, for

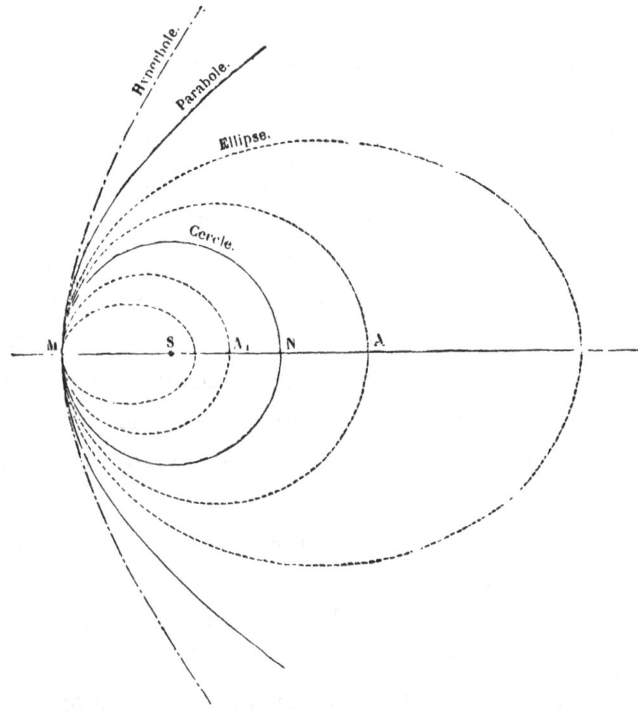

Fig. 8.—Relation between the velocities and forms of Orbits.

the sake of greater simplicity, that *M* is situated at the point where the planet is moving in a direction perpendicular to the radius vector joining the planet and the sun.

The geometrical form of the orbit described by the planet about the sun will depend solely upon the relation between the initial velocity of the planet and the distance of the latter from the sun. For a certain value determined by this relation the curve described becomes a circle of which the sun occupies the centre, and the planet traverses with uniform velocity every part of the circumference. The velocity which for a given distance compels a mass subject to the law of gravitation to describe a circle is known as *circular* velocity. A less velocity would give rise to an elliptic orbit; in which case, the sun, instead of occupying the centre of the curve, would be situated at one of the foci, namely, that which is the further removed from M; and the point M would be the aphelion of the planet.

If the velocity be greater than circular velocity the orbit would still be an ellipse, having the sun in the focus; but in this case M is the perihelion, and the planet attains its greatest distance from the focus of attraction at the opposite extremity of the diameter MS.

The greater the initial velocity the more elongated will be the orbit, and the greater the eccentricity* of the ellipse. But if the velocity should attain a certain value—viz., should be equal to circular velocity multiplied by the number 1·414 (or by the square root of 2)—at this moment the ellipse, the major axis of which has been continually lengthening, and has at last increased in the most rapid manner, changes into a curve with endless branches, called a parabola. A planet animated by this velocity, or, let us say, by parabolic velocity, at the moment when it is at its least distance from the sun—*i.e.* when it is at its perihelion—is a body which comes to us out of infinite

* The *eccentricity* is the distance from the centre of the ellipse to one of its foci, measured in parts of the semi-major axis, which is taken as unity. In an elliptic orbit the eccentricity is always less than unity, and is usually expressed in decimal fractions. Amongst the orbits of the eight principal planets that of Mercury has the greatest eccentricity, 0·2056; Neptune and Venus have the smallest, 0·0087 and 0·0068. Both these orbits differ very slightly from a circle. In a parabola the eccentricity is equal to 1. In a hyperbola it is greater than unity.

space and returns into infinite space; such a body, supposing one to exist, before arriving at that region of the heavens where the action of the sun preponderates, could form no part of the solar system. After passing its perihelion it would depart to an infinite distance; and unless the form of its orbit should become changed by the disturbing influence of the planets, it would again become alien to the solar group.

Lastly, to omit no case that can possibly occur, we must consider that of a planet moving with a velocity greater than parabolic velocity; the orbit now described will continue to be a curve of endless branches, but it will be an hyperbola, of which the sun is, as before, situated at one of the foci.

These preliminary notions understood, we are in a position to consider the question of the geometrical determination of cometary orbits.

These orbits are, in general, very long ellipses, of considerable eccentricity, that is, of eccentricity very nearly equal to unity. And this explains why comets remain visible during comparatively so short a time, as the arc which they describe is only a very limited portion of the entire orbit. During the remainder of their journey they are too far distant from the earth to be perceived either by the naked eye or by the aid of the most powerful telescope.

The orbit of a comet being thus a very long ellipse, and the portion of the arc observed being of very limited extent as compared with the dimensions of the whole orbit, it follows that it is generally very difficult to determine to what ellipse this arc belongs, or even to decide whether it may not form part of a parabola or hyperbola of the same perihelion distance.

These different curves are, so to speak, blended into each other, and only become sensibly distinct at a distance too remote for the comet to be within our range of vision. In these different orbits the positions of the comet obtained by calculation would not be distinguishable from the positions obtained by direct observation, or would differ by quantities so small as to be liable to be confounded with the errors made in the observations themselves.

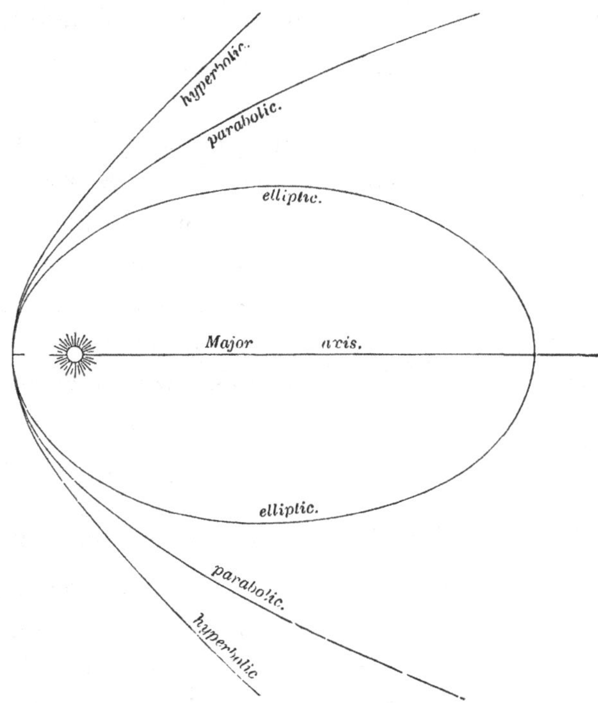

Fig. 9.—Cometary Orbits, elliptic, parabolic, and hyperbolic.

This was recognised by Newton, who at once conceived the idea of simplifying the problem involved in the determination of cometary orbits. He assumed the orbit, in the first instance, whatever might be its real form, to be parabolic, because the elements of a parabola, or the conditions which determine its position in space, its form, dimensions, &c., are less numerous and more simple than the elements of an ellipse.

Let us, then, consider what are the elements of a parabolic orbit. A parabola is a plane curve, that is to say, a curve all the points of which are situated in the same plane, which in our case passes through the centre of the sun. The first thing, therefore, is to define the true position which this plane occupies in space. This will be accomplished by determining first the line of intersection in which it cuts the plane of the earth's orbit, or the ecliptic; and, secondly, the inclination or the angle which the two planes make with one another.

The comet in its movement necessarily cuts the ecliptic in two diametrically opposite points, called the two *nodes*; the line which joins these two points and passes through the centre of the sun is called the line of nodes. It is sufficient to know one of the nodes—for example, the ascending node—that is to say, the node which corresponds to the passage of the comet from the region south of the ecliptic to the region north of the ecliptic. Let N (fig. 11) be this point, which can be obtained by calculation from observations of the comet; its position will be determined if we know in degrees, minutes, and seconds the value of the arc $O☊$ or of the angle OSN measured from the zero of the ecliptic, in the direction in which the celestial longitudes are reckoned.

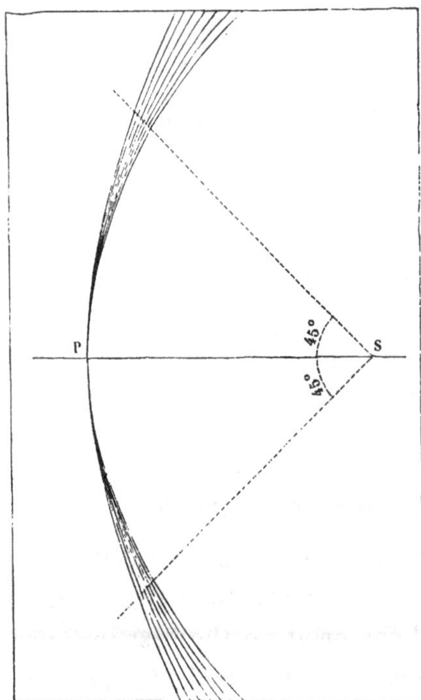

Fig. 10.—Confusion of the arcs of Orbits of different eccentricities in the neighbourhood of the perihelion.

This first element is called the longitude of the ascending node, or, more simply, the longitude of the node. But the plane of the orbit remains undetermined, unless we add to it a second element, viz., its inclination.

If, through the centre of the sun, we imagine two straight lines drawn perpendicularly to the line of the nodes, the one in the ecliptic, and the other in the plane of the comet's orbit, these two lines will make between them two angles, the smaller of which measures the angle between the two planes. The angle i is the inclination.

It next becomes necessary to define and fix accurately the actual curve described by the comet in this plane, determined by the longitude of its node and its inclination. In the first place, we must know the position of the planet at its perihelion, or least distance from the sun. Let A be this point. SA is then the axis of the parabola, the direction of which will be known, if we determine the longitude of the point A, or of the point π, obtained by projecting SA upon the ecliptic. If to the longitude of the perihelion we add another element, the length SA, or the perihelion distance—which, like all celestial distances, is measured in parts of the sun's mean distance from the earth—the vertex of the parabola will be completely fixed.

The parabolic curve described by the comet is now entirely defined, both as regards its position in space and its dimensions. It remains, however, to find the direction of the comet's movement, and to determine at what epoch the comet will occupy any given position in its orbit. For the purpose of obtaining the direction we will suppose the parabola laid or pressed down upon that side of the ecliptic where the inclination is least, or more simply, in the language of geometers, projected upon the plane of the earth's orbit. The direction of movement will be called *direct*, if, when estimated from above the ecliptic or from the north region of the heavens, it takes place in a direction from right to left,

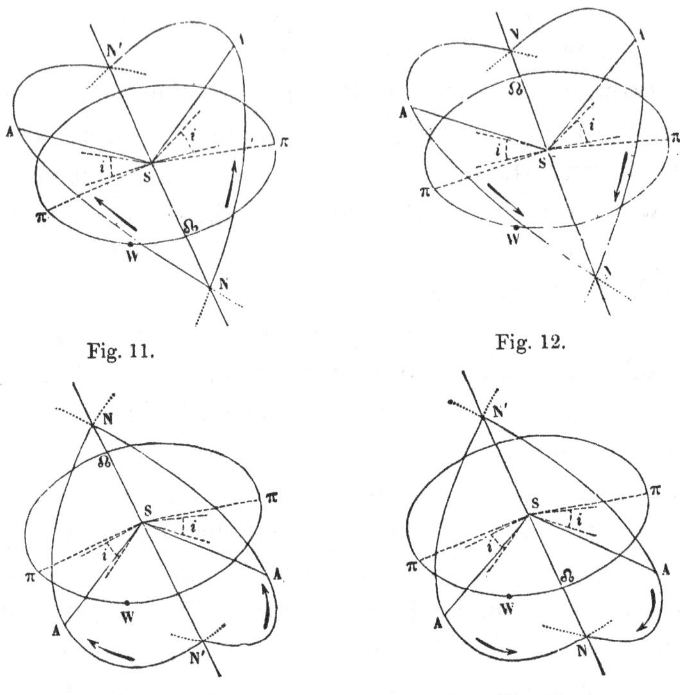

Fig. 11. Fig. 12.

Fig. 13. Fig. 14.

Determination of a cometary orbit: parabolic elements.*

or from west to east, as is the case with the earth and all the planets, and *retrograde* when it takes place in the contrary direction.

* 1. Inclination, 20°. Direction in longitude of the line of Nodes, 35° to 215°.

Fig. 11.—Movement retrograde.		Movement direct.	
Node	35°	Node	35°
Perihelion .	318°	Perihelion . .	112°
Fig. 12.—Movement direct.		**Movement retrograde.**	
Node	215°	Node . . .	215°
Perihelion . .	318°	Perihelion . .	112°
Fig. 13.—Movement retrograde.		**Movement direct.**	
Node	215°	Node	215°
Perihelion . . .	318°	Perihelion . .	112°
Fig. 14.—Movement direct.		**Movement retrograde.**	
Node	35°	Node	35°
Perihelion . .	318°	Perihelion . .	112°

Lastly, the exact date of the perihelion passage of the comet completes the determination of the orbit both in time and space, so that all other positions are deducible by calculation from the elements we have mentioned. Figs. 11, 12, 13, and 14 show the different cases that may arise, that is to say, the different positions the same parabolic orbit may occupy with respect to the plane of the ecliptic, when the inclination, the line of the nodes and the perihelion distance remaining the same, the direction of movement only is varied. It will be seen that eight distinct paths are open to the comet in space.*

Briefly to recapitulate, we subjoin in the following table these various elements, in the order usually adopted by astronomers, taking for examples the two great comets which appeared in 1744 and 1858 :—

T,	Epoch of perihelion passage, 1744, March 1.	7h. 55m. 39s. Paris mean time.	
π,	longitude of perihelion . .	197° 13′ 58″	Mean equinox, 1744·0
☊,	longitude of node . . .	45 47 54	
i,	inclination	47 7 41	
q,	perihelion distance . . .	0·222209	
	Movement direct, D.		

T,	Epoch of perihelion passage, 1858, September 29.	23h. 8m. 51s.
π,	longitude of perihelion . .	36° 12′ 31″
☊,	longitude of node . . .	165 19 13
i,	inclination	63 1 49
q,	perihelion distance . . .	0·57847
	Movement retrograde, R.	

Such are the elements the determination of which is necessary to enable us to find the orbit of a comet supposed to be parabolic. These elements are not determined directly, but

* [There are two planes (N A N) each having the same inclination i, and, the perihelion distance remaining the same, there are therefore four positions of the vertex (A) of the parabolic orbit, viz., two in each plane, one above and the other below the plane of the ecliptic, as shown in the four figs. 11–14. As, also, the direction of motion of the comet in the parabolic orbit may be either direct or retrograde, we have, in all, eight cases.—Ed.]

are derived by mathematical calculation from a certain number of observations of the comet, at least three accurately observed positions of the comet being required. Three complete observations are strictly indispensable; but in order to deduce from them the true curve of the orbit it is necessary that they should have been made with the utmost precision. One or two positions of the comet would leave the problem indeterminate. If we have more than three, they are of great value for verifying the results given by calculation. Of course all the observed positions should correspond to points lying on the orbit which has been determined, or, in other words, the calculated ephemeris should agree with the apparent path obtained from direct observations of the comet.

But if, all these considerations being fulfilled, the difference between the observations and the calculated results should nevertheless prove too great to be attributed to errors in the observations themselves, it is then proper to conclude that the comet is not describing a parabola, and that the hypothesis of a parabolic orbit must be rejected, in which case there remains no other alternative than that of a hyperbolic or elliptic orbit. The latter are much the more common; and it is thus that we have been led to recognise the periodicity of certain comets. We are, in this case, concerned with a body which forms a part of the solar system, and whose movements are regulated in the same manner as those of the planets.

SECTION V.

THE ORBITS OF COMETS COMPARED WITH THE ORBITS OF THE PLANETS.

Differences of inclination, eccentricity, and direction of motion.

If, then, periodical comets, calculated as such, and known to be periodical by their return, are governed by the same laws as the planets, why is a distinction made between these two kinds of celestial bodies? This is a question of high importance, and one which we cannot completely answer at the present moment. A full reply would necessitate some definite knowledge concerning the origin of the bodies which compose the solar world. It would be necessary to have studied and compared the physical constitution of comets with that of planets. Both in origin and constitution we shall see further on that they appear to be essentially different. Surveying the question, however, from a single point of view, regarding it as a question of movement only, we can already show differences which separate these two classes of celestial bodies, and justify the double denomination by which they are distinguished.

Comets, as we have already seen, appear in any quarter of the heavens, instead of moving, like the planets, in the narrow zone of the zodiac. This difference arises from the inclinations of their orbits to the plane of the ecliptic. Among the principal planets Mercury alone has an inclination as great as

7 degrees; and among 115 telescopic planets 29 only have an inclination greater than 10 degrees, and very few exceed 30 degrees;* but we see, on the contrary, the planes of cometary orbits admit of all inclinations. Out of 242 comets which have been catalogued 59 have an inclination included between 0 and 30 degrees, 93 have inclinations between 30 and 60 degrees, and 90 an inclination amounting to between 60 and 90 degrees.

This first characteristic is important. When we add to it the second distinction, that, whilst the movement of the planets is without exception direct, out of 242 comets 123 have a motion that is retrograde, it is impossible not to recognise a difference of origin in the two classes of bodies. It is nevertheless curious to remark, that out of nine comets whose return has been established there are eight whose movement is direct; one alone, the great comet of Halley, which is a comet of long period, moves in a direction contrary to that of the planets; and one alone, that of Tuttle, a comet of mean period, moves in a plane whose inclination to the ecliptic is considerable (54 degrees). The inclinations of each of the eight others are less than 30 degrees.

Let us proceed to another distinctive feature of cometary and planetary orbits. We have already seen that of the eight principal planets Mercury is that which describes an orbit which differs most from a circle. The distance, however, between its aphelion and perihelion distances does not amount to half its mean distance. Its mean velocity is 29·2 miles per second; at the aphelion it is not less than 24·9 miles; at the perihelion it attains 37 miles per second. The orbits of the other principal planets differ much less from the

* Felicitas has an inclination of 31½°, Pallas of 34°. The very great inclinations of some of the small planets, belonging to the group comprised between Jupiter and Mars, have obtained for them the appellation of *extra-zodiacal* planets.

figure of a circle. But in the group of small planets there are orbits the eccentricity of which markedly exceeds the orbit of Mercury; twenty-six of these ellipses have greater eccentricities ; but one in particular, that of the planet Polyhymnia, has an eccentricity comparable to that of some elliptic cometary orbits. Fig. 15, in which the orbit of Faye's comet and the orbit of the planet Polyhymnia are represented, as regards their forms and relative dimensions, clearly shows how close is sometimes the degree of resemblance in point of eccentricity between cometary and planetary orbits.

The divergence may be of any amount; the eccentricity of the great majority of cometary orbits is so great that it may be considered equal to unity, and this is expressed, let us repeat, by assimilating them to parabolas. Is this assimilation to be considered absolute, or are we to suppose that all comets belong to the solar world? It appears certain that some orbits at least are hyperbolic. As regards these there can be no doubt. But if so, it may be regarded as not improbable that amongst observed comets there are some which describe true parabolas ; so that, after having once arrived within the sphere of the solar gravitation, like those which describe hyperbolic orbits, they take their leave of us for ever.

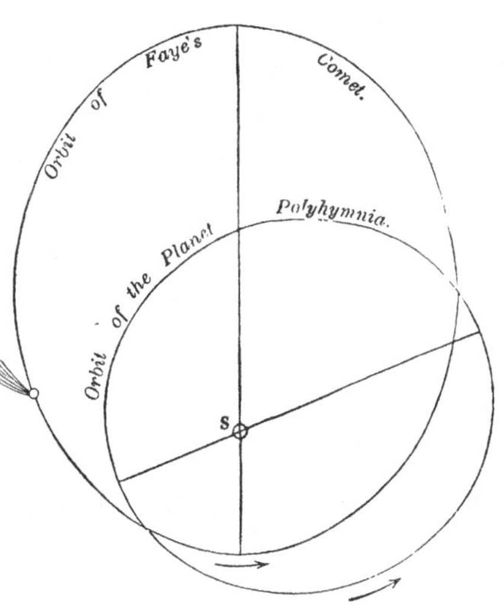

Fig. 15.—Comparison of the eccentricities of the orbit of Faye's Comet with that of the planet Polyhymnia.

Amongst the comets whose periodicity has been calculated there are some which describe ellipses of such great eccentricity that, as far as we or our descendants are concerned, it is almost the same as if they were non-periodic. The great comet of 1769 (eccentricity 0·9992) has a period of about twenty-one centuries; at its aphelion it will reach a point in space the distance of which from the earth will be 327 times the distance of the earth from the sun. The comets of 1811 and 1680 have periods respectively of 3,065 and 8,814 years (eccentricities 0·9951 and 0·9999). The first comet of 1780 and that of July 1844 will only return to their perihelia after journeys the respective durations of which will be 75,840 years and about a thousand centuries. These comets will penetrate so far into the depths of space that at the time of their aphelion they will be distant from our world about 4,000 times the distance of the sun.

If the calculations upon which these necessarily uncertain values depend are not rigorously exact, they nevertheless show that the comets to which they relate always remain an integral part of our system. Their greatest distance is still fifty times less than that of the nearest fixed star. The action of the sun upon these bodies will, therefore, always preponderate over that of any other body, and their masses will be incessantly drawn towards those regions of the heavens traversed by our earth, unless, indeed, the perturbations which the planets can exercise upon them should interfere so as to divert them from their course and modify the elements of their orbits.

SECTION VI.

DETERMINATION OF THE PARABOLIC ORBIT OF A COMET.

Three observations are necessary for the calculation of a parabolic orbit—Cometary ephemerides; what is meant by an ephemeris; control afforded by the ulterior observations—Elements of an elliptic orbit—Can the apparition or return of a comet be predicted?—State of the question—Refutation by Arago of a current prejudice.

THREE observations of a comet—that is to say, three different positions (in right ascension and declination) of the nucleus of a comet, or, in a word, three points of its trajectory or apparent orbit sufficiently distant from each other—are required, as we have said, for the calculation of the parabolic elements of the true orbit.

In the last century this determination was not only a long and laborious operation, but involved much tentative and uncertain work. Before engaging in the difficult calculation of the elements of an orbit, astronomers made trial graphically and even mechanically of different parabolas, and only began the calculation after satisfying themselves that one of these curves nearly represented the positions furnished by observation. Great improvements were introduced into these methods during the last century by Lalande, Laplace, and Gauss. But the calculation of a cometary orbit is always a sufficiently complex operation, even if it be simply parabolic, and it still takes a skilful computer accustomed to this kind of work, several hours to find approximate values of the different elements. This is not the place for us, of course, to attempt an explanation of the work itself.

A first orbit having been found, what astronomers call an ephemeris is then deduced from it. This term is applied to the calculated positions which the comet must have occupied or will occupy day by day during the period of its visibility. These calculated positions should agree with the observed positions, that is to say, with the positions obtained by direct observations with instruments. This comparison furnishes a means of control from which it will result either that the elements are correct, or, on the contrary, that the parabola is unfitted to explain the movements of the comet. In the latter case it remains to examine whether this movement might not be better represented by an hyperbolic orbit, or, as most frequently happens, by an ellipse. In this way a certain number of comets have been found to describe ellipses round the sun, and have been accordingly classed amongst periodical comets of the solar system.

We may remark, while speaking of elliptic orbits, that two more elements must be added to the elements of parabolic orbits for the purpose of determining an elliptic orbit: firstly, the eccentricity above defined, and which, in conjunction with the perihelion distance, enables us to calculate the major axis of the orbit ; secondly, the duration of the revolution, a duration connected with the value found for the major axis by the third law of Kepler.*

* Take for example, the following table of the elliptic elements of Tempel's short period comet, 1867. II., for its return in May 1873 ; e is the eccentricity, a the semi-axis major :—

Perihelion Passage, May 9·74218.

π,	longitude of perihelion .	238° 2′ 34″	Duration of the revolution, 5 years 97 days.
☊,	longitude of node . .	101 12 50	
i,	inclination	9 12 6	
e,	eccentricity	0·5076428	
a,	semi-axis major	3·1721	

Movement direct.

This leads us to say a few words on a question which has nearly always been imperfectly understood by the public, notwithstanding the repeated explanations of astronomers: we mean the possibility of predicting the advent of a comet.

Can the apparition of any comet whatever be predicted?

In these terms the preceding question has been invariably asked. As regards the public which has faith in astronomical science, but very little knowledge of astronomy, an answer in the affirmative is not for a moment doubted; and, in their opinion, astronomers who allow themselves to be surprised by the apparition of a comet have certainly failed in their duty—in their duty as observers, if the discovery of this new comet rests with an amateur, and in their duty as mathematicians, if they have not foretold it.

As a rule these reproaches are unjust. They are founded upon a false idea of the power of astronomy and the nature of cometary orbits. Arago, who never lost an opportunity of endeavouring to destroy popular misconceptions on scientific matters, has given a perfect refutation of this error, which, nevertheless, is still widely spread. The opportunity was furnished by the brilliant comet of 1843, which appeared unexpectedly, its arrival not having been announced by astronomers, and with reason. Let us, therefore, endeavour, following the example of the late well-known Secretary of the Academy of Sciences, to dissipate this generally received error as far as lies in our power.

On referring to the preceding sections of this chapter we perceive that the greater number of comets which have been seen and observed * from the earliest times to the present day

* It should be borne in mind that to *see* a star and to *observe* it are two very different things. In the long list of comets mentioned in history, from the earliest times to the eighteenth century, when Pingré lived, the indefatigable author of the *Cométographie* is unable to find more than sixty-seven comets observed with sufficient accuracy to allow of their orbits being calculated.

describe parabolas, or, at all events, ellipses so elongated that we may be certain either that these comets have never visited our world before, or that their visits have been made in prehistoric times. For this reason they will never return, or if they should return it will be at an epoch so far distant from our own that it need not for a moment occupy our attention. It is, therefore, evident that a comet which thus appears for the first time within sight of the earth could not have been announced before it was perceived: no prediction of its apparition was possible.

Here, then, is a first point established, which, I repeat, applies not only to the great majority of recorded comets, but also to comets which have been catalogued; that is to say, to comets whose orbits have been calculated with more or less precision. Out of the 262 comets in the catalogue that we publish at the end of the present volume nine only are periodical comets whose return has been verified by observation; sixty others have elliptic orbits, but the greater number of these are so eccentric that for our present purpose they practically fall within the category of comets with infinite orbits.

Arago was, then, perfectly justified in the following remarks in reference to the above question so incessantly repeated by persons who are not astronomers. 'Is it reasonable to hope,' said he, 'that a time will come when we shall be able to predict the arrival within our sphere of vision of comets which have remained for ages as if lost in the furthest regions of space, which no one has ever seen, whose action upon the bodies of the solar system is too small to be appreciable, both in consequence of the excessive rarity of the vaporous matter of which they are composed, and of their prodigious distance? A comet is revealed to man when it becomes visible or produces some perceptible effect. That which has never been beheld, and has never produced any observed displacement,* is for us as if it had never existed. The announcement of the apparition of a new and totally unknown comet would belong to the domain of sorcery, and not to that of science. Astrology itself never pushed its pretensions so far even in the day of its greatest favour.'—*Annuaire de* 1844.

* Theoretical astronomy has attained, in fact, to such perfection that the perturbations of unknown bodies have led to the discovery of new planets, as in the case of Neptune. Arago, who wrote the above passage in 1844, ten years before the discovery of the planet Neptune, has thus, as it were, foreshadowed the possibility of such a prediction.

CHAPTER IV.

PERIODICAL COMETS.

SECTION I.

COMETS WHOSE RETURN HAS BEEN OBSERVED.

How to discover the periodicity of an observed Comet and predict its return—First method: comparison of the elements of the orbit with those of comets that have been catalogued—Resemblance or identity of these elements; presumed period deduced from it—Second method: direct calculation of elliptic elements—Third method.

THERE are, however, a certain number of comets of whose return astronomers are certain, and the time of whose apparition they can calculate. The prediction of the probable epoch at which these comets will be situated in regions of the heavens where they will be visible from the earth, and the determination of their perihelion passage, can be effected more or less accurately. These are the comets whose orbits, when calculated from a sufficient number of observations, prove to be neither parabolas nor hyperbolas, but, on the contrary, are closed and elliptic, and such that the comet thenceforth continues to describe them in regular periods; in a word, they are *periodical comets.** Newton treated the orbits of comets as parabolic, merely in order to so represent the arc, always very short, described in the neighbourhood of the perihelion, when the comparatively small distance of the comet from the sun

* [It may be stated here that the duration of revolution of a body, that is, the time occupied by it in a complete revolution round the sun, is called its 'period.' And, in general, the period of any periodical phenomenon is the interval of time between two of its successive returns to the same position.—ED.]

renders observations possible. In his opinion comets were bodies of regular periods, and which described ellipses, certainly very elongated, but in all respects similar to the planetary orbits. The first certain proof of the periodicity of a comet, the indisputable return of a comet in the same orbit, was, therefore, a confirmation and a brilliant triumph for the Newtonian theory. Neither Halley, who had the glory of the first prediction, nor Newton, who made it possible, lived long enough to see the event justify the theory. Since then, as we are about to see, facts of the same kind have been multiplied, and the number of comets whose return can be calculated, and which, moreover, have actually reappeared, is already considerable, and is gradually increasing. Side by side with the planetary system, therefore, another system was being founded, and the history of this part of our solar world is sufficiently interesting and instructive to be given with some detail.

But first let us endeavour to explain by what methods astronomers discover the periodicity of a comet.

When a new comet, or one supposed to be new, makes its appearance, can we tell if it has been seen or observed at any previous epoch? The reply to this question serves as a foundation to the first method employed for the resolution of the problem. But the reply is not easy if the apparition or previous apparitions of the comet (supposing it to have appeared to us before) have not been observed with some degree of precision, and if the tradition or record is limited to a vague mention of the size, the brilliancy of the nucleus, the form or the dimensions of the tail. The outward appearance of a comet, its physical aspect, would be in almost all cases insufficient. We shall see as we proceed that these are variable features, that the aspect of a comet changes in the course of a single apparition. But even if it remained the same, the different circumstances of its visibility and distance from the earth would suffice to prevent the identification of the two comets. A comet formerly of extreme brilliancy might reappear as a feeble nebulosity. It would have been difficult to recognise the same body in the comet of 1607, whose light appeared to Kepler pale and weak; in that of 1682, which Lahire and Picard compared to a star of the second magnitude; in that of 1759, which appeared to Messier like a star of the first magnitude; and, lastly, in the famous comet of 1456, 'which all historians (except two Poles),' says Pingré, 'agree in describing as great, terrible, and of an extraordinary size, drawing after it a long tail which covered two celestial signs, or 60 degrees.' These were, nevertheless, one and the same comet. Astronomers, it is true, mistrust, and justly, the nearly always exaggerated expressions of the ancient chroniclers; but precisely for that reason a resemblance of aspect is not to be relied upon for establishing the identity, and consequently the periodicity, of two comets. We must have more precise elements of comparison. These elements are those of the parabolic orbit, when records have been left of observations—that is to say, of positions and dates sufficient for the calculation of the orbit—when, in a word, the comet instead of having been simply seen has been observed. A catalogue of ancient comets is therefore necessary, and it was whilst consulting the table of twenty-four comets which he had calculated that Halley made the prediction, the history of which we are about to give.

If the longitudes of the ascending node and of the perihelion, the inclination of the plane of the orbit, the perihelion distance, and the direction of movement, are all the same, or nearly the same, in two cometary orbits, in all probability we have two successive, if not consecutive, apparitions of the same comet. Taking the interval between the apparitions for the period itself, we are enabled by the third law of Kepler to calculate the dimensions of the major axis of the corresponding elliptic orbit, and to assure ourselves that the new orbit is in accordance with the whole of the known observations. If this be so, we can calculate more or less exactly the comet's next return; that is to say, its perihelion passage, and all the circumstances of its future apparition.

The second method consists in the direct calculation of the elliptic elements. It requires, as a rule, exact observations, especially if the orbit be greatly elongated, since there is then but little difference between the apparent path followed by a comet, whether it be a parabola, a very long ellipse, or an hyperbola slightly flattened. The first attempts by this method—a very legitimate one in theory—prove that it is subject to many difficulties and uncertainties. Euler, on first applying it to the comet of 1744, obtained a hyperbolic orbit from the observations made at Berlin. But afterwards, having received the observations made by Cassini, he found the orbit to be a very long ellipse, with a period of many centuries.

The first example of an elliptic orbit calculated with precision by this second method is, we believe, that of Lexell's comet (or comet of 1770), a comet of short period (five years and a half), and having an orbit of comparatively slight elongation, but which, unfortunately—we shall come to its history further on—has undergone enormous perturbations, and has not again been seen. Since then the direct calculation of the elliptic movement, without reference to previous observations, has been employed for various comets, and with success in several instances, as the return of the periodical comets of Faye, Brorsen, d'Arrest, and Winnecke (1819) has been rendered certain by numerous and careful observations.

The above two methods both require observed positions of the comet, whose periodicity is to be discovered, and also that these observations should possess a certain degree of accuracy.

In the absence of these conditions the end may, however, be attained, but the result is, in that case, as conjectural as the method itself. This third method consists in making a comparison of the different historical comets, in noting the resemblance of their aspect, and in ascertaining if the intervals of their successive apparitions agree with the hypothesis of a certain period, whose duration, in this case, must be necessarily contained nearly an exact number of times in these intervals. The elements calculated for one apparition may then suffice to render probable the identity of several comets. In this way M. Laugier is of opinion that he has identified the comets of 1299, 1468, and 1799 by assuming a period of one hundred and sixty-nine years, which is twice included between the two last dates. In the same manner the comets of 1301, 1152, 760, and several others (which we shall mention presently) have been identified as former apparitions of Halley's comet, the true period of which has long been calculated and known.

SECTION II.

HALLEY'S COMET.

Discovery of the identity of the comets of 1682, 1607, and 1531; Halley announces the next return for the year 1758—Clairaut undertakes the calculation of the disturbing influence exercised by Jupiter and Saturn upon the comet of 1682; collaboration of Lalande and Mdlle. Hortense Lepaute—The return of the comet to its perihelion is fixed for the middle of April 1759; the comet returns on the 13th of March—Return of Halley's comet in 1835; calculation of the perturbations by Damoiseau and Pontécoulant; progress of theory—The comet will return to its perihelion in May 1910.

LET us recal the memorable words of Seneca in his *Quæstiones Naturales*: 'Why should we be surprised that comets, phenomena so seldom presented to the world, are for us not yet submitted to fixed laws, and that it is still unknown from whence come and where remain these bodies, *whose return takes place only at immense intervals?* ... An age will come when that which is mysterious for us will have been made clear by time and by the accumulated studies of centuries. ... Some day there will arise a man who will demonstrate in what region of the heavens the comets take their way, why they journey so far apart from other planets, what their size, their nature.' Eighteen centuries have elapsed, and not one man, but the accumulated efforts of many men have raised a corner of the veil spoken of by Seneca. As far as the laws of cometary movement are concerned Newton has realised his prediction; whilst that which relates to the return of comets and their calculated periodicity has been fulfilled by Halley.

This learned man, modest as he was laborious, published in 1705 his catalogue of twenty-four comets. On comparing their elements he remarked that three comets—namely, those of 1531, of 1607, and of 1682—had orbits nearly identical. He at once suspected the identity of the comets themselves; and more than that, he announced the next return of the comet for the year 1758. Let us subjoin the elements which Halley calculated, and leave him afterwards to speak for himself:—

	Comet of 1531.	Comet of 1607.	Comet of 1682.
Longitude of node . . .	49° 25′	50° 21′	51° 16′
Inclination of orbit . .	17° 56′	17° 2′	17° 56′
Longitude of perihelion . .	301° 39′	302° 16′	302° 53′
Perihelion distance . .	0·56700	0·58680	0·58328
Direction of movement . .	Retrograde.	Retrograde.	Retrograde.

The following is the passage in Halley's memoir * concerning the periodicity of the comet which at the present day bears his name:—

'Now, many things lead me to believe that the comet of the year 1531, observed by Apian, is the same as that which, in the year 1607, was described by Kepler and Longomontanus, and which I saw and observed myself, at its return, in 1682. All the elements agree, except that there is an inequality in the times of revolution; but this is not so great that it cannot be attributed to physical causes. For example, the motion of Saturn is so disturbed by the other planets, and especially by Jupiter, that his periodic time is uncertain, to the extent of several days. How much more liable to such perturbations is a comet which recedes to a distance nearly four times greater than Saturn, and a slight increase in whose velocity could change its orbit from an ellipse to a parabola? The identity of these comets is confirmed by the fact that in

* [The title of Halley's memoir is *Astronomiæ Cometicæ Synopsis*, and it was published in the *Philosophical Transactions*, vol. xxiv. (1704–5), pp. 1882–1899.—Ed.]

the summer of the year 1456 a comet was seen, which passed in a retrograde direction between the earth and the sun, in nearly the same manner; and although it was not observed astronomically, yet, from its period and path, I infer that it was the same comet as that of the years 1531, 1607, and 1682. I may, therefore, with confidence predict its return in the year 1758. If this prediction be fulfilled, there is no reason to doubt that the other comets will return.'

Later, in his *Astronomical Tables*, published in 1749, ten years before the return of the comet, Halley recurs again to his prediction in the most decided terms. 'You see, therefore,' he says, 'an agreement of all the elements in these three, which would be next to a miracle if they were three different comets; or, if it was not the approach of the same comet towards the sun and earth in three different revolutions, in an ellipsis around them. Wherefore, if according to what we have already said, it should return again about the year 1758, candid posterity will not refuse to acknowledge that this was first discovered by an Englishman.' *

Posterity has remembered and science recognised the claim of the English astronomer, by giving his name to the first comet whose periodical return, announced beforehand, was confirmed by the event. But the same posterity will not be unjust: it will give a legitimate share of honour to the French astronomers Clairaut and Lalande, who completed the work of Halley by calculating the retardation the comet would be subjected to in its voyage of seventy-six years. This second part of the history of a great discovery is perhaps still more surprising and instructive than the first.

As the epoch of the return predicted by Halley drew near,

* [Halley died on January 14, 1741–2, and his *Tabulæ Astronomicæ* were published seven years after his death, in 1749. In 1752 a second edition appeared, and to it was appended an English translation, from which the passage cited in e text is extracted.—Ed.]

all astronomers in France and Europe, occupied with this great event in the annals of science, held themselves in readiness to make observations of the comet. The time of its reappearance was uncertain. The known periods, as Halley had himself remarked, were unequal. Between 1531 and 1607 the interval was 27,811 days; from 1607 to 1682, 27,352 days, with a difference of 459 days between the perihelion passages. Would the new period be still shorter, or, on the contrary, after having been diminished by fifteen months and a half, would it return to its old value, or even exceed it? Several savants made calculations and offered various hypotheses respecting the path of the comet on its return and the date of its apparition, which was watched for from 1757.

It was then that Clairaut, a great mathematician, undertook the rigorous solution of the problem which Halley had only indicated—viz., the calculation of the perturbations which the comet of 1682 would experience whilst passing in the vicinity of the planets, especially of Jupiter and Saturn. It was a work of immense difficulty, and Clairaut, pressed for time, sought the assistance of Lalande, one of the most illustrious of French astronomers. Mdlle. Hortense Lepaute, the lady who has given her name to the *Hortensia*, undertook part of this laborious work. Thanks to the devotion to science of these three worthy *collaborateurs*, the work was brought to a close in November 1758, and Clairaut presented to the Academy of Sciences a memoir from which the following is a short extract:—

'The comet which has been expected for more than a year has become the subject of a curiosity much more lively than that which the public generally bestows upon questions of astronomy. True lovers of science desire its return because it would afford striking confirmation of a system in favour of which nearly all phenomena furnish conclusive evidence.

Those, on the contrary, who would like to see the philosophers embarrassed and at fault hope that it will not return, and that the discoveries of Newton and his partisans may prove to be on a level with the hypotheses which are purely the result of imagination. Several people of this class are already triumphing, and consider the delay of a year, which is due entirely to announcements destitute of all foundation, sufficient reason for condemning the Newtonians.

'I here undertake to show that this delay, far from invalidating the system of universal gravitation, is a necessary consequence arising from it; that it will continue yet longer, and I endeavour to assign its limit.'

Let us say at once that Clairaut found that the perihelion passage of the comet would be delayed 618 days, and that it would take place in 1759, a hundred days being due to the action of Saturn, and 518 days to that of Jupiter, bringing the perihelion passage to the middle of the month of April. Nevertheless, he made reservations with a modesty not less to his honour than his immense work, reservations necessitated by the terms omitted from the calculations, such as unknown causes of perturbation, and the fear that some errors might have been committed in the numerous and delicate operations performed. All these accumulated uncertainties might, according to Clairaut, make the difference of a month in the appointed time. The comet was actually seen on the 25th of December, 1758, by a Saxon peasant of the name of Palitsh in the environs of Dresden. Observations were made of the comet, and astronomers were soon able to prove that the perihelion passage would take place on the 13th of March, 1759, thirty-two days before the epoch calculated by Clairaut. Such a triumphant success of the theory produced in the scientific world a deep impression, and Lalande said with very legitimate enthusiasm:—

'The universe beholds this year the most satisfactory phenomenon ever presented to us by astronomy; an event which, unique until this day, changes our doubts to certainty, and our hypotheses to demonstration. . . . M. Clairaut asked one month's grace for the theory; the month's grace was just sufficient, and the comet has appeared, after a period of 586 days longer than the previous time of revolution, and thirty-two days before the time fixed; but what are thirty-two days to an interval of more than 150 years, during only one two-hundredth part of which observations were made, the comet being out of sight all the rest of the time? What are thirty-two days for all the other attractions of the solar system which have not been included; for all the comets, the situation and masses of which are unknown to us; for the resistance of the ethereal medium, which we are unable even to estimate, and for all those quantities which of necessity have been neglected in the approximations of the calculation? . . . A difference of 586 days between the revolutions of the same comet, a difference produced by the disturbing action of Jupiter and Saturn, affords a more striking demonstration of the great principle of attraction than we could have dared to hope for, and places this law amongst the number of the fundamental truths of physics, the reality of which it is no more possible to doubt than the existence of the bodies which produce it.'

Another return of Halley's comet took place in 1835. It furnished an opportunity of testing the progress made by theoretical astronomy during the period of seventy-six years occupied by the comet in once more performing its revolution. Taking the perihelion passage of 1759 as the point of departure, and following in the steps of Clairaut, two French astronomers, Damoiseau and Pontécoulant, independently undertook the laborious task of determining the epoch of the perihelion passage of the comet, taking into account the perturbating action of the planets. Amongst the unknown disturbing causes which Clairaut had been unable to take into account, but which entered into the researches of the two above-mentioned savants, was the planet Uranus, discovered by Sir William Herschel in 1781. According to Damoiseau the comet should have passed its perihelion on November 4; according to Pontécoulant not till November 13, 1835. Two other astronomers, Lehmann and Rosenberger, had fixed the dates of November 11 and 26. On August 5 the comet was seen at Rome. Observations gave for the exact date of the perihelion

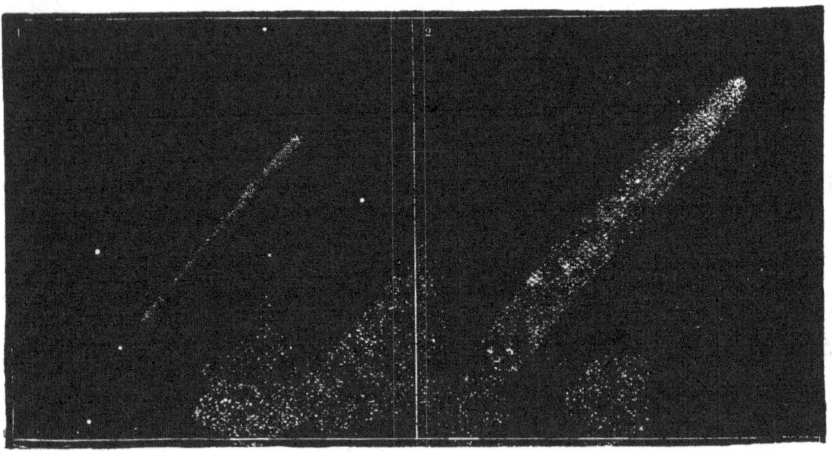

Fig. 16.—Halley's Comet in 1835. 1. As seen by the naked eye October 24. 2. As seen in the telescope the same day.

passage November 16, at half-past ten in the morning, the difference between the observed date and the mean of the calculated dates being less than *three days.** The result showed an increase of sixty-nine days above the length of the preced-

* On re-computing the disturbing influence of the planets Pontécoulant calculated that a period of 28,006$\frac{2}{3}$ days should have elapsed between the perihelion passages of 1835 and 1759. Observations proved it to be 20,006 days. The difference, which is only two-thirds of a day, shows what progress had been made both in theoretical and practical astronomy.

ing period, the new period amounting to 27,937 days; this increase arose from two antagonistic causes—

1. An increase of 135 days, 34 being due to the action of Jupiter.

2. A diminution of 66 days, 30 being due to the action of Saturn, Uranus, and of the Earth.

The duration of this last period was found equal to seventy-six years and 135 days, or seventy-six years and four months and a half. An equal period would bring the next time of perihelion passage to March 29, 1912. But this date will be subject to modification by the perturbations incident to the journey. Jupiter will exercise a considerable retarding influence, and the revolution which the comet is now accomplishing will be shorter than any yet observed—it will be 27,217 days; that is to say, hardly seventy-four years and six months. This brings the next apparition, according to the calculations of Pontécoulant, to May 24, 1910, about nine o'clock in the morning.* If, on the contrary, we look back into the past and consult old chronicles and records, we shall find several apparitions of Halley's comet, the dates of which are as follows, some nearly certain, others somewhat doubtful:—

June 8,	1456.	Halley had already given notice of this apparition.
November 9,	1378.	
In December	1301.	According to the researches of E. Biot and Laugier.
In September	1152.	According to the researches of E. Biot and Laugier.
In May	1066.	
In September	989.	
In June	760.	According to the researches of E. Biot and Laugier.
In October	684.	According to Hind.
In July	451.	According to the researches of E. Biot and Laugier.
In March	141.	
In January	66.	
In October	12 B.C.	

* The elements of the orbit calculated for this epoch by the same mathematician give the 16th of May, 1910, about 11 P.M.

In addition to these dates Mr. Hind gives the following as corresponding probably to former apparitions of the same comet: 1223, 912, 837, 608, 530, 373, 295, 218.

The period which these apparitions lead us to infer (notably those of 1456, 1378, 760, 451) amounts to about seventy-seven years and a quarter, which is longer in a marked degree than that of the three or four last revolutions.* M. Laugier asks if this diminution which we are obliged to admit is not due to the same cause as that which has been assigned to account for the similar diminution undergone by Encke's comet; that is to say, the resistance of the ether; or if, as Bessel thought, it was due to a dispersion or loss of matter abandoned by the comet in the course of its successive revolutions. These are questions of high interest, and we shall recur to them again.

* From the year 12 B.C. to the year 1835, 1,847 years have elapsed, a period comprising twenty-four revolutions of Halley's comet. The mean duration would thus be 76 years 350 days.

SECTION III.

ENCKE'S COMET; OR, THE SHORT PERIOD COMET.

Discovery of the identity and periodicity of the comets of 1818, 1805, 1795, and 1786; Arago and Olbers—Encke calculates the ellipse described by the comet—Dates of twenty returns up to 1873—Successive diminution of the period of Encke's comet.

In 1818 Pons, one of the most indefatigable of observers and comet-seekers, discovered at Marseilles a comet which passed its perihelion on the 27th of the following January. The elements of the new comet, when compared with those of comets already catalogued, gave reason to suppose that it had been observed in 1805. Arago remarked this to the Board of Longitude when Bouvard presented the parabolic elements of the new comet; and Olbers, on making the same remark in Germany, added that it was doubtless the same comet which had been observed in 1795 and 1786. We subjoin the elements of the comets of 1818 and 1805:

	1818.	1805.
Longitude of perihelion	144° 15′	147° 51′
Longitude of node	329° 5′	340° 11′
Inclination	14° 48′	15° 36′
Perihelion distance	0·353	0·378
Direction of movement	Direct.	Direct.

The resemblance was too striking not to produce attempts to determine the periodicity of the new comet. The elliptic elements of the orbit were calculated by Encke, astronomer at the Observatory of Gotha, and for this reason the comet received the name of Encke's comet; but it is also sometimes called the short-period comet, in contradistinction to Halley's comet, whose period of revolution is so much longer. The comet of Encke, in fact, describes its orbit in about 1,210 days, or three years 114 days. 'If we only consider,' says Poisson, 'the rapidity of its successive revolutions, this body might be regarded as a planet, but it continues to have a place assigned to it amongst comets, because of the *appearances which it presents*, and because *it is not visible to us throughout its entire orbit.*' In fact, at the time when Poisson wrote his report the belief in the extreme elongation of all cometary orbits still existed, and it seemed improbable that a comet should have so short a period of revolution. But successive observations of its return removed all doubt, and soon new periodical comets were discovered, which justified the possibility of cometary orbits, comparable in point of their relatively slight eccentricity with the orbits of the planets themselves.

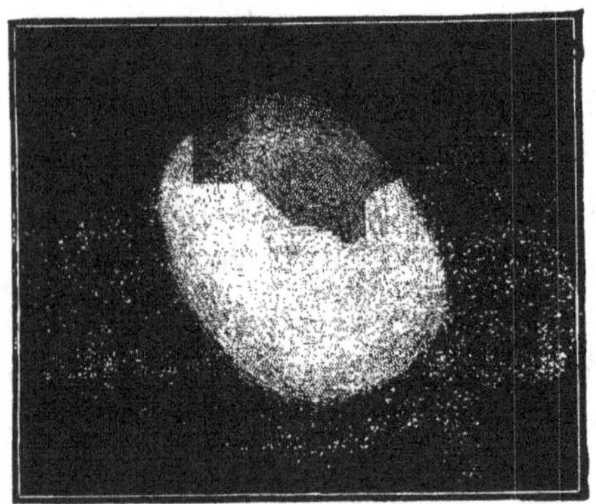
Fig. 17.—Encke's Comet, at its passage in 1838, August 13.

The first return of the short-period comet to its perihelion took place towards the end of May 1822. Encke calculated the epoch of its return, and computed an ephemeris; then, taking into account the perturbations which must have been experienced by the comet in the course of its preceding revolution, owing to its passage near to Jupiter, he showed that its period would be lengthened about nine days, and that the comet would be invisible in Europe; and in fact it was only observed in the southern hemisphere (in Australia).

We extract from the *Annuaire du Bureau des Longitudes* for 1872 the epochs of the perihelion passages of the comet from its discovery in 1818 up to its last and next passage :—

January 27	1819	November 26	1848
May 23	1822	March 15	1852
September 16	1825	July 1	1855
January 10	1829	October 18	1858
May 4	1832	February 6	1862
August 26	1835	May 28	1865
December 19	1838	September 14	1868
April 12	1842	December 29	1871
August 10	1845	April 12	1875

By adding to the preceding the previous apparitions of 1786, 1795, and 1805 we have in all twenty observed returns; but since the first date twenty-seven consecutive revolutions have really taken place. Now, if the exact intervals between the perihelion passages be noted, and the durations of the corresponding periods deduced from them, we have the following table, which was calculated by Encke:—

	D.	H.	M.
From 1786 to 1795, three times	1212	15	7
„ 1795 „ 1805 „ „	1212	12	0
„ 1805 „ 1819 four „	1212	0	29
„ 1819 „ 1822	1211	15	50
„ 1822 „ 1825	1211	13	12
„ 1825 „ 1829	1211	10	34
„ 1829 „ 1832	1211	7	41
„ 1832 „ 1835	1211	5	17
„ 1835 „ 1838	1211	2	38
„ 1838 „ 1842	1210	23	31
„ 1842 „ 1845	1210	21	7
„ 1845 „ 1848	1210	18	29
„ 1848 „ 1852	1210	17	2
„ 1852 „ 1855	1210	11	17
„ 1855 „ 1858	1210	13	41

The above list testifies to a fact of the highest importance: the period of the comet is continually diminishing. Will it continue always to diminish; and if so, what law does this continual alteration of the orbit follow? A diminution in the periodical time of a body cannot take place, according to the laws of Kepler, without a corresponding diminution in the length of the major axis, or mean distance of the comet from the sun. Are we, then, to suppose that this comet is continually approaching the focus of our world, and some day will be precipitated upon its mass? These interesting questions and this hypothesis have been studied from different points of view by several astronomers, who have endeavoured to find the physical cause of this diminution. We shall return to this point further on; it is one that concerns the whole solar world.

SECTION IV.

BIELA'S OR GAMBART'S COMET.

History of its discovery; its identification with the comet of 1805—Calculation of its elliptic elements by Gambart—Apparitions previous to 1826—Peculiarities in the apparitions of 1832, 1846, and 1872.

SEVEN YEARS elapsed between the discovery of Encke's comet and that of Biela or Gambart, which likewise may be called a comet of short period, since it performs its revolution in less than seven years.

The comet was first observed by an Austrian major of the name of Biela, at Johannisberg, February 27, 1826; it was seen ten days after at Marseilles by the French astronomer Gambart. The latter, after having calculated the elements of the parabolic orbit, immediately recognised their resemblance to those of a comet which had been observed in 1805 and in 1772. The following table affords a comparison between the elements of these three orbits:—

	1772.	1805.	1826.
Longitude of perihelion . .	108° 6′	109° 23′	104° 20′
Longitude of ascending node .	252° 43′	250° 33′	247° 54′
Inclination	19° 0′	16° 31′	14° 39′
Perihelion distance . .	1·018	0·89	0·95
Direction of movement . .	Direct.	Direct.	Direct.

I especially wish to direct attention to these comparisons as examples of the employment of the most simple method for determining the periodical orbit of a comet, a method merely suggestive and provisional, and for which direct calculations are immediately substituted. These calculations for the comet of 1826 were performed by Gambart and Clausen,* who both obtained accordant results, and assigned to the duration of the comet's revolution a period of six years and three quarters. Damoiseau, then, taking account of the perturbations, was able to predict its next return, which he fixed for the 27th of November, 1832. The comet made its appearance on the 26th, only

* It is customary to give to a periodical comet the name, not of the observer by whom it was first seen or observed, but that of the astronomer by whom the elliptic character of its orbit was first recognised. This comet that astronomers of both hemispheres persist in calling the comet of Biela ought, therefore, to bear the name of Gambart. It is not the only instance of injustice in the history of astronomy. Non periodical comets generally receive the name of their first observers—thus we speak of Donati's comet, Coggia's comet, &c. But in our opinion the best system of denomination is that of designating comets by the year in which they have effected their perihelion passage, and affixing to them a numeral, according to the order of their discovery. Thus, we say comet I., 1858; comet II., 1858, &c. This method leaves no opening for small rivalries of the kind above alluded to.

[It seems natural, and is, in fact, unavoidable, that a comet should be known by the name of the astronomer with whom it is chiefly associated, whether as calculator or observer, without there being any fixed rule in the matter. Astronomers attach a particular name to a comet, not with the view of honouring the individual, but of having a convenient name for the comet; and although the system of quoting the year and the number is admirable for the majority of comets, still in the case of those that have become celebrated and are frequently referred to, some more distinctive and easily remembered appellation is needed.

But in the present case it seems a matter of justice that the comet should be named after Biela, who not only first discovered it, but who calculated its parabolic elements, remarked their similarity to those of the comets of 1772 and 1805, and thence concluded that the orbit was elliptic, and that the period was six years and nine months. This Biela communicated to the *Astronomische Nachrichten*, in a letter dated March 23, 1826. Gambart also calculated the parabolic elements of the comet, and remarked their resemblance to those of the comets of 1772 and 1805. His letter was dated March 22, and both appear in the same number of the *Astronomische Nachrichten*. Thus Biela and Gambart independently recognised the elliptic motion of the comet, while Biela was in in addition the first discoverer.

If we adopt the rule that a comet should be named after the astronomer who first recognised its periodicity, it is clear that Faye's comet—the subject of the next section—should be named after Goldschmidt.—Ed.]

one day earlier than the date assigned. Thus was perfected more and more the theory of cometary orbits based upon the principle of universal gravitation.

Including the previous apparitions of 1772 and 1805, the comet of six years and three-quarters has been observed on seven of its returns—in 1826, in 1832, in 1846, in 1852, and in 1872. It should have been observed in 1839, 1859, and 1866. 'In 1839,' says M. Delaunay, 'it could not be observed on account of the unfavourable position of its orbit at the time of its perihelion passage.' This passage, in fact, took place during the first days of July, and both before and after the comet was situated in close proximity to the sun, and consequently lost sight of in his rays. Nearly the same thing happened in 1859, the perihelion passage taking place in the first days of June. Lastly, in 1866, although the comet could not have been far distant from the earth about the time of its perihelion (the 26th of January), and notwithstanding the diligent search made for it with powerful instruments, it was not discovered. It was last seen at Madras by Mr. Pogson, on the 2nd and 3rd of December, 1872.

Gambart's comet has furnished some curious events in the history of physical astronomy. In the beginning of 1846 the comet divided into two distinct comets, which appear at the present day in the catalogues, with their respective orbits. Moreover, in 1832, like the comet of 1773, it had the privilege of exciting fears which at that epoch were certainly without foundation. The comet was to come into collision with the earth There was more reason to believe in the possibility of such an encounter at the end of November 1872; and if it is not one of the twin comets that then just grazed the earth, it is at least one of their fragments. I here restrict myself to the simple mention of these events, which further on will receive the development they merit.

SECTION V.

FAYE'S COMET.

First comet whose periodicity, without comparison with previous dates, has been determined by calculation and verified by observation—M. Le Verrier demonstrates that it has nothing in common with the comet of Lexell—Slight eccentricity of Faye's comet and great perihelion distance—Dates of its return—Perturbations in the movements of Faye's comet inexplicable by gravitation alone: a problem to be solved.

A COMMUNICATION by Arago, published in 1844, in the *Comptes Rendus* of the Academy of Sciences, gives an account of the first researches relative to the fourth periodical comet, which we here subjoin:—

'This body was discovered at the Observatory of Paris by M. Faye, on November 22, 1843. This young astronomer hastened to calculate its parabolic elements. But as the number of observations increased M. Faye perceived that a parabola was quite inadequate to represent the series of positions occupied by the comet, and announced that he should determine its elliptic orbit, as soon as the state of the sky should have permitted him to pursue his observations of the new comet in regions so far removed from those in which it had first appeared that no doubt could possibly exist as to the certainty of his results. M. Faye therefore applied himself to the multiplying of observations, which had become extremely difficult to obtain, on account of the indistinctness of the comet. Matters were in this stage when a letter from

Schumacher informed him that Dr. Goldschmidt, a pupil of Gauss, had already calculated an elliptic orbit, having used the observations made at Paris on November 24, and those of December 1 and 9, made at Altona.'

Here, then, is a comet whose periodicity has been at once determined by calculation, and without comparison with the elements of comets previously observed. As the period of its revolution is short, less than seven years and a half (seven years 151 days), it was thought remarkable that the comet had not been seen before. But as at its aphelion it had probably passed within close proximity to the orbit of Jupiter, it was supposed that its orbit had been altered by 'that most powerful member of the solar system,' and that 'from parabolic it had become elliptic and periodic.' But in reality, according to the calculations of M. Goldschmidt, Faye's comet could not have been within sixty millions of miles of Jupiter, and this hypothesis had therefore to be abandoned. M. Valz even thought to identify the new comet with the famous missing comet of Lexell, or that of 1770. But M. Le Verrier showed that there was nothing in common between the two comets. The absence of previous observations appears, therefore, to prove no more than that at the time when its former apparition should have taken place the comet was not favourably situated for observation, a hypothesis which can easily be accepted in the case of a body so faint as to be visible only in the telescope.

Its orbit presents two rather remarkable peculiarities. In the first place, it is the least eccentric of known cometary orbits, as we have already had occasion to remark when calling attention to the ellipses described by it and the small planet Polyhymnia (fig. 15). Besides this, its perihelion distance is somewhat considerable (1·682). When passing the point of its orbit nearest to the sun the comet is twenty-seven millions of miles further removed than Mars at his perihelion, and twenty-two millions of miles further than Mars at his mean distance. At its aphelion it is beyond the orbit of Jupiter—as is the case with all the other periodical comets, with the exception of Encke's—exceeding it by more than half the distance of the sun from the earth, or by fifty millions of miles. What chiefly caused the great perihelion distance of Faye's comet to be remarked is that, in the same year, the great comet of 1843 approached the sun to within 650,000 miles of its centre, so that the nebulous nucleus was not more than 210,000 miles from the surface of the solar photosphere—less than the distance of the earth from the moon.

Faye's comet returned to its perihelion on April 2, 1851, at two o'clock in the morning, about a day after the epoch that M. Le Verrier had calculated for the time of its passage, taking into account the perturbations it had been subjected to from the disturbing influence of Jupiter. It was seen again in 1858, again in August 1865, and lastly in September 1873. The period of its revolution is 7 years 151 days, or 2,708 days; that is to say, 323 days longer than that of the planet Sylvia, the most distant from the sun of the small planets circulating between Mars and Jupiter.

On carefully studying the three first successive apparitions of Faye's comet M. Axel Möller detected variations in the orbit impossible to be explained by gravitation alone. There is need, therefore, as with Encke's comet, for enquiry into the cause of this perturbation. This interesting question will be treated in one of the later chapters which we shall devote to the hypotheses which have been suggested to account for such anomalies.

SECTION VI.

BRORSEN'S COMET.

Discovery of the comet of five years and a half period by Brorsen in 1846—Its supposed identity with the comet of 1532 gives reason to suspect elliptic elements; calculation of these elements—Returns of the comet in 1851, 1868, and 1873.

In the order of their discovery we proceed to pass in review the periodical comets of the solar system—those at least whose return has been confirmed by observation, and which have justified the predictions of calculation.

This brings us to a comet which likewise bears the name of the astronomer who discovered it, at Kiel, on February 26, 1846, viz. to Brorsen's comet, whose period is intermediate to those of Encke and Faye. It performs its revolution round the sun in five years and a half, or, more exactly, in five years 176 days, or 2,002 days.

As soon as the parabolic elements of the new comet were calculated, two astronomers, Goujon and Petersen, suspected its identity with a comet observed in 1532,* and were thus led to the calculation of an elliptic orbit; this orbit was actually determined by Goujon, by Brünnow, and later by Bruhns. The return was predicted for 1851, and the perihelion passage for November 10 of that year. But the comet was not seen till

* And also with the comet of 1661. But Brorsen's comet is now regarded as distinct from both these bodies, whose identity is suspected, but whose period is much longer.

six years later, on its return in 1857, when it made its appearance about three months before its time, for instead of passing its perihelion on June 25, as required by the ephemeris of Dr. Galen, it was observed on March 18 by Bruhns, and eight days later by M. Yvon Villarceau, at Paris, and it was only after a new calculation of its parabolic elements had been made that MM. Villarceau and Pape recognised the comet of Brorsen. The perihelion passage had taken place on March 29, three months before the predicted epoch, as just stated. This error is nothing remarkable in the sum of two entire revolutions of a body observed once only. But it explains why the comet was not seen in 1851, as the place in the heavens to which search was directed and the time of the search were widely different from the place which the comet really occupied and the time at which it passed the perihelion. Instead of November 10 the date ought to have been fixed for September 27, 1851.

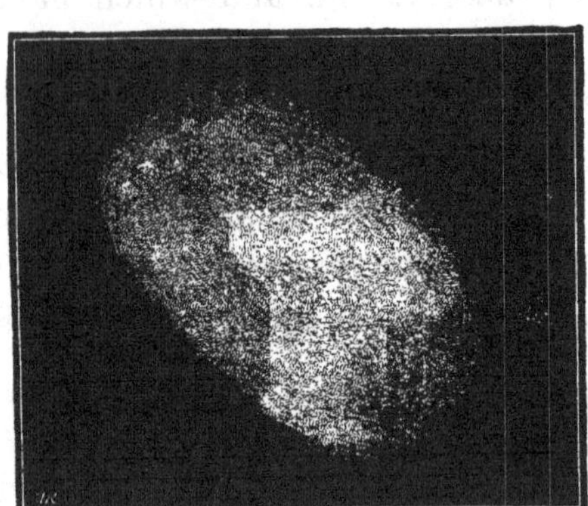

Fig. 18.—Brorsen's comet, as observed May 14, 1868, from a drawing of Bruhns.

Brorsen's comet, which was to have reappeared in 1862, 1868, and 1873, was seen only at its two last returns. In 1868 the return calculated by M. Bruhns, taking into account the perturbations due to Jupiter, for April 18, about midnight, took place instead on the 17th, about ten o'clock in the evening. Theory had reasserted its right. The comet was observed at Marseilles by Mr. Stéphan, and at Twickenham by Mr. Bishop, in the course of September and October 1873.

At its perihelion, Brorsen's comet approaches to within a distance a little greater than half the distance of the earth from the sun, viz. to within 55¼ millions of miles; at its aphelion it is beyond the orbit of Jupiter; and its distance from the sun is then more than nine times as great as its perihelion distance, viz. about 516 millions of miles. Less eccentric than the orbits of the comets of Encke, Tuttle, and Halley, the orbit of Brorsen's comet is more eccentric than that of any other of the known periodical comets.

SECTION VII.

D'ARREST'S COMET.

Discovery of the comet and of its periodicity by D'Arrest—Return predicted by M. Yvon Villarceau for 1857; verification to within half a day—Importance of the perturbations caused by Jupiter—Research of MM. Yvon Villarceau and Leveau—Return of the comet in September 1870.

Here, again, we have a periodical comet whose periodicity has been determined by calculation, and whose returns have been predicted and observed without the help of any comparison with previous comets. It bears the name of the astronomer who discovered it in 1851, and who recognised the periodicity of its orbit. M. Yvon Villarceau had drawn the same conclusion, and calculated the ephemeris for its next return to perihelion, which he announced for the end of 1857, a prediction verified to within twelve hours. The new comet was seen again at the Cape of Good Hope by Sir Thomas Maclear. On its following return, which took place in 1864, astronomers were less fortunate, and were unable to perceive the comet, whose position in the heavens and distance from the earth were very unfavourable. In 1870 the perihelion passage of the comet took place on September 23; it was observed three weeks before by M. Winnecke, thanks to the ephemeris calculated by MM. Yvon Villarceau and Leveau.

'Of all the comets which have not failed to return to us,' says M. Yvon Villarceau, 'the comet of D'Arrest is perhaps the most interesting in regard to its perturbations. I do not think that any other comet has been so closely followed by Jupiter.' These perturbations, which the above-named astronomer had calculated for 1864, had increased by more than two months the duration of the first revolution, the comet being situated in 1862 at a distance from Jupiter equal to three-tenths of the distance of the sun from the earth, or at a distance of about twenty-seven millions of miles. They were calculated with great care by M. Leveau for the ensuing period, and it is doubtless owing to this great work, the labour of three years, that observations of the comet were rendered possible at its apparition in 1870. The comet, in fact, passed its perihelion on September 23 of that year. We enter into these details to show the difficulties of cometary astronomy, and how science is able, if not always to surmount them, at least to diminish them considerably.*

D'Arrest's comet describes its orbit in a little less than six years and a half (6·567 years), or in 2,398 days, only thirteen days more than the planet Sylvia. Next to Faye's comet, its orbit has the smallest eccentricity, or, in other words, the least elongated figure.

* [M. Leveau has since performed the calculations for the next revolution of the comet, and has given an ephemeris for every twentieth day throughout the year 1877. The perihelion passage is found to occur 1877, May 10·339 Paris mean time, and the comet will attain its maximum intensity of light about a fortnight later. It will be nearest to the earth in the middle of October, when its distance from us will be about one hundred and thirty millions of miles. It will probably be a very faint object.—ED.]

SECTION VIII.

TUTTLE'S COMET.

The period of Tuttle's comet is intermediate to that of Halley's comet and those of other periodical comets that have returned—Very elongated orbit of the comet of $13\frac{2}{3}$ years period—Previous observation in 1790; five passages not since observed—Next return in September 1885.

THE periodical comets of which we have just given an account, and that of Winnecke, which we shall next describe, may be considered, that of Halley excepted, as comets of short periods. Tuttle's comet, discovered sixteen years ago by the American astronomer of that name, is intermediate to Halley's comet of long period and the others. It performs its revolution in $13\frac{2}{3}$ years, or more exactly in 13 years 296 days, or about 5,044 days—a period nearly two years longer than that which Jupiter occupies in his revolution. But it describes a very elongated orbit, so that at its aphelion it is removed from the sun a distance exceeding ten times its distance at its perihelion; it penetrates depths of space that are even beyond the orbit of Saturn; in fact, it attains the distance of nearly 955 millions of miles; at the perihelion it is about as far distant from the sun as is the earth.

Tuttle's comet was first observed in 1790 by Méchain, who discovered it, and by Messier, and it was the comparison of the parabolic elements of the comets that caused their identity to be recognised. From 1790 to 1858 there is an interval of sixty-eight years; that is to say, five times the duration of the periodic revolution of the comet, which must, therefore, have returned to its perihelion, without having been seen, in 1803, 1817, 1830, and 1844.

To the calculation of the elliptic elements, performed by M. Bruhns, we owe our knowledge of the exact period and the prediction of the return of the comet in the year 1871. It was, in fact, observed at Marseilles on October 13 of that year, and afterwards at Carlsruhe, at Paris, and at the Cape of Good Hope. It passed its perihelion on November 30. Leaving out of consideration the perturbations the comet may have to experience in the course of its present revolution, the next return of Tuttle's comet may be expected in the middle of September 1885. But, as with all other periodical comets, the date may be somewhat modified* under the influence of the planetary attractions, and the consequent disturbance of the orbit.

* Tuttle's comet, it should be observed, moves in an orbit the inclination of which is considerable—it exceeds 54°; consequently, in withdrawing from the sun and penetrating to the distances of the larger planets, Jupiter and Saturn, the comet moves further and further from the paths which they pursue. The disturbing influence of the masses of these planets upon the comet would, therefore, in any case, be inconsiderable.

SECTION IX.

WINNECKE'S PERIODICAL COMET.

Discovery of the periodicity of the third comet of 1819; calculation of its elliptic elements by Encke—Discovery of Winnecke's comet in 1858; its identity with the comet discovered by Pons—Return of the star to its perihelion in 1869; probable date of its next return in 1875.

In 1819 Pons discovered, at Marseilles, a comet the elliptic elements of which were afterwards calculated by Encke; these elements assigned to it a period of $5\frac{6}{10}$ years. Now, in March 1858 M. Winnecke discovered, at the Observatory at Bonn, a new comet, whose parabolic elements, it was soon ascertained, bore considerable resemblance to those of the comet discovered by Pons. To determine if this identity were real, it was necessary to wait for the comet's return in 1863 and 1869. It was actually seen in the month of April of the latter year by M. Winnecke himself, and passed its perihelion on June 30. The date of its next return is, therefore, approximately fixed for the month of April 1875.* It will be requisite, however, as with all comets liable to approach Jupiter or the other planets, to allow for the perturbations it may have to undergo.

From its first apparition, in 1819, to its last, in June 1869, is an interval of fifty years, corresponding to nine revolutions of the comet. Three passages only, as we have seen, have been

* [It was detected by M. Borrelly, at Marseilles, on the morning of February 2, 1875.—Ed.]

observed; seven have taken place unperceived. But the elliptic orbit is now determined with precision. Observers are numerous and vigilant, and the comet will doubtless no longer escape the researches of astronomers, except when its apparent proximity to the sun and its distance from the earth are such as not to admit of its being seen.

The period of the revolution of Winnecke's comet is 2,042 days, only forty days more than that of Brorsen's comet; the eccentricity of its orbit is somewhat less. In perihelion the comet is situated at a distance from the sun equal to four-fifths of the distance of the earth; in aphelion it exceeds the orbit of Jupiter by about one-fifth of this distance.

SECTION X.

TEMPEL'S SHORT PERIOD COMET.

Calculation of the elliptic elements of the second comet of 1867, discovered by Tempel—Perturbations due to Jupiter, and consequent delay in the return of the comet to its perihelion in 1873—Remarkable agreement of observation and calculation.

THE second comet of 1867, discovered by M. Tempel, was found by several astronomers to have elliptic elements. It passed its perihelion on May 23, 1867, and its period had been calculated at 2,064 days. But Dr. Söllinger, taking into account the perturbations its passage in the vicinity of Jupiter would produce in the elements of its orbit, assigned a retardation of 117 days in the date of its return to perihelion in 1873. It was, in fact, seen again in the course of that year, and its perihelion passage took place on May 9, which gives for the duration of the revolution performed in the interval between the two apparitions a value of very nearly six years, or 2,178 days, three days less than the number determined by calculation.

Tempel's comet of short period is, therefore, the ninth periodical comet whose return has been verified by observation; that is to say, which really forms an integral part of our solar system. Observed in May 1873, at Greenwich, by Messrs. Christie and Carpenter, it appeared in the telescope like an elongated nebulosity, about 40″ in diameter, with a central nucleus, which shone like a star of the twelfth or thirteenth magnitude.

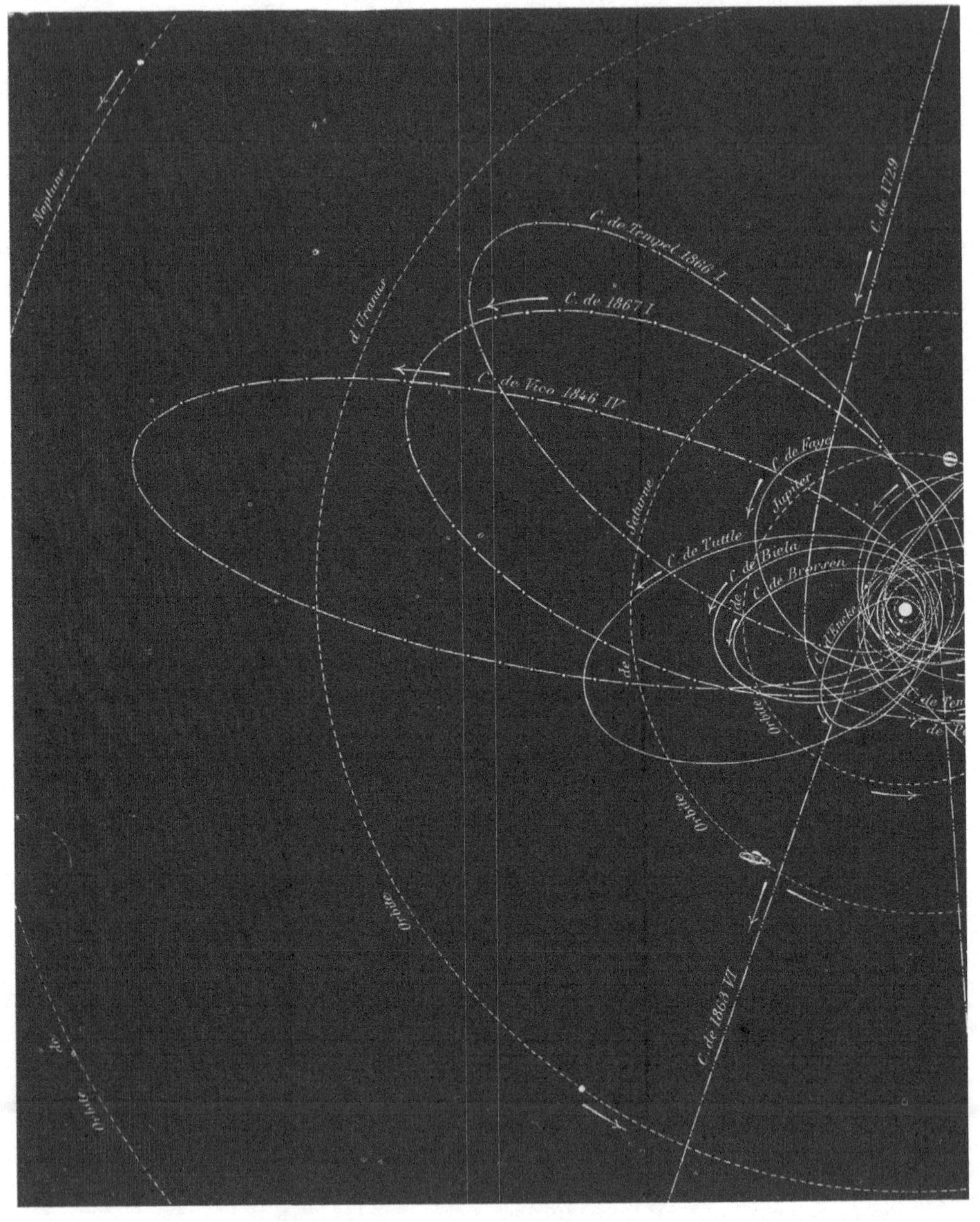

ORBITS OF PERIODICAL COMETS.

Plain line ———— Nine recognised Periodical Comets, whose returns have been observed.

Dotted line - - - - - - The Plane's.

" " —.—.— Periodical Comets whose returns have not yet taken place.

" " —..—..— Comets with Hyberbolic Orbits.

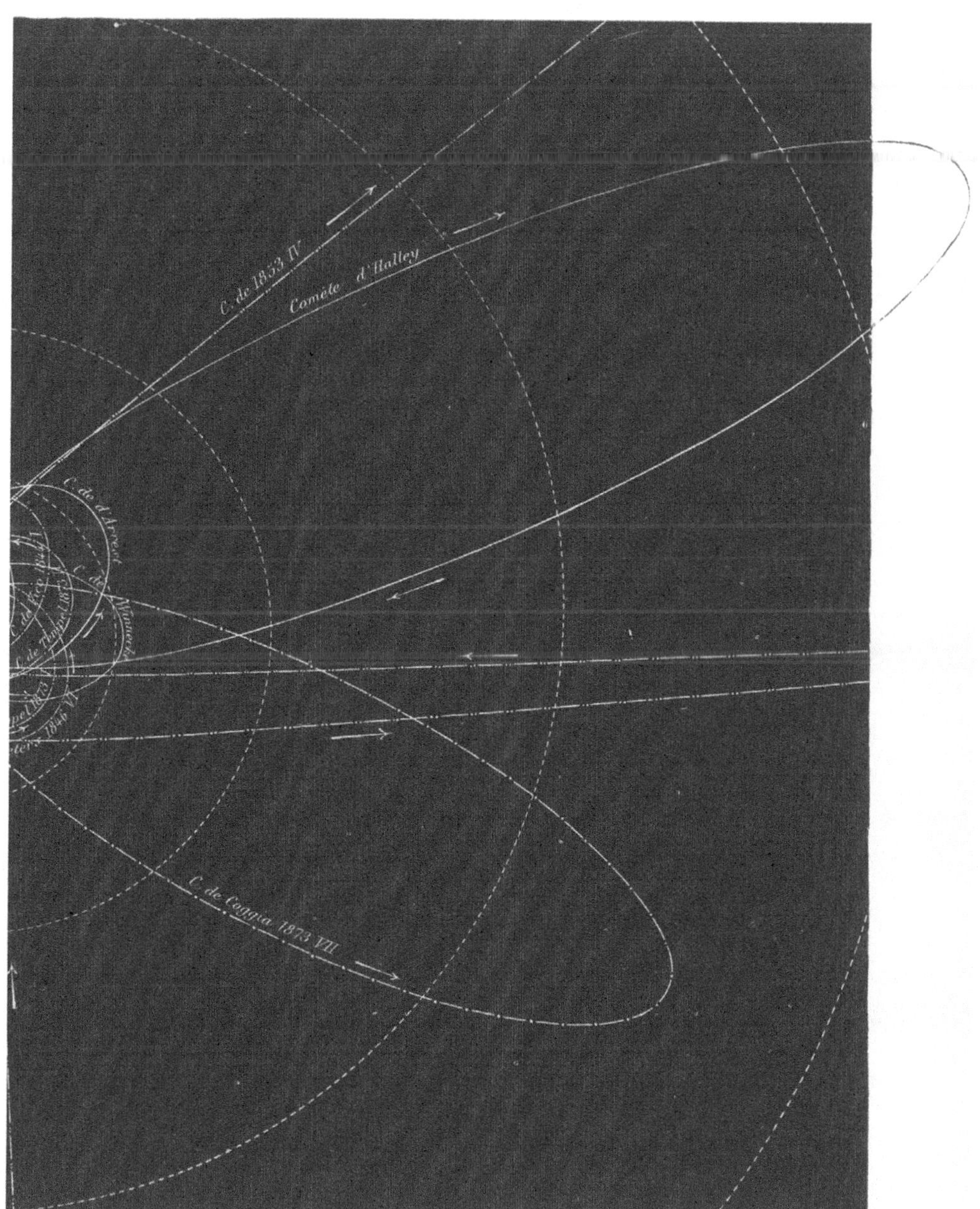
C. de 1853 IV
Comète d'Halley
C. de d'Arrest
C. de Coggia 1873 VII

CHAPTER V.

PERIODICAL COMETS.

SECTION I.

COMETS WHOSE RETURN HAS NOT BEEN VERIFIED BY OBSERVATION.

Periodical comets which have not been seen again; long periods; circumstances unfavourable to observation; motions possibly disturbed by perturbations—Elliptic orbits determined by calculation—Uncertainty of return under these different hypotheses.

THE nine comets of which we have just given an account are up to the present time the only comets which can be considered as certainly belonging to our system. But they are not the only comets which regularly perform their revolutions round the sun. Of the numerous comets moving in apparently elliptic orbits some, we shall now see, have been regarded as new apparitions of comets previously observed, the great resemblance of their parabolic elements having caused their periodicity to be suspected. But either their return to perihelion has not yet taken place, or circumstances favourable to observation have not occurred; or, an equally likely hypothesis, they may have been disturbed in their courses by the vicinity of the planetary masses, producing perturbations powerful enough to change their periods, or even to cast them out of the sphere of the sun's attraction, of which perhaps until then they had formed for a time a part.

Other comets, which have not been assimilated to comets already observed, have elliptic orbits determined by calculation;

but for the reasons that we have just enumerated they have not been seen again; that is to say, they have periods much too long, or they have been subjected to disturbing causes.

We will now pass in review the principal comets of these two categories, and for greater clearness we will divide them into three classes:—

1. *Comets of short period*; that is to say, those which perform their revolution in a few years, like the eight periodical comets above described. All comets of this first class are interior comets, because their orbits do not exceed the known boundaries of the planetary orbits; in other words, because at their aphelia they are at a less distance from the sun than Neptune.

2. *Comets of mean period*; that is to say, those which describe their orbits in less than two centuries, like Halley's comet. Comets of this class are not interior comets.

3. *Comets of long period*, whose revolution exceeds two centuries, and may amount to hundreds of thousands and even to millions of years. Comets of this class penetrate to remote depths of space far exceeding the limits of the solar system.

SECTION II.

INTERIOR COMETS, OR COMETS OF SHORT PERIOD, THAT HAVE NOT YET RETURNED.

Comets lost or strayed: the comet of 1743; the comet of Lexell, or 1770; perturbations caused by Jupiter; in 1767 the action of Jupiter shortens the period, and in 1779 produces an opposite effect—Comet of De Vico; short period comets of 1783, 1846, and 1873.

DURING the month of February 1743 a comet was observed at Paris, Bologna, Vienna, and Berlin whose parabolic elements were calculated by Struyct and Lacaille. A mathematician of our time, M. Clausen, identified it as a comet of short period, performing its revolution in five years and five months. Is this, as has been supposed, the same comet that was seen in November 1819? If so, its period must have greatly changed, since the calculations of Encke assign to the latter a period of about four years and ten months.

The comet of which we are now about to speak is celebrated in the history of astronomy. The following extract from a memoir by M. Le Verrier gives an account of the circumstances of its first apparition:—

'Messier perceived, during the night of the 14–15th of June, 1770, a nebulosity situated amongst the stars of Sagittarius, but not discernible by the naked eye; it was a comet first coming into view. On the 17th of June it appeared surrounded by an atmosphere the diameter of which was about

5′ 23″. In the centre appeared a nucleus; its light was bright, like that of the stars. Messier estimated its diameter at 22 seconds.

'The comet rapidly approached the earth. On the 21st of June it was visible to the naked eye, and three days after it shone like a star of the second magnitude. The diameter of the nebulosity, which was not more than 27′, gradually increased, and by the night of the 1st–2nd of July had attained 2° 23′. But, whilst the diameter of the nebulosity thus increased, according to the laws of optics, in the inverse ratio of the distance of the body from the earth, the diameter of the supposed nucleus remained, on the contrary, nearly invariable.

'From the 4th of July the comet became lost in the rays of the sun and remained for a short time invisible. Pingré, from the observations of Messier, calculated a parabolic orbit. The comet, it was found, would again become visible in the month of August, and Messier was once more able to observe it on the 4th of that month. From this date it was seen almost without interruption; but as it gradually drew further away from the sun and the earth it faded out of sight, and was lost to view in the first days of October.

'Before the time of the perihelion passage no indication of a tail had been perceived. But from the 20th of August to the 1st of September the comet was provided with a faint tail, the length of which was about one degree.

'The parabolic elements given by Pingré were in accordance with the first observations, but they differed greatly from the last. Nor were other elements calculated by Slop, Lambert, Prosperin, and Widder more exact. The discrepancies were generally referred to the derangement of the orbit caused in June by the action of the earth. Prosperin, however, suspected that the orbit of the comet might be elliptic, but he contented himself with the hypothesis and did not verify it. Lexell at last discovered that the comet was moving in an elliptic orbit, which it described in 5·585 years (a little more than five years and a half): and rejecting, with Dionis du Séjour, the supposition that the disturbing action of the earth could have considerably changed this orbit, he proved, first, that the observations were all satisfied by a period of five years and a half; second, that it was impossible to admit a period of five or six years without introducing differences into the theory incompatible with the observations.

'"But," said Messier, "if the duration of the comet's revolution is only five years and a half, how is it that it has only once been observed? This is a very strong objection to the researches of M. Lexell."

'Lexell replied, "As the aphelion distance of the comet from the sun is nearly equal to the distance of Jupiter from the sun, a suspicion arises that the comet may have been disturbed in its movement by the action of Jupiter, and that at one time it described an orbit altogether different from that in which it moves at present. It is found by calculation that this comet was in conjunction with Jupiter on the 27th of May, 1767, and that the distance between the comet and the planet was only $\frac{1}{580}$th of the distance of the comet from the sun; hence we may conclude, bearing in mind the masses of Jupiter and the sun, that the action of Jupiter has been powerful enough to change the movement of the comet in a sensible degree." Lexell further pointed out that the comet would be in proximity to Jupiter about the 23rd of August, 1779, and that this circumstance might perhaps prevent the comet returning to its perihelion in 1781, as it would do if undisturbed by perturbations. And, in fact, the astronomical world vainly awaited the return of the comet in 1781 and 1782.'

Since the end of the last century Lexell's comet, or the

comet of 1770, has not been seen. It is, therefore, a comet lost or gone astray, and it is easy to conceive the fascination of the problem offered to astronomers by the fact of its disappearance. Several mathematicians following Lexell have attempted its solution—Burckhardt, Laplace, and M. Le Verrier himself. According to Laplace the action of Jupiter in 1767 rendered the comet visible by diminishing its perihelion distance, and the same action in 1779, by increasing the same distance, has rendered it invisible for ever.

M. Le Verrier has discussed the question anew; he subjected to enquiry the amount of the perturbations the earth would have caused in 1770, when the comet approached it to within a distance not exceeding seven times that of the moon. Proceeding next to estimate the disturbing influence of the mass of Jupiter upon the comet at its aphelion passage in 1779 and during the next two years, he has shown that the perturbations produced during the twenty-eight last months were considerable and must have completely changed the orbit of the comet.

'From the 28th of May,' he proceeds, 'the comet was rapidly approaching Jupiter in a hyperbolic orbit, and it is impossible that the comet should have become a satellite of Jupiter, as has been supposed.' Did it strike against that powerful mass, or, at all events, did it traverse the regions where the four satellites describe their orbits? To these questions M. Le Verrier replies that 'it is not absolutely impossible, but that it is very improbable, and that conclusions in the affirmative, based upon the diminutive size of the comet's mass, are very hazardous.'

According to his first memoir there is every reason to believe that the comet of 1770 has not been carried away from our solar system.

But if so, would it not have been seen again? Among the

number of comets which have been seen since 1780 are there none identical with Lexell's comet? It is not surprising that no comet has appeared with elements similar to those of Lexell's comet, if we bear in mind the perturbations which it underwent during its return to aphelion, in 1779. A first examination seemed to indicate a similarity between the comets of Lexell and Faye. But M. Le Verrier has demonstrated the contrary, by tracing back the course of the latter comet before its discovery in 1843, and has shown that 'we must go back to the year 1747 for the time when the comet of Faye began to describe (under the influence of Jupiter) the contracted ellipse in which it moves at the present day.'

Here, then, is a comet which is as lost to our world, or at least to astronomers, for even if it should return how would it be possible to recognise its identity?

A comet was discovered on August 22, 1844, at Rome, by the astronomer De Vico. The elliptic elements calculated by Faye and Brunnow proved that its period of revolution was five years and a half only (or more exactly, 1,996 days); so that, but for perturbations, the comet would return to its perihelion in February 1850, again in August 1855, in January 1861, July 1866, December 1871, and to its next perihelion in June 1877. It has not been seen, however, either on the first of its probable returns or subsequently. It is, therefore, at all events, a comet that has gone astray. It has also been supposed that the comet of De Vico is a new apparition of that of Lexell, with which it has points of vague resemblance. Acccording to M. Le Verrier the two comets are entirely distinct. Nor does he admit the conclusions of Mauvais and Laugier, who consider the comet of 1844 identical with that of 1585. But he considers it very probable that the comet observed in 1678 was De Vico's comet. The following are his conclusions on this point, written in 1847, before the time fixed for the first return of De Vico's comet:—

'The comet of 1844 might have come to us, as others have come, from the furthest regions of space, and have been attached to our system by the powerful influence of Jupiter. The date of its arrival may doubtless be referred back many centuries. Since that epoch it has often passed in the neighbourhood of the earth, but in all this time it has only once been observed—166 years before the apparition of 1844 (viz. at the apparition of 1678, mentioned above). This comet will for a length of time move in the restricted orbit it now describes. In a certain number of ages, however, it will again reach the orbit of Jupiter, in a direction opposite to that in which it may have first entered the solar system, and its course will once more be altered. Perhaps Jupiter himself will restore it to the regions of space from which he had previously appropriated it.'

At the epoch of its greatest visibility, which took place in September, the comet of De Vico was for several days perceptible to the naked eye. In the telescope it presented a remarkable peculiarity: the nebulosity, which was fan-shaped, contained a circular nucleus, pretty well defined; it had a tail—of bluish tint, but of no great length—which pointed from the sun.

Amongst the comets of short period whose orbits have been calculated, but which have not returned, we have yet to mention the comet of 1766, which, according to Burckhardt, performs its revolution in five years, and is perhaps a previous apparition of the comet discovered by Pons in June 1819. This latter, according to Encke, would have a period of 5·6 years, the orbit having in the interval been changed by the planetary perturbations. Next comes the comet discovered by Peters on the 26th of June, 1846. Its time of revolution is about sixteen years, but it was not seen in 1862, and will have to be again looked for in 1878. Lastly, a comet was seen at Marseilles in 1873 by M. Stéphan, moving in an elliptic orbit, and with a period of 1,850 days, or a little more than five years. In the course of 1878 this comet may, therefore, be expected; and should it reappear it will be the tenth periodical comet belonging to the solar system, whose return has been observed, or even the eleventh, if the comet discovered by Peters should also return.

Four other periodical comets (one of which we shall refer to again when we explain the connexion existing between comets and shooting stars) must be placed amongst the number of interior comets—we can hardly say of short period comets—not yet seen again. One is Tempel's comet I., 1866, which performs its revolution in a period of 33·176 years, or thirty-three years and sixty-four days. It passed its perihelion on January 11, 1866, and will consequently return in the spring of 1899. This comet approaches the sun to within a distance rather less than that of the earth; but at its aphelion it is far beyond the orbit of Uranus. The comets of 868 and 1366 are very probably anterior apparitions of this thirty-three years comet. Since the first of these dates it has, therefore, returned to its perihelion twenty-nine times without having been perceived, and has thus effected at least thirty entire revolutions round the sun. A second comet—the first of the year 1867—has also a period of more than thirty-three years (33·62 years, or rather more than thirty-three years and a half). At its aphelion distance, which is equal to nineteen and one-third times the distance of the sun from the earth, it is far beyond the orbit of Uranus, but at its descending node (which it passes through about 5,800 days after perihelion passage) the two orbits are situated very near together, the distance being scarcely 2,237,000 miles. In 1817, but more especially in

1649, the comet in passing through its node was situated in the close vicinity of Uranus, and hence must have been produced very considerable perturbations in the movement of the former. It should be looked for again in the year 1900. Lastly, there are two comets with nearly the same period of fifty-five years. The one was discovered in 1846 by De Vico, and should effect its first return in the year 1902; at its aphelion it withdraws to a distance nearly equal to the radius of the orbit of Neptune; it is thus an interior comet. The other comet of fifty-five years, which likewise does not pass beyond the orbit of Neptune, was discovered in 1873, by M. Coggia, at Marseilles; it is suspected to be identical with a comet observed by Pons in 1818.

SECTION III.

COMETS OF MEAN PERIOD.

Periodical comets exterior to the solar system; the type of this class is Halley's comet, which is the only comet of mean period whose return has been verified by observation—Enumeration of comets with periods between 69 and 200 years—Periods; aphelion and perihelion distances.

Of the comets belonging to this class Halley's comet is the type; but it is the only one of which we have several undisputed apparitions. When a comet is suspected to be identical with some other comet that has been previously observed, from the similarity of the parabolic elements, its return is probable; but as a rule great uncertainty attaches to the length of the period, even if, assuming the identity of the two comets, the perturbations be left out of the question. A third apparition is, therefore, generally necessary before the identity and real periodicity of a comet can be affirmed. And this third element up to the present time is wanting in the comets we are now engaged upon. But it will evidently suffice to prove a second apparition, when the elliptic elements have been calculated solely from observations of the first apparition.

The following, in the order of their discovery, are the nine comets of mean period which we have to mention:—

The first on the list is the comet of 1532, observed by Apian and by Fracastoro, 'whose head,' says the latter observer, 'was three times larger than Jupiter, with a beard two fathoms long.' According to the calculation of Olbers this comet has a period of 129 years, and is identical with a comet which appeared in 1661, and at several other remarkable epochs. Sir John Herschel mentions it in the following manner: 'In 1661, 1532, 1402, 1145, 891, and 243 great comets appeared, that of 1402 being bright enough to be seen at noonday. A period of 129 years would reconcile all these appearances, and should have brought back the comet in 1789 or 1790. That no such comet was observed about that time is no proof that it did not return, since, owing to the situation of its orbit, had the perihelion passage taken place in July it might have escaped observation.' Its next return should take place between 1918 and 1920.

A comet observed by the English astronomer Flamsteed, from the end of July to the commencement of September 1683, has, according to the elliptic elements calculated by Clausen, a period of about 190 years. This comet, whose return ought to have been observed about 1870, has failed to reappear; but the perturbation caused by the larger planets might occasion a delay of several years, and its re-appearance may still be expected. At its aphelion it recedes far beyond the orbit of Neptune, surpassing it by 497 millions of miles. A new discussion of the observations, however, by Mr. Plummer affords a presumption that the comet of 1683 describes a parabolic orbit, in which case it should be removed from the list of periodical comets.

About the years 1882 and 1887 search will have to be made for two comets, the first of which, discovered by Pons, in July 1812, has a period of about seventy-one years, and the second, discovered by Olbers, in March 1815, has a period of about seventy-four years, as calculated by Bessel. The return of the comet of 1815 to its perihelion would be accelerated two years by the perturbations of the planets. Then comes the comet discovered in February 1846 by De Vico and Bond; a period of seventy-three years would bring it back to perihelion about the middle of the year 1919. Next we have the comet discovered by Brorsen (July 1847), with a period of seventy-five years, to return in 1922; that of Westphal (July 1852), with a period of sixty-one years, and next return in 1913; that of Secchi (comet I., 1853), whose period would be 188 years, and which, according to Mr. Hind, has a great resemblance to the comet of 1664. Lastly, the third comet of 1862, of whose connexion with the meteor stream of August 10 we shall have to speak in a later chapter. This comet has a period of about 120 years; its next appearance, therefore, should be expected about 1982.

We now append in order, according to the duration of their periodical revolutions, a list of the nine comets above enumerated. They would be ten in number, if Halley's comet, which we have taken as a type of this class (and be it noted that this division into classes is quite arbitrary), had not been placed amongst comets of verified return. We also add to this enumeration their greatest and least distances from the sun, expressed both in mean radii of the earth's orbit and in miles:—

Comet.	Period.	Perihelion distance.		Aphelion distance.	
		(radii.)	(miles.)	(radii.)	(miles.)
1852 II.	61 years	1·25	115,800,000	29·61	2,713,200,000
1812	71 ,,	0·78	70,800,000	33·40	3,060,900,000
1846 III.	73 ,,	0·66	60,500,000	34·40	3,152,900,000
1815	74 ,,	1·21	110,600,000	34·10	3,125,600,000
1847 V.	75 ,,	0·49	46,600,000	35·10	3,217,500,000
1862 III.	120 ,,	1·01	92,600,000	48·70	4,463,300,000
1532	129 ,,	0·61	55,900,000	48·05	4,403,800,000
1853 I.	188 ,,	1·03	94,400,000	65·02	5,959,100,000
1683	190 ,,	0·55	50,400,000	65·50	6,003,900,000

SECTION IV.

COMETS OF LONG PERIOD.

Periodical comets exterior to the known limits of the solar system—Distance to which the comet of longest calculated period recedes from the sun—The so-called comet of Charles V.: its apparitions in 1264 and 1556; its return predicted for the middle of the nineteenth century, between 1848 and 1860—Calculation of the perturbations; another comet lost or strayed—The great comet of 1680: the Deluge and the end of the world—Magnificent comets of 1811, 1825, and 1843.

Of the comets we are now about to mention, the periods of which have been calculated approximately, none certainly will be seen by anyone now living. One alone was expected about fifteen years ago; and if it really did return to its perihelion, in spite of all researches it was not observed, and it will not be visible again until after the lapse of three centuries.

We will begin by enumerating the comets, and will afterwards give some details about the most remarkable of them. The following tables contain the durations of their revolutions and their distances from the sun, expressed in radii of the terrestrial orbit:—

	Probable previous appari-tions	Period of revolutions	Distances from the Sun	
			Perihelion	Aphelion
Comet of 1845 III.	1596	249 years	0·401	78·38
„ 1556	1264	292 „	0·500	87·53
„ 1840 IV.	—	344 „	1·481	96·76
„ 1843 I.	—	376 „	0·006	104·28
„ 1846 VI.	—	401 „	0·633	108·21
„ 1861 I.	—	415 „	0·921	110·40
„ 1861 II.	—	422 „	0·822	111·70
„ 1793 II.	—	422 „	1·495	111·03
„ 1746	1231	515 „	0·950	127·55
„ 1840 III.	1097	743 „	0·742	163·20
„ 1811 II.	—	875 „	1·582	181·44
„ 1860 III.	—	1,000 „	0·292	211·30
„ 1807	—	1,714 „	0·646	286·07
„ 1858 III.	—	1,950 „	0·578	311·40
„ 1769	—	2,090 „	0·123	326·80
„ 1827 III.	—	2,611 „	0·138	379·10
„ 1846 I.	—	2,721 „	1·481	388·32
„ 1811 I.	—	3,065 „	1·035	421·02
„ 1763	—	3,196 „	0·498	434·32
„ 1825 III.	—	4,386 „	1·241	534·64
„ 1864 II.	—	4,738 „	0·909	563·30
„ 1822 III.	—	5,649 „	1·145	618·15
„ 1849 III.	—	8,375 „	0·895	812·73
„ 1680	—	8,813 „	0·006	855·28
„ 1840 II.	—	13,866 „	1·221	1,053·00
„ 1847 IV.	—	43,954 „	1·767	2,489·03
„ 1780 I.	—	75,838 „	0·096	3,974·88
„ 1844 II.	—	102,050 „	0·855	4,366·74
„ 1863 I.	—	1,840,000 „	0·795	29,989·00
„ 1864 II.	—	2,800,000 „	0·931	40,485·00

We scarcely need warn the reader that the periods of the comets in this division are far from being well determined. In some cases the periodicity has been determined from the similarity presented by the elements of the orbits to those of preceding comets, and in others by a direct calculation of the elliptic elements. But even when the calculation rests upon a very sure basis, as is the case with several, it must be remembered that the next returns deduced from the periods given are subject to modification from the casualties of the voyage; that is to say, from perturbations which may be experienced on the journey from known and unknown planets of the solar system.

The second comet of the foregoing table is interesting, not only from an astronomical but also from an historical point of view. The following are some details concerning the history of its apparitions.

About the middle of July, after sunset, there appeared in France, in the year 1264, a comet, which Pingré, in his *Cométographie*, calls a '*great* and celebrated comet.' Several causes contributed to its celebrity At the epoch of its first apparition superstitious beliefs in cometary influences were still rife, and, as we may well believe, these were not diminished by this apparition, for after exhibiting itself in Europe for two months and a half it disappeared on October 3, 'the very day on which Pope Urban IV. died.' Eye-witnesses who attest this fact did not fail to conclude 'that it had only appeared to announce this death.' In the last century Dunthorne and Pingré calculated the parabolic elements of the comet's orbit, which they found bore great resemblance to those of the comet of 1556. 'The comet of 1264,' says Pingré, 'is very probably the same as that of 1556;

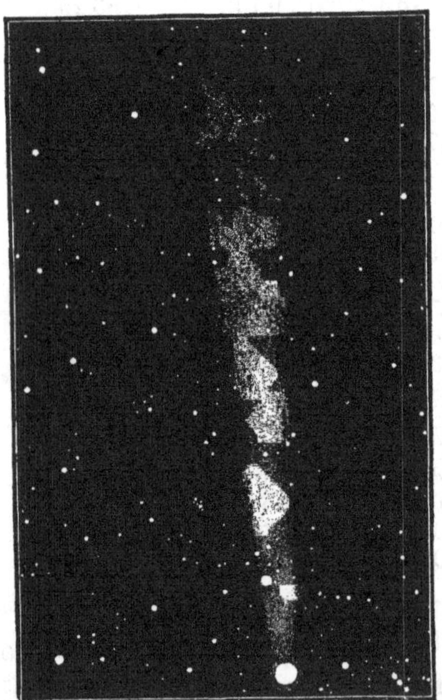

Fig. 19. Great comet of 1264, from Theatrum Cometicum of Lubienietzki.

its periodical revolution is about 292 years, and its return may consequently be expected about 1848.'

This identity, which till quite recently was considered beyond dispute, would make the comet in question a most formidable body, since, after presiding at the death of a Pope, it came to decide the abdication of a famous sovereign. It is known in history as the comet of Charles V., and is thus mentioned in the *Cométographie*:—

'The apparition of this comet produced, according to several writers, a very singular effect. It struck terror into the Emperor Charles V.; this prince doubted not his death was at hand, and is said to have exclaimed—

His ergo indiciis me mea fata vocant.

This verse has been translated into French:—

Par la triste comète,
Qui brille sur ma tête,
Je connois que les cieux
M'appellent de ces lieux.

For this translation, which is open to improvement, Pingré has proposed to substitute the following:—

Dans ce signe éclatant je lis ma fin prochaine.

Be this as it may, 'if the historians I have quoted,' Pingré continues, 'are to be believed, the panic contributed not a little to the design which Charles V. formed, and executed a few months later, of ceding the imperial crown to his brother Ferdinand; he had already renounced the crown of Spain in favour of his son Philip. If this account be true, the fact deserves to be added to the numbers of great events produced by very little causes.'

But is it true? The tradition, it is certain, was current a few years ago, as the following passage will testify, taken from a lecture given by M. Babinet at one of the public *séances* of the Institute: 'In 1556 a great and beautiful comet appeared. Charles V., who had hitherto delayed his abdication, hesitated no longer. To him, as the greatest of living sovereigns, the comet was addressed. The influence which menaced the emperor would, he hoped, be divested of evil for the private individual—would fall harmless upon the monk.' Was this, we may ask, the decisive reason that determined the famous emperor to retire to the cloisters of St. Justus? Such is not the opinion of M. Mignet, who has established the fact that Charles V. abdicated in 1555, and that consequently 'it was not the fear of the hairy star of 1556 which caused him to descend from the throne.'

Leaving history, however, let us return to science, and to the scientific reasons which have drawn the attention of savants to the comet of 1264 and 1556. It has just been seen that in the eighteenth century its return was predicted for the year 1848. Encke believed its return was possible in 1844, and that comet III. of 1844 was identical with that of Charles V. Mr. Hind, on the contrary, having calculated the elliptic elements of comet III., assigned to it a period of 41¾ years. But this period may not be incompatible with the three dates of 1264, 1556, and 1844; for the first interval, 1264–1556, would give seven periods of 41¾ years; and the second, 1556–1844, would give, pretty nearly, the same number of revolutions. An acceleration of four years, in so long a time, distributed moreover amongst several successive revolutions, is not at all inadmissible. Sir John Herschel, in the sixth edition of his *Outlines of Astronomy*, published in 1858, expresses himself in these terms on the subject of the supposed identity of the comets of 1264 and 1556: 'Mr. Hind,' he says, 'has entered into many elaborate calculations, the result of which is strongly in favour of the supposed identity. This probability is further increased by the fact of a comet, with a tail of 40° and a head bright enough to be visible after sunrise, having appeared in 975; and two others having been recorded by the Chinese in 395 and 104. It is true that, if these be the same, the mean period would be somewhat short of 292 years. But the effect of planetary perturbation might reconcile even greater differences; and though even to the time of our writing (1858) no such comet has yet been observed, two or three years must yet elapse, in the opinion of those best competent to judge, before its return must be considered hopeless.'

Let us finish the history of this celebrated comet and the efforts that have been made to re-discover it. Mr. Hind began by calculating the amount of perturbation the comet would be subjected to in 1556 by its passage in the vicinity of the earth. Its return was first expected in 1848. 'But 1849, 1850, 1851, and 1852 have passed, and the great comet has failed to appear! Here, however, is news of it at last'—these lines were written by M. Babinet, in March 1853—'which I take from Mr. Hind's excellent treatise, that I have just received. It is due to M. Bomme, a learned mathematician of Middleburg (Zealand), who appears to have resolved the question completely. Dissatisfied, like all astronomers, at the non-arrival of the comet, M. Bomme has performed *de novo* the whole of the calculations, and estimated the separate action of each of the planets upon the comet for 300 years of its revolution, month by month and day by day, when necessary. M. Bomme, aided by the preparatory work of Mr. Hind, with a patience characteristic of his countrymen, has re-calculated, at a great expenditure of time and labour, the entire path of the comet.'

The result gave for the epoch of its return to perihelion the month of August 1858, with an uncertainty of two years either way. But in vain was it looked for; astronomers swept with their telescopes every region of the heavens. Splendid comets appeared in 1858, 1861, and 1862, but the comet of Charles V.

never returned. Mr. Hind and M. Bomme had not the same good fortune as Clairaut, Lalande, and Mdlle. Lepaute, in the last century. Like Lexell's comet and De Vico's comet of short period, the comet of 1264 and 1556 must be considered lost; and if in reality merely accidental causes prevented its being observed, and it should appear again, it will be our descendants in the twenty-second century who will have the satisfaction of celebrating its return.

Amongst other comets of long period must be mentioned the great comet of 1680, made famous by the hypothesis of Whiston, who assigned to it a revolution of only 575 years, and thus made one of its previous apparitions coincide with the date of the Deluge. The Deluge, according to Whiston, was caused by a rencontre between the earth and this formidable comet, which, being destined to destroy our globe by fire, after having first drowned it, is to bring about the end of the world. Further on we shall return to these fancies. According to the calculations of Encke the comet of 1680 has a period of more than eighty-eight centuries. At its aphelion it would be distant from the sun and the earth 850 radii of the earth's orbit, or about 77,673 millions of miles. 'At this enormous distance,' says Humboldt, 'the comet of 1680, which at its perihelion has a velocity of 244 miles per second—that is to say, thirteen times greater than that of the earth—moves at a rate hardly greater than ten feet per second; that is, scarcely more than triple the speed of our European rivers, and only the half of that which I have myself observed in the Cassiquiare, a branch of the Orinoco.' It was the comet of 1680 which furnished Newton with the elements of his theory of cometary movements. Of all known comets it is, after the one which we mention next, that which approaches most nearly to the sun; its perihelion distance being 0·0062. The perihelion distance of the great comet of 1843 is 0·0055, which is equivalent to only about 506,000 miles measured from the centre of the solar sphere. Thus, the nuclei of these two famous comets have passed respectively, the one to within 143,000 and the other to within 78,000 miles of the surface of the sun, and have, therefore, certainly passed through that hydrogenous atmosphere the existence of which the corona in total eclipses has revealed to us.

The great comet of 1769, which was observed in Europe, in the island of La Réunion, and at sea, near the Canaries, has, according to the calculations of Euler, Lexell, and Pingré, an elliptic orbit, but there is an uncertainty in the period of from 450 to 1,230 years. Bessel, after a profound discussion, fixed its

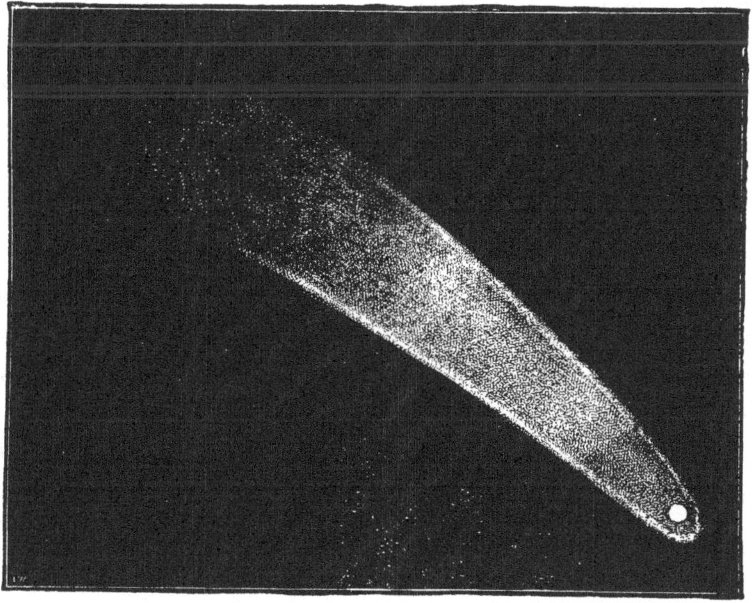

Fig. 20.—Great comet of 1811.

most probable period at 2,090 years; but there remains an uncertainty at least of 500 years. Similar uncertainty exists with regard to the period of the comet of 1843, which, if identical, as has been believed, with that of 1668, would have a period of 175 years instead of 376; and also with regard to the comet of 1793, whose period was first calculated at twelve years. That of 422 years, which we have given, following D'Arrest, is the result of a more careful investigation. The period of De Vico's comet (1846)—viz. 2,721 years—is not more certain; the approximation is only to within 400 or 500 years either way. The comet of 1840, whose period we have given as nearly 14,000 years, has, according to Mr. Loomis, a period of only 2,423 years. We have seen above a similar kind of difference in regard to the great comet of 1680. But the most recent discussions of cometary elements of all kinds should inspire more confidence, and for this reason we have used them in preference to the older determinations.

Two comets amongst those in the preceding table still remain to be noticed. The first is the great comet of 1825, or the comet of Taurus, which was visible for nearly a year—from July 15, 1825, the day of its discovery by Pons, to July 5, 1826, the last day on which it was seen; the other is that of 1811, the great comet which was also observed in 1812, and which is so well remembered in Europe—in the West, on account of the excellent wine attributed to it, and generally known as the Comet Wine, and in the East, because it was regarded by the Russians as a presage of the great and fatal war of the first Napoleon against Russia.

Comets of long period have nothing to distinguish them from other comets, except the enormous distances to which they recede from the sun at the time of their aphelion. The smallest of their orbits exceeds the known limits of the solar system by more than forty-eight times the mean distance of the sun from the earth. The comet of 1845 recedes to a distance from the sun of two and a half times the distance of Neptune; that is to say, to a distance of 6,260 millions of miles.

The comet of 102,000 years period penetrates to a distance fifty-five times greater still. Finally, the two last comets of

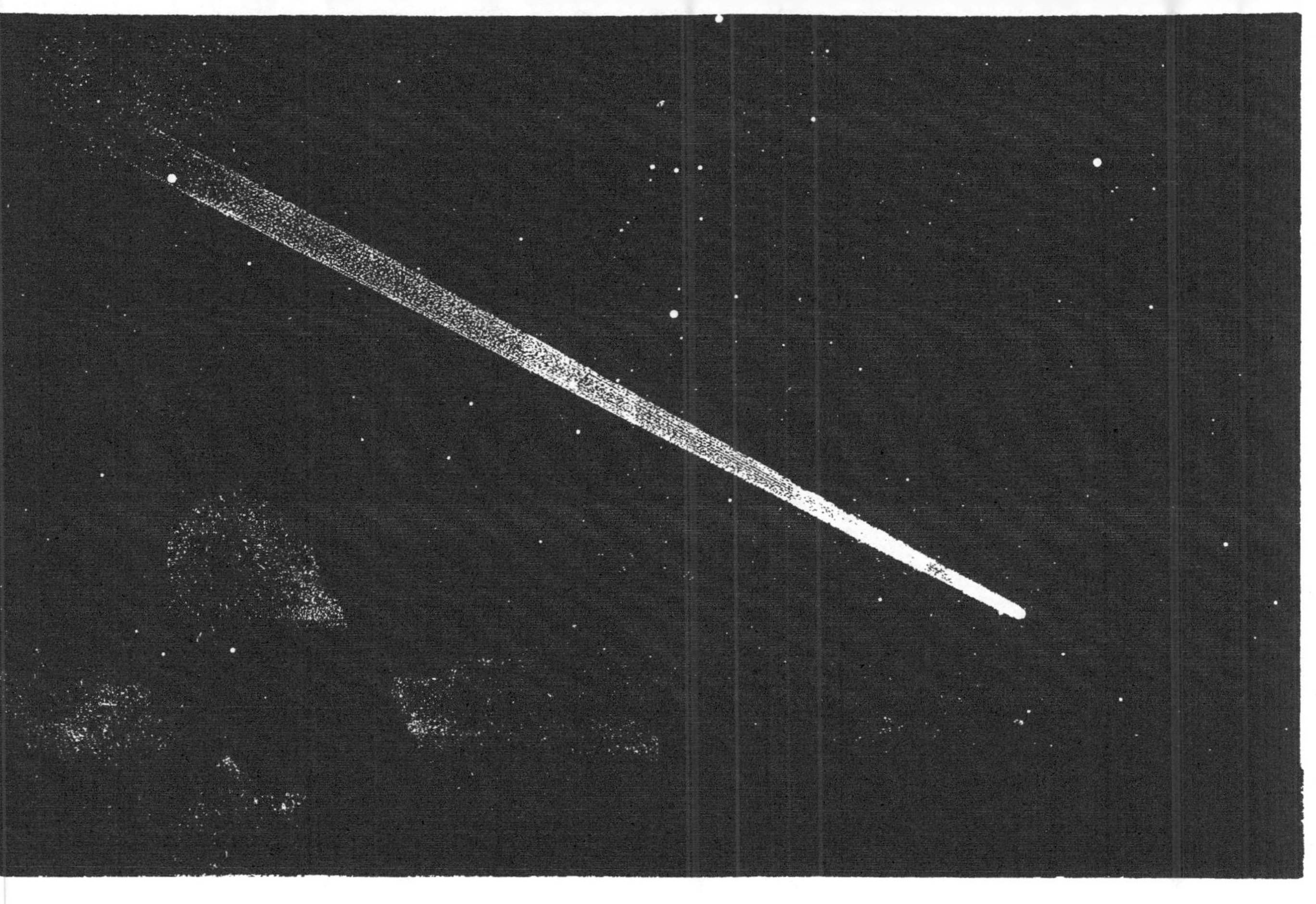

THE LARGE COMET OF 1843

as seen at Paris during the night of the 19th of March.

our table which perform their revolutions, the one in 18,400 centuries, the other in 28,000 centuries, reach at their aphelia to regions of space so remote that their light would require 171 days and 230 days respectively to reach our earth. The comet of 1864 (the last on the list) attains a distance from the sun equal to one-fifth of the distance of the star Alpha Centauri from our system, the distance of this star being assumed to be equal to 200,000 times the mean distance of the sun from the earth. The voyage outward, it is true, takes 1,400,000 years, and the return also 1,400,000 years. It has been said of the comet of 1844 (that of 102,000 years period) that it has left us for depths of space more remote than Vega, Capella, or Sirius. This is not true; it would not be true even for comets with periods measured by millions of years; but there is nothing to prevent it being so for comets which have hyperbolic or parabolic orbits.

[Reference is made to certain points in this chapter and in the next, in the editor's "note upon the designation of comets and the catalogue of comets," which will be found at the end of the volume.—Ed.]

CHAPTER VI.

THE WORLD OF COMETS AND COMETARY SYSTEMS.

SECTION I.

THE NUMBER OF COMETS.

Kepler's remark upon the number of comets—Comets observed—Comets calculated and catalogued—Conjecture as to the number of comets which traverse the solar system or belong to it; calculations and estimates of Lambert and Arago—Calculation of the probable number of comets from the actual data; Kepler's remark verified.

'COMETS are as numerous in the heavens,' said Kepler, 'as fishes in the ocean, *ut pisces in oceano.*' In quoting this comparison of the great astronomer we only follow the invariable custom of all the authors who have hitherto treated the question of the number of comets; but we remark that the expression employed by Kepler is only the result of an opinion which is little more than a conjecture, and that the words ought to be taken in their poetical rather than in their literal sense; but, making allowance for some exaggeration in the expression, we shall see that Kepler was justified in considering the number of comets as very great.

Our inquiry, it is evident, must be confined to comets which are liable temporarily to traverse our system, or to revolve for ever about the sun as an integral part of the solar system. Any attempted estimate of comets situated outside this sphere, beyond our range of vision, and exterior to the planets which belong to our group, could not rest upon any certain data. Our calculations and conjectures must be limited to the domain of that which admits of proof, and is strictly within the power of observation. Beyond this limit number fails us—we lose ourselves in the infinite.

Let us, in the first place, speak of comets which have been observed or at least of those which have been noted by history and tradition. The following is a passage from Lalande, acquainting us with their number as it was known in the last century: 'Riccioli,' says he, 'in his enumeration of comets, reckons only 154 as mentioned by historians up to the year 1651, in which year he composed his *Almagest*, the last having appeared in 1618. But in the great work of Lubienietzki, in which all the passages to be found in any author having the slightest reference to a comet are scrupulously recorded, we find 415 apparitions, up to the comet which appeared from the 6th to the 20th of April, in the year 1665. Since that time forty-one have been observed, including those which appeared in the year 1781.' This makes in all, therefore, up to this last date, 461 comets.

This number has been much increased since, partly by the apparition of new comets and partly by the researches of scholars and the study of the Chinese annals, which have brought to light many apparitions of comets forgotten or not observed in Europe. The following is a table based upon that published by Mr. Hind in 1860, and completed to the present time:—

	Comets observed	Comets calculated	Comets recognised as re-apparitions
Before our era	68	4	1
First century	21	1	1
Second "	24	2	1
Third "	40	3	2
Fourth "	25	0	1
Fifth "	18	1	1
Sixth "	25	4	1
Seventh "	31	0	2
Eighth "	15	2	1
Ninth "	35	1	1
Tenth "	24	2	3

	Comets observed	Comets calculated	Comets recognised as re-apparitions
Eleventh century .	31	3	2
Twelfth " .	26	0	1
Thirteenth " .	27	3	3
Fourteenth " .	31	8	3
Fifteenth " .	35	6	1
Sixteenth " .	31	13	5
Seventeenth " .	25	20	5
Eighteenth " .	69	64	8
Nineteenth " .	189	189	42
Total .	790	326	85

Deducting from the total number of 790 comets observed, those which have returned, some several times, we find altogether 705 distinct comets. With regard to this number we must bear in mind that up to the sixteenth century all comets were observed with the naked eye, and that since the invention of the telescope a great number have been discovered by its means. The preceding table gives, therefore, up to about the year 1600, the most brilliant comets only; but subsequently telescopic comets, too faint or too far from the earth to be visible to the naked eye, have outnumbered the others. In the sixteenth century thirty-one comets were observed, of which eight were telescopic. In the seventeenth century the number of telescopic comets amounted to thirteen out of twenty-five; to sixty-one out of sixty-nine in the eighteenth century; and in the three-quarters which have elapsed of the nineteenth century, out of 189 comets fifteen only have been visible to the naked eye; 174 have been discovered or recognised by the help of instruments, thanks to the zeal of numerous astronomers who have occupied themselves with such researches in both hemispheres.

The progress, however, accomplished by astronomers in our own and the last century consists not so much in the number of discoveries as in the determination of orbits, the precision of the observations, and the study of the physical constitution of these bodies. Before the time of Newton we must distinguish between the number of comets observed or simply seen and the number catalogued. These last are not very numerous before the sixteenth century, because the ancient historians have left but very inexact records of the positions and movements of the comets of their time; and indeed the documents which have since rendered possible the calculation of the orbits are chiefly those of the Chinese annalists. On the other hand, for the last three and a half centuries nearly all comets observed have been catalogued. In our century this is true of all. Except in the unusual case of the apparition being so brief that three observations, separated by the requisite intervals, cannot be obtained, when a comet is seen and observed its elements are now promptly calculated.

But to return to the number of comets that have been catalogued. It amounts to 326, a number, it is true, which must be reduced to 241, if we desire to take into consideration distinct comets only. As regards the probable number of comets which have made their appearance in the solar system in historic times, it is clear that it must be considerably greater than the number given above, even if we take into consideration those comets only which have crossed our system under favourable conditions of visibility for an observer situated on the surface of the earth. To explain this let us take as a basis the number afforded by the present century. It amounts, in the three-quarters of a century, to 147 comets, which gives, therefore, 185 comets for the entire century. This number is well within the truth, and we might even assume without exaggeration an average of two new comets per year. But, restricting ourselves to the number 185, and confining our calculation only to the last twenty centuries, we thus obtain a total of 3,700 comets—an enormous number, but one which must be still further augmented, for the following reasons.

By an examination of the monthly distribution of the comets according to the dates of their perihelion passages—that is, of the times when they occupy that portion of their orbit in the vicinity of which they are most visible—Arago has found that out of 226 comets the distribution is 130 for the winter months, and 96 only for the summer months. Out of 301 apparitions of comets which have been catalogued we find that 165 have passed their perihelion between September and March. The proportion is thus in both cases about fifty-four or fifty-five to one hundred. Now, such a difference can only arise from one cause, and that a very natural one—viz. that the nights in winter are long and favourable to a lengthened observation of a much greater portion of the heavens, whilst in summer, the nights being much shorter, and further diminished by the long twilight, it necessarily follows that a greater number of comets escape detection. The difference is one-seventh, a fraction which may be added to the preceding number, to include comets which in this way escape observation, making in all 4,228 comets.

To this reason for increasing the number may be added the fact that observers and comet-seekers are more numerous in the northern than in the southern hemisphere of the earth, and in consequence a certain number of comets whose orbits are so inclined that they are only visible about the time of their perihelion in the southern hemisphere are unobserved. But it is clear that we have already partially allowed for this, by making the number of comets which pass their perihelion in the summer equal to those which make their passage in the winter. In fact, those portions of the heavens which cannot be examined, on account of the shorter nights in the northern hemisphere of the earth, are precisely those which at the same epoch are in view in the southern hemisphere. If the number of observers were the same in both hemispheres, it is clear that there would be no correction whatever to be made.

Moreover, it is far from being the fact that all comets coming within sight of the earth are observed; or that the astronomers who devote themselves to the laborious search for new bodies can explore continually every region of the heavens. There is a great difference between searching for telescopic comets and small planets; the latter must cut the ecliptic at some time during their revolution, without ceasing to be visible, and observers have only to examine a comparatively small region of the heavens situated north and south of the ecliptic. Comets appear and disappear in all regions of the heavens alike, and thus a great number must escape the researches of astronomers. Again, the weather is not always favourable; and if the comet should be one of those which pass close to our globe, and from the rapidity of its motion should be visible only for a few weeks or days, a cloudy sky may very easily veil the whole of the apparition. The light of the moon is also another obstacle which may cause a comet to escape the detection of observers. The following passage from the *Quæstiones Naturales* of Seneca proves that the ancients even suspected that comets existed in greater numbers than the frequency of their apparition seemed to indicate: 'Many comets,' he says, 'are invisible, because of the far greater brightness of the sun.' Posidonius relates 'that during an eclipse of the sun a comet became visible which had been hidden through his vicinity.' *

Thus, already we must reckon the comets by thousands,

* [Mr. Ranyard has remarked a structure upon the photographs of the solar eclipse of December 12, 1871, which may possibly be a faint, though large comet near to perihelion. See Monthly Notices of the Royal Astronomical Society, vol. xxxiv., p. 365 (June 1874).—Ed.]

confining ourselves to those which have appeared in the course of 2,000 years,—a minute in the probable duration of the solar system! But our calculations have reference only to those which approach the earth near enough to become visible. It remains to estimate the probable number which traverse our system at all possible distances, when we shall arrive at numbers so great that they will justify the expression used by Kepler. We shall follow as our authorities Lambert and Arago, modifying according to the state of the facts in the present day the figures employed by them in their estimation of the number of comets within our system. Lambert relies for his values upon the elements of the twenty-four comets of Halley's table. In the first place, he reduces their number to twenty-one, on account of the two re-appearances amongst them; in the next, he considers the position of the perihelia, two of which exceed the orbit of the earth; two are situated between the earth and Venus, twelve between Venus and Mercury, and, lastly, six between Mercury and the sun. These numbers, according to him, are in accord with the hypothesis that comets are uniformly distributed throughout the interplanetary spaces. But, we may ask, according to what law is the number of known comets found to increase? At first sight it would seem that it should increase in the ratio of the spaces included within the spheres of the different planets; that is to say, proportionally to the cube of the distance. But Lambert assumes 'that comets are disposed in such a manner that they never meet or disturb each other in their movements. To effect this their orbits must not intersect each other anywhere; further, these orbits are not to be regarded as geometric lines, but are to include as much of the sphere of activity of each comet as will prevent incursions into the spheres of others, and avoid the disorders which would result from these incursions.' From this restriction—deduced from the principle of final causes, and which at the present time seems to us quite unjustifiable, as, observations having proved the intersection of the orbit, the resulting perturbations are perfectly possible—the celebrated mathematician reduces the increase of the number of comets to the ratio of the square of the distance. The numbers six and seventeen, which in Halley's table give the comets comprised within the spheres of Mercury and Venus respectively—numbers which are nearly in the relation of one to three—in his opinion justified this hypothesis. Taking, then, as our basis the comet of 1680, whose perihelion was more than sixty times nearer the sun than Mercury, Lambert came to the conclusion that the sphere of the orbit of this planet may contain sixty times sixty, or 3,600 comets. Considering, then, the orbit of Saturn, whose radius is equal to twenty-four times the radius of the orbit of Mercury, he multiplies by 600 the preceding number, and finds there would be more than two millions of comets moving within this sphere.

If at the present time we were to make a similar calculation, it would be, in the first place, necessary to double our fundamental number, since in addition to the comet of 1680 we have the comet of 1843, whose perihelion distance is equally small, and in the next place, we should have to extend the limits so as to include the planet Neptune. Under these conditions we should find not less than 45,500,000 comets!

Arago, after discussing the elements of the comets contained in the catalogues at the time when he wrote his *Astronomie Populaire*, adopted the same fundamental principle as Lambert; that is to say, he assumed the uniform distribution of comets within the space included by the solar system. At all events, he says, 'no physical reason can be advanced for assuming the contrary.' But, with reason, he rejects Lambert's second principle, which restricts the increase of comets to the ratio of the squares of the distances; he adopts the hypothesis that the increase is as the cube. Now, in the catalogue of 1853 there are thirty-seven perihelia whose distances from the sun are less than the radius of the orbit of Mercury. It will be necessary, therefore, he says, to make this proportion:

1^3 is to 78^3 as 37 is to the number required;

or, performing the operations indicated,

1 is to 474,552 as 37 is to 17,558,424.

Thus, within the orbit of Neptune the solar system would be traversed by seventeen and a half millions of comets. A similar calculation, now that the number of comets whose perihelion distances are inferior to the distance of Mercury amounts to forty-three, would give more than twenty millions of comets.

We are unwilling to leave the subject without appending to the values we have just recorded a few reflections which may enable the reader to better appreciate their import.

Everyone will readily comprehend that the question is indeterminate, and that its approximate solution can never do more than assign an inferior limit of the number required. Admitting the uniform distribution of comets in space as a probable fact, it will be seen at once that the result of the calculation will depend solely upon our fundamental number, as, for example, on the number of comets that pass between the sun and Mercury. Now, the number we have taken is evidently much inferior to the number of comets which in reality have penetrated this region of space in historic times. If for 2,000 years observation and research had been carried on as during the last two centuries, what numbers of distinct comets would not our catalogues contain! More than that, in each century the number, not counting re-apparitions, would continue to increase, and the preceding values would rise in a similar proportion.

Besides, why limit the space by the orbit of Neptune? Is it not evident that the sphere of the comets which have gravitated at least once round the sun must extend to all those regions of the heavens where the attraction of his mass preponderates? Let us suppose that stars of the first magnitude have masses nearly equal, on the average, to that of the sun, and that they are nearly equally distributed over the sphere whose radius is equal to the mean distance of Alpha Centauri; the action of the sun would extend to the half of this distance; that is, to about 100,000 times the radius of the earth's orbit. Every comet penetrating within this distance would fall under the dominion of our system and gravitate around its central luminary in an orbit whose elements would depend upon its initial velocity.

If we extend to a sphere of these dimensions the calculation we have made for a sphere extending to Neptune, does the reader foresee in what enormous proportion the results of our previous calculation will be multiplied? It would be in the proportion of the cube of 30 to the cube of 100,000 ; that is to say, it would be multiplied by thirty-seven thousand millions. So that instead of obtaining the already great number of twenty millions of comets we should arrive at the stupendous number of 74,000,000,000,000,000, or seventy-four thousand billions of comets, as the minimum number of those which have each been submitted for one at least of their periods to the empire of the sun!

In presence of such considerations the comparison of Kepler is no longer a metaphor, and we are permitted to say literally with the great astronomer of the sixteenth century: 'Comets are as numerous in the heavens as fishes in the ocean.'

SECTION II.

COMETS WITH HYPERBOLIC ORBITS.

Do all comets belong to the solar system?—Orbits which are clearly hyperbolic—Opinion of Laplace with regard to the rarity of hyperbolic comets—Are there any comets which really describe parabolas?—First glance at the origin of comets.

Do all the comets which have been observed up to the present time belong to the solar system? Or, as we have already suggested, are there comets which visit the sun but once, and which before penetrating to the sphere of his activity and submitting to the influence of his attraction were altogether strangers to our system?

Theoretically speaking the reply is not doubtful. A celestial body, describing under the influence of gravitation an orbit of which the sun is the focus, may move in a parabola, an ellipse, or an hyperbola. All depends upon its velocity at any one given point of its course, that is, upon the relation existing between the velocity and the intensity of gravitation at that point. The better to explain this let us take a point whose distance from the sun is equal to the mean distance of the earth, and let us suppose the body to have arrived at this point. For certain velocities, which we may call elliptic or planetary velocities, the orbit described will be either a circle or an ellipse; for a greater velocity (equal to the mean velocity of the earth multiplied by the number 1.414, that is by the square root of 2), the curve will be a parabola, with endless branches; for a velocity greater still the orbit will be an hyperbola, which also is a curve with branches extending to infinity.

The question is then reduced to this: Are there any known comets with parabolic or hyperbolic orbits? In regard to the parabolic orbits there may be a doubt, because we may always suppose that apparently parabolic orbits are really ellipses of extreme length and considerable eccentricity; but if there be orbits of manifestly hyperbolic character, that is to say, whose eccentricity exceeds unity by an amount greater than we can attribute to errors of observation, no doubt can exist, because the curve described cannot be mistaken for a closed or elliptic orbit. Now, among the comets whose elements have been calculated there are a certain number whose orbits manifestly present this character. We give a list of those which, according to M. Hoek, merit in this respect a certain amount of confidence:—

Comets with hyperbolic Orbits.

			Eccentricity.	Perihelion distance.
Comet	1824	II.	1·00173	1·05
„	1840	I.	1·00021	0·62
„	1843	II.	1·00035	1·62
„	1844	III.	1·00035	0·25
„	1847	VI.	1·00013	0·33
„	1849	I.	1·00002	0·96
„	1849	II.	1·00071	1·16
„	1853	IV.	1·00123	0·17
„	1863	VI.	1·00090	1·30

In the catalogue of comets given by Mr. Watson there are also several comets with hyperbolic orbits—viz. those of 1729, 1771, 1773, 1774, 1806 II., 1826 II., 1852 II., and 1853 IV.—whose respective eccentricities are 1·00503, 1·00937, 1·00249, 1·02830, 1·01018, 1·00896, 1·05250, and 1·00123. As the number of comets catalogued is about 311, it will be seen that one out of every twenty, or nearly so, is certainly foreign to the solar system. It is, therefore, possible that a certain number of the non-periodical comets describe hyperbolas, the visible portion of which is for us confounded with the arc of a parabola; all others would have for their orbits very elongated ellipses, and thus would be confirmed the hypothesis advanced by Laplace in the last chapter of his *Exposition du Système du Monde*:—

'If we connect the formation of comets with that of nebulæ, we may regard the former as small nebulæ wandering from one solar system to another, and formed by the condensation of the nebulous matter scattered with such profusion throughout the universe. Comets would thus be in respect to our system what the aërolites are to the earth, to which they appear to be foreign. When these bodies become visible to us they offer so strong a resemblance to nebulæ that they are frequently mistaken for them, and it is only by their motion, or by our knowledge of all the nebulæ belonging to the part of the heavens in which they are moving, that we are able to distinguish them.'

In the celebrated passage which closes the *Exposition du Système du Monde*, in which the illustrious mathematician expresses his views on the formation of the planets and the sun, he has added this remark in regard to comets: 'We see that when they reach those regions of space in which the influence of the sun is predominant he compels them to describe either elliptic or hyperbolic orbits. But there being no reason why they should have a velocity in one direction rather than in any other, all directions are equally likely, and they may move indifferently in any direction and at any inclination to the ecliptic, which is in accord with what has been observed.'

Laplace next examines the cause of the rarity of hyperbolic orbits, and, in fact, at the time when he wrote no orbits were known that could with certainty be said to possess this character, and he concludes that it is owing to the conditions of visibility, by which it happens that comets are observable only when their perihelion distances are inconsiderable. 'We may imagine,' he proceeds, 'that, to approach so near the sun, their velocity at the moment of their entrance into the sphere of his activity must have an amount and direction comprised within very narrow limits. Determining by the theory of probabilities the ratio of the chance that, within these limits, the orbit should be an appreciable hyperbola, to the chance that it should be an orbit which could be confounded with a parabola, I find that the odds are at least six thousand to one that a nebula penetrating into the sphere of the sun's activity in such a manner as to admit of its being observed should describe either a very long ellipse or an hyperbola, which through the magnitude of its greater axis would sensibly coincide with a parabola in the part of its orbit where it is observed. It is, therefore, not surprising,' concludes Laplace, 'that up to the present time hyperbolic movements have not been recognised.' But, in the last three-quarters of a century, the progress of theoretical and practical astronomy, by rendering the determination of cometary orbits more exact, has altered the chances whose ratio was calculated by Laplace.

As regards truly parabolic orbits they can only be rare exceptions. If we imagine a parabolic comet entering into the sphere of the planetary system, the least perturbation modifying its velocity in one direction or the other will transform the orbit either into an hyperbola or an ellipse, either casting the comet thenceforth from our system or, on the contrary, compelling it to become a periodical satellite of the sun.*

* [It is especially to be noticed that while for an elliptic orbit, the eccentricity may have any value less than 1, and for a hyperbolic orbit any value greater than 1, yet for a parabolic orbit the eccentricity must be *exactly equal* to 1: so that parabolic orbits are infinitely less likely to occur than elliptic or hyperbolic orbits, as the least deviation from the exact value, 1, would make the orbit fall within the two latter categories. Of course no orbit is accurately an ellipse, parabola or hyperbola, as the planetary perturbations must produce some modification of form; but ignoring these deviations, an absolutely parabolic orbit is all but an impossibility.—Ed.]

SECTION III.

REMARKS ON THE ORIGIN OF COMETS.

Have all the known comets of the solar world always belonged to it?—Probable modification of their original orbits through the planetary perturbations—Cause of the gradual diminution of the periods of certain comets.

THE origin of comets is a question equally interesting and difficult.

On comparing all the orbits that have been calculated we find that they pass by almost imperceptible gradations from comets of short period to comets of periods of immense length, and thence to others the major axes of which are of infinite dimensions. If we suppose the latter to be strangers to our solar system, have the former, we may ask, always formed a part of it? In which case why should periodical comets in the elements of their orbits and their physical constitution differ so essentially from planets? Why do they cut the plane of the ecliptic at all inclinations, and why are their movements sometimes direct and sometimes retrograde? Why are their masses so small, and why do they exhibit such vaporous appearances, such rapid changes of aspect, and the phenomenon of tails?

On the other hand, if comets are all of extra-solar origin, why have not all cometary orbits a major axis equal at least to the radius of the sphere of the sun's activity?

The reply to the first questions would be difficult on the hypothesis of comets having the same origin as the planets.

If, on the contrary, we admit that comets come from the depths of the sidereal universe, the comparatively slight eccentricity of certain orbits can be explained by the modifying action of the planetary masses upon the original orbit. We have seen that perturbations exerted in the opposite direction have been able to eject certain comets from the system, and that the disappearance of some periodical comets is thus explained. Moreover, independently of this cause, there is another which also depends upon the insignificance of cometary masses. We refer to the cause which diminishes continually the periods of revolution of the comets of Encke and D'Arrest. Whether it arises from a resisting medium or a repulsive force radiating from the sun, the result is the same—a progressive diminution of the mean distances of the two comets from the sun, and the probability that in the lapse of time these two vaporous masses will become blended with the solar globe itself.

In our second volume, which will form a continuation of the present work, and which will be devoted to the subject of *Shooting Stars*, we shall give new proofs with regard to the origin of comets. It will be there seen how they arrive from all points of space in their peregrinations from world to world, and wheel around the sun like moths about the flame of a candle, some to be there consumed and feed the incandescent ruler of our system, others to be dispersed in long trains and shed the dust of their atoms in the interplanetary spaces. The shooting stars, which vary the sublime, but never-changing, spectacle of the starry nights of the earth, are but fragments of dispersed comets. Who knows but that the incessant rencontres of the planets with these cosmical atoms may be a means of increasing the planetary masses? Who knows but that comets play an important part in the formation and evolution of planetary systems? This point, insignificant as it may appear, in contrast with our enormous planet on the one side and our brief existence on the other, may in the course of time exercise a considerable influence upon the earth's mass. For its operation, this influence has time—millions of years,—and the nebulous matter 'scattered,' as Laplace has said, 'with such profusion throughout the universe.'

SECTION IV.

SYSTEMS OF COMETS.

Comets which have or seem to have a common origin—Double comets—Systems of comets according to M. Hoek—Distribution of aphelia over the celestial vault; region of the heavens particularly rich in aphelia.

When, in accordance with the actual facts of science, we endeavour to form an idea of the constitution of the visible universe, we see that the celestial bodies which compose this whole are everywhere distributed into groups and associations united by the common bond of universal gravitation.

There are the *planetary systems*. In the centre of each group is a star or central sun, whose preponderating mass retains near him, circulating in regular orbits, other stars or planets, to which this central sun distributes heat and light. Our planetary system is the type of associations of this kind.

There are the *stellar systems*, groups of two, three, or more suns gravitating about one another, probably in accordance with the same laws. These systems are themselves the elements of greater associations, which, like the resolvable nebulæ known under the name of *stellar masses*, are composed of myriads of suns. The Milky Way is one of the most splendid examples of these immense agglomerations.

In certain regions of the heavens the nebulæ are themselves to all appearance grouped into systems, so that the general plan of the universe is one vast synthesis of associations of different orders encompassing each other without end. Nor can any individual star escape the necessity of forming a part of one of these groups.

Are there likewise *systems of comets?*

It is certain, in the first place, that there are some comets which belong to the solar system. Originally strangers, they have become drawn into it by the action of the planetary masses, and have since contributed to form an integral part of the group. We have seen that it is possible for comets, through the effect of perturbations, to escape from the power of the sun's attraction; others, on the contrary, owing to the insignificance of their masses, unable to resist the causes that tend to precipitate them into the focus of their movement, may possibly become blended with the central mass; or perhaps, shattered and scattered throughout the interplanetary spaces by the successive perturbations of the planets, they may constitute a sort of resisting medium, the elements of which in the course of time may be a source of increase to the planetary masses themselves.

Besides, we are already in a position to answer the question. We have seen Biela's comet divide into two; and the twin bodies into which it separated, performing their voyage in concert, may be said to constitute an embryo cometary system. The comet observed by M. Liais in 1860 was an example of another kind, since, if the two comets of which it is formed should withdraw from the sun, and, still maintaining their relative position, should leave the system, they would constitute in space a group of two independent comets.

But are all the other comets—I mean the non-periodical comets which describe parabolas or hyperbolas—are they to be regarded as independent voyagers journeying from one solar system to another, and never staying their vagrant course? Are there not amongst these some which move in groups and make the circuit of their long orbits in company together?

This question appears capable of direct solution through the researches of a Dutch astronomer, M. Hoek. By comparing and studying the elements of different comets M. Hoek has discovered that several of their number appear to have had a common origin, and that before entering the sphere of the sun's attraction they formed groups or systems, in proof of which he shows that at some former epoch these bodies were near together, and had each an initial movement in the same direction and of the same velocity. Moreover, in his opinion, comets of elliptic or periodic orbits form the exception, the immense majority of comets moving in curves with endless branches. Arriving singly or in groups from the sidereal depths, they enter our system, sent thither by some star from which they have receded so far as to be beyond the preponderance of its attraction, and to fall temporarily under the attraction of our own sun. But in what manner has M. Hoek discovered that certain comets have emanated from the same focus and have probably a common origin?

To solve this difficult question the Dutch astronomer has compared the elements of the comets which are determined with sufficient accuracy to admit of comparison, those—for example, of the comets calculated since 1556. He has determined the positions of their aphelia, collecting first in a separate group the comets whose apparitions were not separated more than ten years, and whose aphelia were included within a circle of about ten degrees diameter. And further, he has investigated whether the orbits of comets thus grouped three and three or in greater numbers have not points of intersection in common.

Let us, following M. Hoek, take an example, selecting in the first place the comets of 1672, 1677, and 1683, and in the next place the comets 1860 III., 1863 I., and 1863 VI. The positions of the aphelia of these six comets are as follows:—

	Longitudes.	Latitudes.
	°	°
1672	279·4 . .	69·4
1677	286·4 . .	75·7
1683	290·8 . .	83·0
1860 III.	303·1 . .	73·2
1863 I.	313·2 . .	73·9
1863 VI.	313·9 . .	76·4

Now ten degrees of longitude, at a latitude of 73°, represent an angular distance of $3\frac{1}{3}$°, so that the differences of longitude measured upon the arc of a great circle are equivalent in each group to a little more than three degrees. This of itself is a remarkable coincidence. But if we investigate the points of intersection of the different orbits a still more surprising coincidence appears, for we find that these points are grouped together in a region of the heavens the extent of which is not more than two degrees in diameter, and which has its centre at about 319° of longitude, and 78° of south latitude. By drawing a straight line joining the sun and γ Hydræ we obtain nearly the common intersection of the orbits of the last five comets.

On calculating the distances between the comets and the sun at different epochs in past ages M. Hoek has obtained the results which are given in the following tables, the unit of distance being the mean radius of the terrestrial orbit:—

Date.	Distances from sun. Comet 1677,	1683.
573·9 . .	600	601·9
837·8 . .	500	502·2
1076·5 . .	400	402·4
1286·9 . .	300	302·9
1464·7 . .	200	203·6
1602·0 . .	100	105·1

Date.	Distances from sun. Comet 1860 III.	1863 I.	1863 VI.
757·0 .	600	600·4	600·2
1020·9 .	500	500·6	500·4
1259·6 .	400	400·7	400·5
1470·0 .	300	300·9	300·8
1647·8 .	200	201·1	201·2
1785·1 .	100	101·8	102·1
1833·7 .	50	52·8	53·3
1853·6 .	20	24·4	25·5
1858·0 .	10	15·9	17·4

These tables show that the further back we go the more nearly the comets of 1677 and 1683, and the three comets of 1860 III., 1863 I., and 1863 VI., are found respectively to approach each other. Have they started simultaneously on their course, or has each had a separate epoch of departure? M. Hoek gives no opinion in favour of either of these hypotheses. Only, he shows that the extremely small difference of 26 inches per second between the initial velocities of the comets of 1677 and 1860 (supposing them to have started together from a distance so great as to be practically infinite, *i.e.* to have been originally fragments of the same body) would suffice to produce a difference of 200 years in the times of their arrival into our system; it is, therefore, not impossible that the two comets of 1677 and 1860 may have quitted at the same time the focus from which they emanated.

Let us take another example from M. Hoek—comets III. and V. of 1857 and comet III. of 1867. These three comets, in fact, described orbits with elements so similar, and the intervals separating their apparitions were so short, as to point to the probability of a common origin. At first M. Hoek only regarded the two former comets as forming a system, but the comparison of the third with the other two removed all doubt from his mind.

Speaking of the two comets III. and V., 1857, M. Hoek proceeds, 'I did not hesitate to attribute to these two bodies a common origin, considering the extreme resemblance of all the elements of their orbits, and the short interval between their appearance. The comet 1867 III. has just given an unexpected confirmation to this view. The circle which is the intersection of its orbit with the sphere passes through almost the same point of the sky. The planes of the three orbits intersect therefore in the same line, which is necessarily parallel to the direction of the initial motion of the comets.'

The radiant point of their orbits—that point in which their planes intersect each other—is situated in the southern hemisphere, upon the confines of the constellation of Piscis Australis.

This cometary system is not the only one. In the first place, the three comets mentioned above are not the only members of the group, to which must be added the following comets: 1596, 1781 I., 1790 III., 1825 I., 1843 II., and 1863 III., and even 1785 II., 1818 II., 1845 III. The subjoined table sums up the conclusions of the learned astronomer:—

	Comets.	Longitudes and latitudes of the radiant point.
I. First system.	1677 1683 1860 III. 1863 I. 1863 VI.	319°, – 78°·5
II. Second system.	1739 1793 II. 1810 1863 V.	267°, – 52°
III. Third system.	1764 1774 1787 1840 III.	175°·5, – 46°·5
IV. Fourth system.	1596 1781 I. 1790 III. 1825 I. 1843 II. 1863 III. 1785 II. 1818 II. 1845 III. 1857 III. 1857 V. 1867 III.	75°·5, – 51°·7
V. Fifth system.	1773 1808 I. 1826 II. 1850 II.	274°·6, + 38°·7
VI. Sixth system.	1689 1698 1822 IV. 1850 I.	92°·9, + 0°·6

	Comets.	Longitudes and latitudes of the radiant point.
VII. Seventh system.	1618 II. 1723 1798 II. 1811 II. 1849 I.	217° 8, + 26°·6

In the preceding section we have already said a few words on the origin of comets, a question still involved in much obscurity. We here merely quote from the *Monthly Notices of the Royal Astronomical Society* * the following summary of the views to which M. Hoek's researches lead:—'Every star is associated with a cometary system of its own; but owing to the attraction of planetary or other cosmical matter, these bodies continually leave their proper primaries, and revolve either permanently in ellipses, or temporarily in parabolas or hyperbolas, round other suns.'

On studying the distribution throughout the celestial sphere of the aphelia of 190 cometary orbits M. Hoek discovered a somewhat curious fact. If we suppose a circle drawn through three points, the respective longitudes of which are 95°, 169°, and 243°, and the latitudes 0°, 32°, and 0°, the sector comprised between this circle and the ecliptic will be found particularly poor in aphelions. Instead of including fifteen, as it would were the distribution uniform, it contains only one, that of the comet of 1585, situated at a distance of three degrees only from the ecliptic. How is this peculiarity to be explained? To this question M. Hoek replies, 'If we knew that the solar system was removing from the point situated in the middle of that sector, I should be inclined to attribute the phenomenon to a difficulty comets might experience in overtaking the sun. But the direction of the solar motion, such as it was given by Mädler's investigations, does not allow of such

* Vol. xxvi., p. 147 (February 1866).

an explanation.* Therefore we may ask if the phenomenon is a real one, and there is in that direction of the heavens a scarcity of centres of cometary emanations; or rather, if it depends on the circumstances under which comets are ordinarily detected, the sector in question being so near the part of the ecliptic occupied by the sun from July to December.'†

The first of these two hypotheses is not in our opinion at all improbable; the labours of Sir John Herschel on the distribution of nebulæ prove that they are disposed very unequally in the different regions of the sky. A similar inequality in the distribution of the nebulous centres from whence the comets emanate would be a fact of the same kind, and one perhaps not without physical connexion with the first. If future observations should establish this connexion, it would add one more gleam of light to those which astronomy has already thrown on the constitution of the universe.

* See on this subject two interesting letters from M. Hoek to M. Delaunay. *Comptes rendus de l'Académie des Sciences*, 1868, I.

† *Monthly Notices of the Royal Astronomical Society*, vol. xxvi., p. 207. M. Hoek's other papers are published in vol. xxv., p. 243 (June 1865), vol. xxvi., p. 1 (November 1865), and vol. xxviii., p. 129 (March 1868).—Ed.

SECTION V.

COMETARY STATISTICS.

Comparison of the elements of cometary orbits—Eccentricities; numbers of elliptic, parabolic, and hyperbolic comets—Distribution of comets according to their nodes and perihelion distances—Equality of the numbers of direct and retrograde orbits.

If we arrange in the order of date the various apparitions of comets that have been recorded, and note how these bodies appear in different regions of the heavens, and how some pursue a direct and others a retrograde course; or, better, if we study their elements in a catalogue, our attention is at once arrested by the diversity of these elements, which seem connected by no relation.

It may, however, be instructive to examine, by comparing these materials, whether any law presides over the distribution of comets in time and space. We shall, therefore, give a rapid *résumé* of the analysis we have made with this object. We have taken the catalogue published by Mr. Watson at the end of his work on Theoretical Astronomy as the basis of our investigation.

In this catalogue, which we reproduce at the end of this work, we find 279 comets arranged in the order of their successive apparitions, from the most ancient times to the commencement of the year 1867; we have ourselves completed it for the seven following years, including also the first half of the year 1874; so that the total number of comets in the catalogue is by this means increased to 311, a number very inferior, not only to the actual number of comets, but to the number of those which have received mention in history. Pingré, in his *Cométographie*, enumerates 400 comets whose apparition he considers almost certain, and many others which he has registered as doubtful. His list, however, ends with the year 1781. Since that epoch 212 comets have appeared. The catalogue that we are about to study includes only those comets whose elements astronomers have found means to calculate. Up to the end of the sixteenth century these calculations are in general founded upon observations oftentimes uncertain and leaving much to be desired on the score of accuracy; since then, under the twofold influence of improved observation and theory, a greater and steadily increasing degree of accuracy has been obtained.

Let us first consider the form or geometrical nature of cometary orbits. This form is determined by the element *eccentricity*. If the eccentricity is equal to 1 (unity), the orbit is a parabola, or an ellipse so elongated as to be indistinguishable from a parabola of the same perihelion distance and direction of axis. If it be less than 1, the orbit is an ellipse; in this case the comet is periodical, and the duration of its revolution round the sun may be more or less approximately determined. Lastly, if the eccentricity be greater than 1, the orbit is hyperbolic.

This being premised, out of 311 comets in the catalogue we find that 177 have parabolic orbits, 120 elliptic, and only fourteen hyperbolic. But these numbers require modification, because they apply, not to distinct comets, but to all observed apparitions, and consequently to comets which, having reappeared, are included more than once in the enumeration. Taking into account, then, these multiple apparitions, we have in all 264 distinct comets, the orbits of which are thus distributed:—

Parabolic orbits	177
Elliptic orbits	73
Hyperbolic orbits	14

This proves that of known comets the most numerous are those which really perform their revolution round the sun, and, but for unknown perturbations, would remain members of the solar system. If we confine ourselves to the eighty-seven comets whose orbits have been really determined, we find that about one in six are foreign to our system. With respect to the 177 comets which describe parabolic orbits it is still a matter of doubt whether in reality they move in very long ellipses or in hyperbolas differing but little from parabolas.

If the 177 comets which seem to be parabolic were divided in the same proportion between the really elliptic and decidedly hyperbolic, we should then find that, out of 264 distinct comets, the distribution would be as follows:—

222 elliptic orbits, or periodical comets.
42 hyperbolic orbits, or comets foreign to the solar system.

But in respect to elliptic orbits we must remember that, out of the seventy-three comets whose orbits have been calculated, nine only belong to comets which have actually returned, or, what comes to the same thing, which have been observed on two of their successive revolutions.

Let us now proceed to an element of great importance as regards the study of the distribution of comets in space, viz. the inclination of the planes of their orbits. The inclination, however, does not suffice of itself to determine the nature of this distribution; it is necessary to add to it the other elements which fix the position of the curve traced by the comet in the plane of its motion; the position of this plane itself being given, in the first place, by the longitude of the node, and in the second place by that of the axis of the orbit, or the longitude of the perihelion.

We will begin by the study of the inclinations.

These, as we are aware, vary from 0° to 90°. In other words, a certain number of comets move in the ecliptic, or deviate but little from it, and might be called *zodiacal comets;* others describe orbits which have a moderate inclination to that of the earth, and others again move in curves which cut nearly at right angles the paths pursued by our earth and by the other planets of the solar system.

The following table, in which distinct comets only are included, shows this distribution:—

Inclinations between ° °	Number of comets.	
0 and 10	21	
10 ,, 20	20	62
20 ,, 30	21	
30 ,, 40	23	
40 ,, 50	39	97
50 ,, 60	35	
60 ,, 70	31	
70 ,, 80	33	96
80 ,, 90	32	

The inclinations of nine comets are wanting in this table.

These numbers clearly prove that great inclinations occur more frequently than small. The comets, it may be observed, that we have proposed to call zodiacal form only a quarter of the number of distinct comets that have been catalogued. The other three-quarters are pretty evenly distributed between the moderate and great inclinations.

Does not this furnish irrefragable testimony of the extra-solar origin of a great number of comets, since so great a divergence exists between the planes in which they move and the planes of the orbits of the planets? This distinctive feature appears to us all the more striking, because amongst the number of comets of small inclination there are many whose movement is retrograde, a fact which adds another point of difference to those which distinguish the movements of these bodies from the movements of the planets.

We now come to the longitudes of the ascending nodes and those of the perihelia. These will be found in the following table:—

Longitudes of nodes and of perihelia comprised between ° °	Number of comets. Nodes.		Number of comets. Perihelia.	
0 and 30	20	67	17	71
30 „ 60	22		24	
60 „ 90	25		30	
90 „ 120	25	72	25	60
120 „ 150	25		21	
150 „ 180	22		14	
180 „ 210	24	66	16	66
210 „ 240	22		21	
240 „ 270	20		29	
270 „ 300	14	53	30	60
300 „ 330	22		22	
330 „ 360	17		8	

The nodes, as we may perceive by comparison with the table on p. 30, exhibit a greater degree of uniformity in their distribution than the inclinations. Nevertheless, in the last quadrant of the circumference of the ecliptic the number of comets which cross the plane of the earth's orbit, from south to north, is noticeably smaller than in the other three. As regards the perihelia, the differences in the different quadrants are still less. We have seen that M. Hoek, who has studied the question closely, has made a comparison of the opposite points or aphelia of various comets, and has arrived at the important conclusion that a certain number of these bodies are united in groups, and that each of these groups includes comets of probably common origin.

Let us now compare the comets, arranged according to their respective perihelion distances. We will take as unity the mean distance of the earth from the sun and divide it into tenths, each tenth corresponding to 2,320 equatorial radii of our earth, or about 9,200,000 miles. We shall then find the perihelia of the 258 distinct comets distributed as follows:—

Perihelion distances comprised between ° °		Number of comets.	
0·0 and 0·1	53	11	192
0·1 „ 0·2		9	
0·2 „ 0·3		11	
0·3 „ 0·4		22	
0·4 „ 0·5	60	11	
0·5 „ 0·6		29	
0·6 „ 0·7		20	
0·7 „ 0·8	130	28	
0·8 „ 0·9		26	
0·9 „ 1·0		25	
1·0 „ 1·1		16	66
1·1 „ 1·2		12	
1·2 „ 1·3		11	
1·3 „ 1·4		5	
1·4 „ 1·5		7	
1·5 „ 2·0	15	7	
2·0 „ 6·0		8	

This table shows that by far the greater number of comets have their perihelia in the vicinity of the earth, between the planets Venus and Mars, whose mean distances are 0·723 and 1·524 respectively, the earth's mean distance from the sun being taken as unity. There are no fewer than 130 within these limits. Comets, on the contrary, whose perihelion distances are beyond the orbit of Mars, and even beyond that of Jupiter, are few in number—but fifteen in all; fifty-three comets have their perihelia comprised within the mean distance of Mercury, 0·387, and sixty between the orbits of Mercury and Venus. But, as we have already said, in our section upon the number of comets, this distribution is in all probability apparent only, because, being invisible from the earth, except in the neighbourhood of their perihelia, comets which do not make a nearer approach to the sun than the planet Mars are under very unfavourable conditions for observation; unless of exceptional brilliancy they would pass unperceived from the earth. Comets which have a perihelion distance comprised between the orbits of Venus and Mars are, on the contrary, near enough for observation; but, on the other hand, their close vicinity to the earth renders their apparent motion very rapid, and they are only visible for a brief period. In short, the most likely comets to be observed are those which pass between the sun and Venus; and on the hypothesis of an equal distribution in space these ought to be the most numerous, regard being had to the volumes of the spheres in which their perihelion distances are contained.

Lastly, let us consider the movement of comets. All comets whose orbits, projected on the ecliptic, are described in the direction of the earth's movement are direct; all those which move in an opposite direction are retrograde. Now, out of 252 distinct comets 129 are retrograde and 123 direct. Their numbers are, then, nearly equal. How these numbers are divided between parabolic, elliptic, and hyperbolic orbits the following table will show:—

Direct comets	Parabolic	69	123
	Elliptic	44	
	Hyperbolic	10	
Retrograde comets	Parabolic	98	129
	Elliptic	27	
	Hyperbolic	4	

Thus, the comets decidedly elliptic seem to show a greater preference to move in the direction of the planetary movements than comets which are parabolic. However, as the true nature of the curves described by the latter is a matter of doubt, it is hardly possible to draw from this circumstance any certain conclusion as to which direction of movement predominates. It is a more significant fact that, out of nine periodical comets of verified return, one alone (Halley's comet) has a retrograde motion, and that this comet has an aphelion distance exceeding the known limits of the planetary system. If we include the seven other interior periodical comets which have not yet returned, we find that the movement of fourteen of them is direct, and that two only describe orbits in a retrograde direction. These comparisons become still more striking when we observe that the inclinations of the nine first comets are nearly all comprised within the limits of the zodiac. One of them (Brorsen's) has a larger inclination, of about 29½°, which is less, however, than the inclinations of three of the little planets which revolve between Mars and Jupiter. Tuttle's comet forms the sole exception, its inclination exceeding 54°. Of the remaining nine periodical interior comets one alone, the comet 1846 IV., has the large inclination of 85°; two others attain 30°, and six have small inclinations.

Such are the comparisons that have been suggested to us by the study of the elements furnished by existing catalogues of comets. It would be desirable, no doubt, to multiply comparisons of the same nature, and to obtain from them further probable deductions. The work is one that would require long and minute research, and we have only attempted to give our readers some idea of these relations. If, instead of limiting ourselves to points we have considered in this chapter, we were to include all that has reference to the aspect and physical constitution of comets, especially since they have been subjected to rigorous telescopic scrutiny, our field of research would be greatly enlarged, and our results proportionably increased in number and value. We should, perhaps, be enabled by a kind of natural classification to distinguish these bodies into kinds and species and varieties. The physical explanation of the phenomena which they present would be rendered easier, because we should not then be compelled to apply to all a theory which may be suitable for some and not for others. This the reader will better comprehend as he becomes familiar with the subject in this new aspect, the phenomena it includes, and the explanations suggested. Such is the principal object of the following chapters.

CHAPTER VII.

PHYSICAL AND CHEMICAL CONSTITUTION OF COMETS.

to ascend, and depths in the earth to which he has not yet penetrated. The mean density of the earth, its mass and weight, and the relation of its mass to that of the principal members of the solar system, are known.

What are comets from these various points of view? Are they globes similar to our earth, illuminated like it by the sun, or do they shine by their own light? Have they a solid or liquid nucleus, surrounded by a vaporous atmosphere, or are they gaseous masses, collections of particles more or less condensed? Has any certain estimate been formed of their masses, or the density of the matter of which they are composed? As regards their movements we know that they do not differ from other members of the celestial group of which we form a part, and that the same universal force, the same laws govern them. Coming probably from the depths of space, of distinct origin therefore, and of very different aspect to the planets and their satellites, we may not apply to both the lines of Ovid:—

———Facies non omnibus una
Nec diversa tamen qualem decet esse sororum.

Comets are, from all these points of view, their movements alone excepted, conspicuously different from the earth and the rest of the planets. In physical constitution they appear to be quite dissimilar—chemically speaking, are they equally unlike? That is to say, is the matter of which they are composed formed of unknown elements, or of elements identical with those of which the planets themselves are constituted?

All these questions possess a high degree of scientific interest. Nor are they less important if we view them in their relation to the superstitious beliefs which for so long a time made comets formidable to the world—beliefs which, having changed in form perhaps more than in substance, are still to a certain extent current even in our enlightened century. Although not susceptible of proof, the habitability of the planets is a thesis that has long been maintained and is still maintained with very considerable probability. More than this, in the last century it was supposed, and some savants even of our time believe, that comets have likewise their inhabitants. Are comets indeed habitable? We are urged by an instinct of invincible curiosity to put such questions to ourselves; and if it appears next to impossible to return positive replies, at least we are not forbidden to examine the probability of each. But, if we would not abandon ourselves to vain and profitless conjectures, it is clear that we must, in the first place, acquaint ourselves with what science has to communicate, not respecting this problem, which may be considered as extra-scientific, but upon the physical and chemical conditions which observation and experiment show to be compatible with the existence of human beings, as far as they are known to us.

We shall, therefore, examine what is known of the constitution of comets at the present day; and we shall begin with the study of their aspect and external form.

SECTION II.

COMETARY NUCLEI, TAILS, AND COMÆ.

Comæ and tails—Classification of the ancients according to apparent external form; the twelve kinds of comets described by Pliny—The 'Guest-star' of the Chinese—Modern definitions: nucleus, nebulosity or atmosphere; tails.

WHAT is the distinctive sign of a comet by which it is universally known, by which it is distinguished from all other celestial bodies? Everyone answers at once, it is the train of luminous vapour, the nebulosity of more or less length, which accompanies it or at least surrounds it; in other words, the *tail* and the *coma*

This is what the etymology implies, the word *comet* signifying *long-haired* or *hairy*. Armed with its tail, which appears brandished in the heavens like an uplifted sword or a flaming torch, the precursor of some untoward event, a comet is everywhere recognised on the instant of its appearance; it needs no passport signed by astronomers to prove its identity. But should the tail be absent, should no appendage or surrounding nebulosity distinguish the celestial visitor on its apparition, for the world at large it is no comet, but simply an ordinary star like any other.

Nevertheless, there are tailless comets. The comet of 1585 was equal to Jupiter in size, but less brilliant; its light was dull. It had neither beard nor tail, and it might have been compared to the nebula in Cancer (Pingré). Lalande observes that the comets of 1665 II. and 1682 exhibited discs as round, clear, and well defined as that of Jupiter himself, without *tail*, *beard*, or *coma*.' We are here speaking of comets visible only to the naked eye; of telescopic comets a great number are destitute of tail, and it very often happens that they are simple nebulosities, in the midst of which a faint nucleus is but just discernible, sometimes nothing but a luminous condensation at the centre. Moreover, from the presence or absence of a tail at one time of the apparition, we cannot infer that the same is true at another. Thus, the above-mentioned comet of 1682 (no other than Halley's comet), which Cassini observed to be without tail on August 26, had developed one of 30° in length by the 29th of the same month. And as regards the comet of 1585, twelve days after its apparition, 'a slender and hardly perceptible ray was seen to issue from it, a hand's breadth or more in length.' It likewise often happens that the tail which has been invisible to the naked eye is readily perceived in the telescope; instances of this we shall meet with as we proceed. All that we have here to bear in mind is, that the distinctive sign of a comet, astronomically speaking, is not to be sought in the tail, the coma, or in any of the variable appendages which may surround the star during its apparition. The elements of its orbit, its large eccentricity, great inclination, direction (oftentimes retrograde), &c., constitute the true points of difference between a comet and the planets. We have already called attention to these differences, and need, therefore, only allude to them here.

It is clear that up to the sixteenth century, before the employment of the telescope in astronomical observations, the accounts given of cometary apparitions can refer only to comets seen by the naked eye. The strange forms of their tails, their beards and comæ, attracted the attention alike of the multitude and the learned. The ancients, who have not always clearly distinguished them from other luminous meteors, such as bolides and auroræ boreales, applied themselves to a classification of comets according to their appearance. Pliny has distinguished not fewer than twelve kinds, which he describes somewhat obscurely in the following terms:—

'There are,' he observes, 'comets properly so called; they are fearful by reason of their blood-coloured manes and their bristling hair pointing upwards. The Bearded (Pogoniæ) have their long hair hanging down like a majestic beard.' (These first two kinds may be classed together, because they differ only in the direction of their tails.) 'The Javelin (Acontias), which seems to dart forward like an arrow; the effect follows with the utmost speed upon an apparition of this kind. When the tail is short and pointed it is called the Sword (Xiphias); this is the palest of all comets; it shines like a sword, and is without any rays. The Plate or Disc (Disceus) bears a name in accordance with its figure; it is of an amber colour, and emits a few rays from the margin only. The Cask (Pitheus) exhibits the figure of a cask, and appears in the midst of a smoky light. The Horn (Ceratias) has the appearance of a horn, and the Lamp (Lampadias) that of a burning torch. The Horse (Hippeus) resembles a horse's mane, agitated violently by a circular or rather a cylindrical motion. It is also very white, with silver hair, and so bright that it can scarcely be looked at, exhibiting the aspect of a deity in human form. Some there are which are shaggy (*hirti*, and not *hirci*, as several have read); these have the appearance of a fleece, surrounded by a nebulosity. Finally, the hair of a comet has been seen to assume the form of a spear.'

All these denominations are more or less justified by the diversity of aspect which comets are known to exhibit, and by the differences observable in their nebulosities and tails; but they afford us absolutely no information concerning their physical

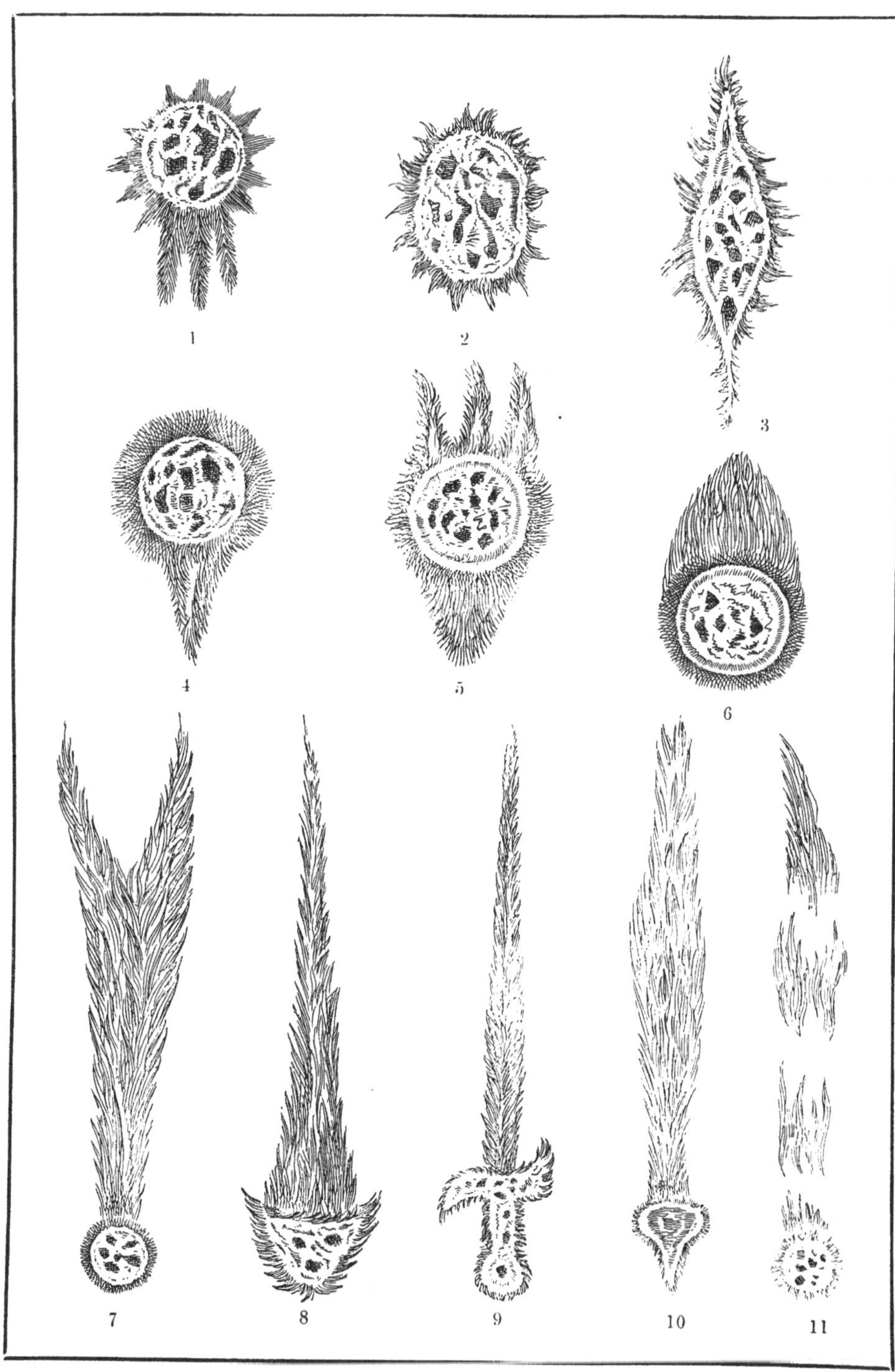

FORMS OF COMETS ACCORDING TO PLINY,

Taken from the *Cométographie* of Hévélius.

Cometæ : 1. Discei, disciformis. — 2. Pithei, doliiformis erectus. — 3. Hippei, equinus barbatus. — 4–5. Lampadiæ, lampadiformis. — 6. Barbatus.—7. Cornutus bicuspidatus.—8. Acontiæ, faculiformis lunatus.—9. Xiphiæ, ensiformis.—10. Longites, hastiformis.—11. Monstriferus.

nature. Nor is Pliny's enumeration complete, if we are to regard as comets the burning torches and beams (*faces* and *trabes*), which he describes separately.

The Chinese, who, fortunately for science, have taken careful note of all cometary apparitions, have given to the tails of these bodies the very prosaic name of *brooms* (*sui* or *soui*).* They likewise acknowledged no comet without a tail. 'If devoid of this appendage,' says Pingré, 'whatever might be its movement, it was spoken of simply as a *star*, or the *new star*, or the *guest-star*, from its visiting the provinces and taking up its abode in different places, as at an inn. Their home was in the vestibules of the celestial palaces; there, under an invisible form, they awaited the order of departure. The order sent, they became visible and commenced their journey. If whilst on their way they put forth a tail, the star was said to have become a comet.'†

* Comets are called in Chinese 'broom stars,' a name derived from the form of their tails. As a rule the records make no distinct mention of the nucleus, and the constellations indicated are generally those over which the tail extended. Thus, in describing the march of the comet of 1301, the text of the records runs as follows: 'It swept the star Thien-ki, the Sankoung, &c.' (Biot and Stanislas Julien, *Comptes rendus de l'Académie des Sciences*, 1842, tome ii. p. 953.)

† This passage will be better understood if we extract from the same author a second paragraph, in which he explains 'the foolish and singular idea that the Chinese had formed of the heavens. The heavens were, according to them, a vast republic, a great empire, composed of kingdoms and provinces; these provinces were the constellations; there was decided all that would happen for good or ill to the great terrestrial empire, that is, to China. The planets were the administrators or superintendents of the celestial republic, the stars were their ministers, and the comets their couriers or messengers. The planets sent their messengers from time to time to visit the provinces for the purpose of restoring or maintaining order; but all that was done in the heavens above was either the cause or the forerunner of what was to happen here below.'

We confess that the ideas of the Chinese appear to us hardly more foolish than the extravagant conceptions of the Europeans in the times of the ancients and in the Middle Ages; they, at all events, give evidence of a higher idea of the disposition of the universe. Nor would it be difficult to find amongst our contemporaries individuals whose views concerning the government of the world differ in no essential respect from those of the Chinese.

But let us return to the definitions accepted by modern astronomers.

A comet consists, generally speaking, of what are invariably termed the *head* and the *tail*.

The head is composed of the star; that is to say, of the nucleus or luminous point in which the brightest light of the star is concentrated, and of the surrounding nebulosity, coma, or atmosphere. All comets do not exhibit a nucleus; but those which appear as simple nebulosities of vaporous appearance are generally telescopic comets. The head of a comet visible to the naked eye is always bright and star-like.

When the nebulosity is of nearly circular form, oval or sometimes irregular—which may arise either from its real configuration or from an effect of perspective—and is devoid of any prolongation or train, the comet is said to have no *tail*, this denomination being reserved for the luminous train, sometimes of no great length, sometimes of immense extent, which escapes from the head in a direction nearly always opposite to that of the sun at the time of observation. It sometimes happens that the train is directed towards the sun, or makes a certain angle with the line joining the head and the sun; it was then called by the ancient astronomers the *beard* of the comet, an expression now discarded. At the present day every luminous appendage or train of vaporous appearance is spoken of as a tail.

[I may here mention that M. E. Biot's 'Catalogue des Comètes observées en Chine depuis l'an 1230 à l'an 1640 de notre ère,' forms a supplement to the *Connaissance des Temps* for 1846; and that in 1871 the late Mr. John Williams published 'Observations of Comets, from B.C. 611 to A.D. 1640, extracted from the Chinese Annals,' which contains a catalogue of the whole of the observations of the comets recorded in the Encyclopædia of Ma Twan Lin, and in the historical work called the She Ke. The catalogue of M. Biot gives notices of 224 comets, and that of Mr. Williams of 373.—ED.]

SECTION III.

COMETS DEVOID OF NUCLEUS AND TAIL.

Gradual condensation of nebulous matter at the centre—Imperceptible transition from comets without apparent tails to the immense luminous trains of great historic comets.

LET us before proceeding further make a few general remarks on the heads and tails of comets. The remaining sections of the chapter we will devote to a more complete examination of their structure.

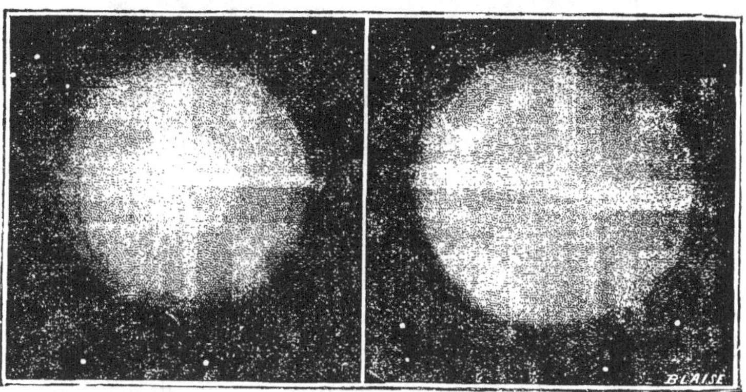

Fig. 21.—Cometary nebulosities; central condensation; absence of tail and nucleus.

Since a systematic search has been made for comets, and powerful instruments have been employed, the number of those discovered has, as might be expected, considerably increased; but the majority are telescopic comets, and amongst them are many nebulosities devoid of nucleus. This fact had been already ascertained by Sir William Herschel in 1807. 'Out of sixteen telescopic comets that I have examined, fourteen,' he observes, 'exhibited nothing remarkable at their centres.'

The following are some examples of comets which were simple nebulosities, and apparently without tail or nucleus. Encke's comet, observed by Mr. J. Tebbutt, June 24, 1865: 'The comet,' he observes, 'was about two minutes in diameter,

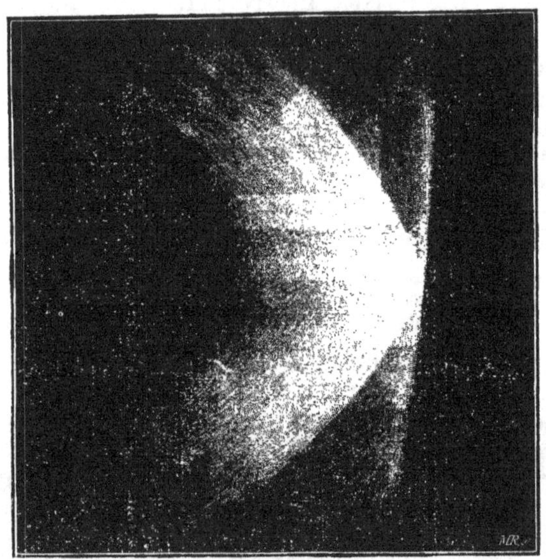

Fig. 22.—Encke's Comet according to Mr. Carpenter.

faint, and without the slightest condensation of light in the centre.' In October 1871 the same comet presented, according to Mr. Hind, when first observed, the aspect of a faint and nearly round nebulosity, without any condensation of its parts. But on the 9th of November the same comet exhibited an appearance anything but globular. According to Mr. Carpenter the nebulosity had expanded like a fan, the apex of which was the most brilliant part; but there was no nucleus. The comet discovered on July 12, 1870, by M.

Winnecke was similar in appearance, and is described as a round nebulosity, of moderate brilliancy, and of 2½ minutes in diameter.

The following is another instance in which the trace of a brilliant nucleus is just discernible. We refer to Brorsen's comet, observed at Marseilles, on September 1, 1873, by M. Stéphan, who thus describes it: 'Nebulosity ovoid, diffuse, and exceedingly faint, with a *trace of condensation* towards its centre.' And likewise Winnecke's comet, seen in April and May 1869: 'It is a faint nebulous patch of some little size,' says Mr.

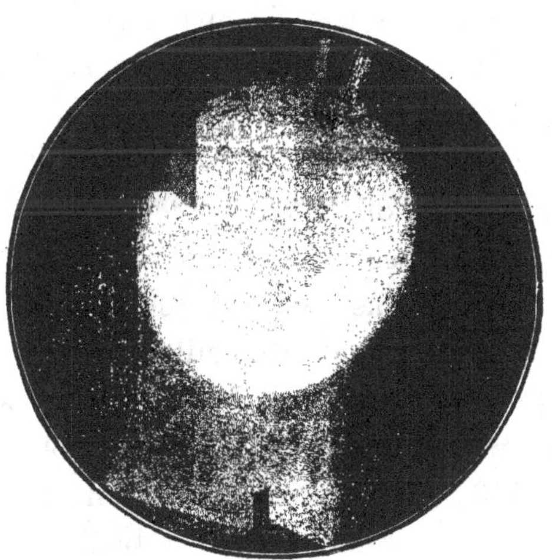

Fig. 23.—Encke's Comet, December 3, 1871, according to Mr. H. Cooper Key.

Hale Wortham, 'appearing occasionally to brighten somewhat to a centre.' According to Father Perry, 'there seems to be a slight condensation towards the centre, but no decided nucleus.' However, we must not forget that the absence of a nucleus may proceed either from the distance of the comet rendering a very slight condensation invisible, or from the position of the comet relatively to the sun. If the nucleus shines by a light which is not its own, its light would increase as the comet draws near to its perihelion. And we see in fact, that in fig.

17, Encke's comet exhibits a visible condensation, while in fig. 23 it has a brilliant and defined nucleus. In like manner Brorsen's comet, observed in October 1873, showed considerable condensation about the centre. On its apparition in 1868 the brightest portion was very eccentric, and there were three or four centres of condensation or brilliant nuclei. (See fig. 18, p. 120.)

The comet of 1867, II., telescopically observed by Mr. Huggins, 'appeared to consist of a slightly oval coma, surrounding a minute and not very bright nucleus.' This bright point was not central, but near to the following (eastern) edge of the coma. The double comet of Biela, as we shall presently see, possesses a well-defined luminous nucleus in the centre of each of the nebulosities which compose its two parts. The same fact is observable in respect to other telescopic comets. In May 1873 Tempel's comet exhibited a head of oval form, with a central nucleus about as bright as a star of the 12th or 13th magnitude. Faye's comet, seen at Marseilles, in September of the same year, although extremely faint, had a small sharply-defined nucleus, which enabled it to be easily observed. Lastly, the comet of 1873, IV., discovered by M. P. Henry at the Observatory of Paris, was round, very brilliant, nearly visible to the naked eye, and had a central condensation. It is shown under this aspect in the left hand drawing of fig. 32.

In some comets, as we have seen in the preceding section, the nuclei have been equal in brilliancy to Jupiter himself; others that we have yet to mention have even exceeded him in the brilliancy of their light. Between simple nebulosities, therefore, devoid of nucleus or luminous condensation, and those comets which have surpassed in lustre the most brilliant of the planets, there is no distinct line of demarcation. The transition from the one extreme to the other is imperceptible. We shall find a similar gradation in respect to cometary tails, from the comets destitute of tail, that we have just described, from hardly visible traces of these appendages in telescopic comets, to the immense luminous trains of the great comets of 1680, 1769, 1811, 1843, 1858, &c., which during their apparition swept the heavens. These differences of aspect the reader will be enabled to follow by the aid of our engravings.

SECTION IV.

DIRECTION OF THE TAILS OF COMETS.

Direction of the tail opposite to the sun; discovered by Apian; the Chinese astronomers were acquainted with this law—Deviations in some comets—Variable aspect of the tail according to the relative positions of the comet, the earth, and the sun.

In respect to the direction of cometary tails let us call attention to an important point—to a general phenomenon which was remarked by the ancients in the very earliest times. Seneca refers to it in the following line:—

Comæ radios solis effugiunt.

The comæ of comets fly the rays of the sun. According to Edward Biot the Chinese astronomers had observed, since the year 837, this constant direction of cometary tails from the sun. 'In Europe,' says Lalande, 'Apian was the first to perceive that the tails of comets were always opposite to the sun; this rule was afterwards confirmed by Gemma Frisius, Cornelius Gemma, Fracastoro, and Cardan. Nevertheless, Tycho Brahé did not believe it to be very general or well demonstrated; but the fact itself is beyond a doubt.'

Pingré observes with truth that the direction of the tail is not always strictly opposite to the sun. He instances the comet of 1577, whose tail was deflected as much as 21° towards the south, and the great comet of 1680, when the deflection was about 4½°. On both these occasions, however, the comet and the earth occupied the same relative positions in the heavens. The deviation is less in proportion as the tail is more inclined to the orbit; viz., considering only the portion of the tail in the neighbourhood of the nucleus, the deviation is less in proportion as the comet draws near to its perihelión; and it takes place towards the region of the heavens last quitted by the comet in its course.

It results, therefore, from this law that the tail of a comet sometimes follows and sometimes precedes that body in its course. It follows the comet before the perihelion

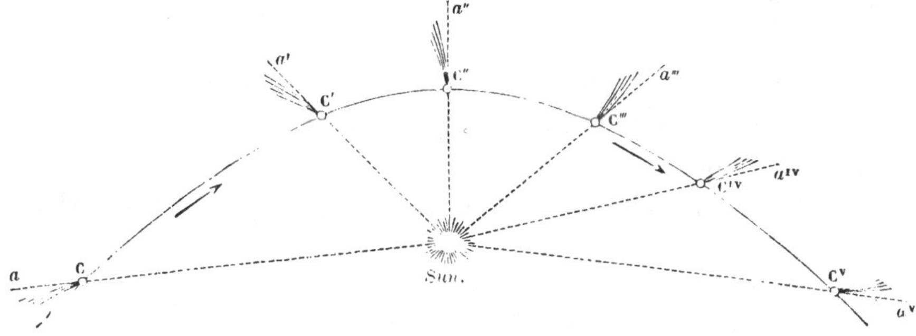

Fig. 24.—General direction of cometary tails.

passage, and precedes it when the perihelion has been passed. Moreover, tails have very frequently a more or less considerable degree of curvature, and this curvature appears more marked in proportion as the earth is removed from the orbit of the comet. If the earth be situated in the plane of the comet's orbit, the curvature is *nil*,* and the tail is rectilinear, or at

* The two tails of the great comet of 1861 at first appeared to offer an exception to this law. On the 30th of June, on which day the earth passed through the plane of the comet's orbit, the two tails, projected the one upon the other, appeared to form but one, wider in the first third of its length, reckoning from the nucleus; but both were rectilinear. But M. Valz and Mr. Bond, from observations made by the former and by Father Secchi, discovered, as they believed, that the two tails presented a slight deviation to the east of the plane of

least appears so; this is no doubt an effect of perspective, and the curvature then takes place in the plane of the orbit. It is more marked in those portions of the tail that are furthest from the nucleus; from which it follows that if we draw radii vectores from the sun to the several positions of the comet, the tail always presents its convex side to these lines, as may be seen in fig. 24.

There is yet another conclusion to be drawn from these facts, which is, that if the earth occupies such a position with reference to the comet and the sun that the comet is in opposition to the latter, its tail, being likewise opposite to the sun, is situated behind the nucleus, and is consequently invisible. It is then only the breadth of the tail that is seen, and it appears to surround the nucleus as a coma, thus increasing only the extent of the cometary atmosphere. This fact may serve to explain the absence of tails in some comets, which, from their nearness to the earth, we should have expected to have been so provided.

the orbit. This would render still more difficult the theory of the formation of tails. But, if we adopt the conclusions of M. Faye in this respect, the deviation existed in appearance only, and this difficulty would be removed. This is a point well deserving the attentive consideration of all future observers of cometary phenomena.

Comets with double tails, one of which is opposite to the sun and the other directed towards that luminary, appear likewise to follow the law stated above. Olbers says of the comet of 1823: 'On the 23rd of January the earth passed through the orbit of the comet; on this day not the least deviation could be discerned between the direction of the abnormal tail and the prolonged axis of the other tail.' 'Thus,' says M. Faye, in citing this passage, 'the two tails of the comet were projected, each on the prolongation of the other, which shows that tails directed towards the sun have, like the others, their axes situated in the plane of their orbit.'

SECTION V.

NUMBER OF TAILS.

Double tails of comets; comets of 1823, 1850, and 1851—Tails multiple, fan-shaped, rectilinear, curved—Variable number of tails belonging to the same comet; comets of Donati, of 1861 and of Chéseaux.

GENERALLY a comet has but one tail, which varies considerably in form or size, or, at all events, appears to do so. Sometimes these changes take place very rapidly, but still, as a rule, the tail consists of one luminous train. Nevertheless, examples may be adduced of double and even multiple tails. The comets of 1807 and 1843 were furnished with double tails, or, what comes to the same thing, single tails formed of two branches of very unequal length. It was the same with the comet of 1823, about which Arago gives the following details:—

'On the 23rd of January, 1824, the comet, in addition to its ordinary tail opposite to the sun, had another which was directed towards the sun, so that it resembled somewhat the great nebula of Andromeda. The first tail appeared to include a space of about 5°, but the length of the second was scarcely 4°. Their axes formed between them a very obtuse angle of nearly 180° (fig. 25). In the close vicinity of the comet the new tail was hardly to be seen. Its maximum brightness occurred at a distance of 2° from the nucleus. During the first few days in February the tail opposite to the sun was alone visible; the other had disappeared, or had become so faint that the best telescopes in the clearest weather failed to show any trace of it.'

The comets of 1850, I., and 1851, IV. (figs. 26 and 27), exhibited the same phenomenon of two unequal tails, the shorter of which was directed towards the sun.

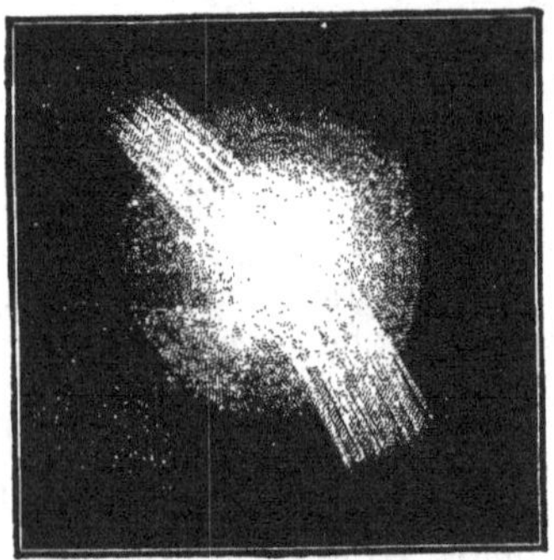

Fig. 25.—Double tail of the comet of 1823.

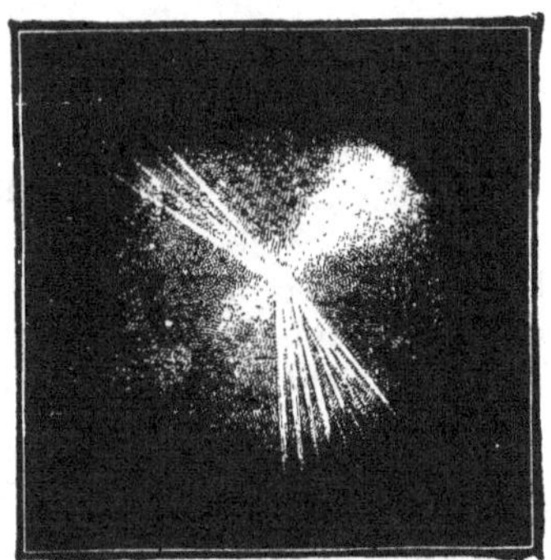

Fig. 26.—Double tail of the comet of 1850.

The observations and the drawings of Messier show that the great comet of 1769 had, if not a multiple tail, at least lateral jets of light, resembling secondary tails, proceeding from the nucleus, but much smaller and less extended than the principal tail, and making unequal angles with the latter: all the tails were rectilinear.

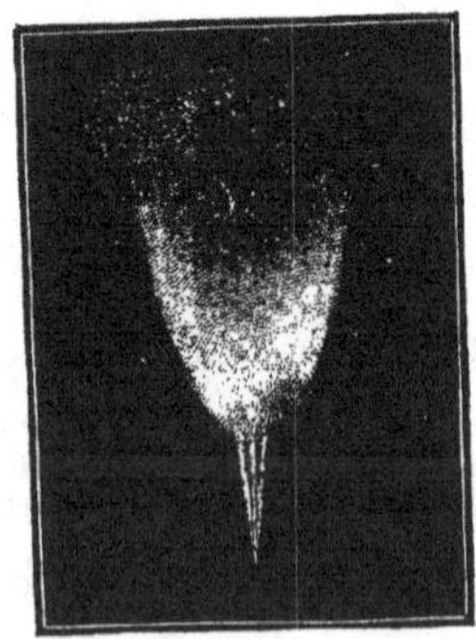

Fig. 27.—Comet of 1851.

Donati's comet exhibited, in 1858, a similar peculiarity. In addition to the principal tail, remarkable for its extent, its curvature, and brilliancy, there appeared first one and then two luminous trains, much fainter, apparently rectilinear, and nearly tangential to the limiting curves of the great tail. The figures of Plates VII. and IX. give a very correct idea of this phenomenon, which was observed in Europe by Schwabe (of Dessau) from the 11th of September, and then by Mr. Hind at London, and by Winnecke and Struve at Pulkowa. In America the secondary tails of the comet were studied and drawn with the utmost care by Professor Bond of the Observatory of Harvard College. On following the development of these remarkable appendages by means of the beautiful plates in the great work* which the American astronomer has devoted to this comet, we obtain the following *résumé* of the changes observed :—

On September 27 a slender rectilinear tail was first per-

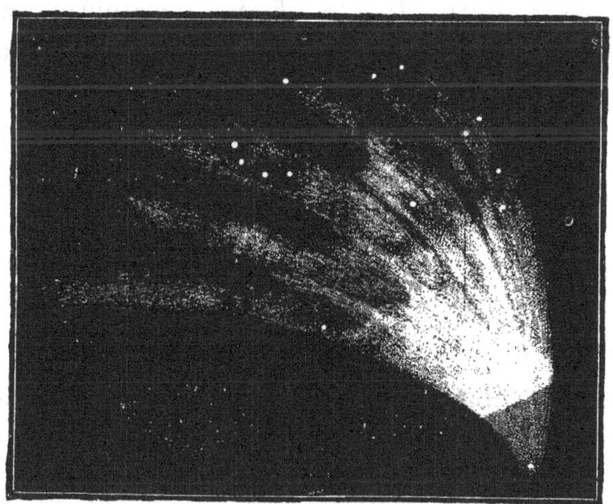

Fig. 28.—Sextuple tail of the comet of 1744, according to Chéseaux.

ceived, in part veiled by the principal tail, and nearly of the same length; it seemed to be tangential to the concave portion of the curve. There was no change on the 28th, but on the 29th it began to approach the nucleus. On the 30th it was

* [The work forms Vol. III. of the Annals of the Observatory of Harvard College (1862). The reader will find in it almost everything that is known about the great comet of 1858, and the plates are so numerous and excellent that all the changes of form and appearance that the comet underwent, both as regards its tails and nucleus, can be easily followed.—Ed.]

hardly visible, but on the following days up to October 3 it became somewhat brighter; it was then half as long again as the principal tail. On the 4th of the same month a second rectilinear tail, not so long as the former, made its appearance, forming with it an angle apparently equal to that enclosed by the two limiting curves of the principal tail at their point of departure from the nucleus. On the 5th the longer was also the brighter. On the 6th, 7th, and 8th of October the longer of the secondary tails alone was seen; but on the 9th the second was seen, and, as it proved, for the last time. The convexity of the principal tail at this date became more marked, and the longer of the rectilinear tails, which had never ceased to form a tangent to the principal tail, was itself somewhat curved near its base, so that, if continued in a straight line, it would no longer have terminated in the nucleus of the comet. These appearances allow us to concede to Donati's comet a triple tail.

In the last century a comet was observed whose tail, which was fan-shaped, presented six distinct branches. This is the famous comet of 1744, known as Chéseaux's comet. Fig. 28 represents, according to the drawing of this astronomer, the sextuple tail in question. On March 8 its remarkable form was most observable. The six divergent branches of the tail proceeded from the nucleus as luminous curves, the outer radii of which included an angle of about 60°, the longest being towards the concave portion. Chéseaux saw the comet rise before the sun, and its large fan appeared above the horizon before the nucleus itself was visible. This curious phenomenon was sketched by Chéseaux at Lausanne, and from his original drawing we have designed Plate V.

Nearly fourteen years ago there was observed in Europe and America a beautiful comet (1861, II.), which is of interest from several points of view. In the first place, it is one of the

CHÉSEAUX'S COMET

as seen at Lausanne during the night of the 8th of March 1744.

comets of long period we have already mentioned—it performs its revolution about the sun in about 422 years. Moreover, as we shall see, the earth, in all probability, passed through its tail on June 30, 1861, an event worthy of notice, if only from the absence of any disastrous consequences to the inhabitants of the earth. Lastly, the comet in question was remarkable at the same date (June 30) for its beautiful fan-shaped tail, the long divergent rays of which gave it some resemblance to the

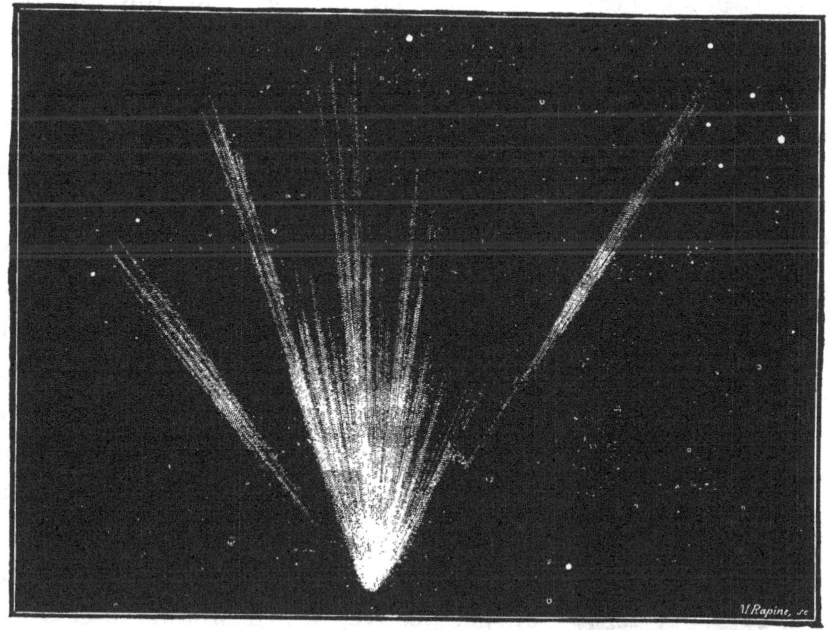

Fig. 29.—Fan-shaped tail of the great comet of 1861, according to the observation of June 30 and the drawing of Mr. G. Williams.

comet of 1744. The drawing which we here reproduce (fig. 29), due to Mr. G. Williams, of Liverpool, shows a striking difference, however, in the form of the appendages of the two comets. The divergent rays which compose the multiple tail of the comet of 1861 are sensibly rectilinear, and emerge from the head of the comet; the extreme or outer rays alone, which include an angle of 75°, are detached from the nucleus, whilst the longer and inner rays are slightly curved, the convexity being outwards.

Before assuming this remarkable form the great comet of 1861 was furnished with two tails of unequal length, making an angle of about 13°. The drawings given by M. Liais, for dates from June 19 to 28, leave no doubt upon this point. Those which we here reproduce (fig. 30) exhibit the comet, according to Father Secchi, as seen on June 30 and July 2. On June 30, the earth being exactly in the plane of the comet's orbit, the two tails, the one long and slender, the other shorter and of greater width, were to all appearance projected the one upon the other. On July 2, the earth being then out of the plane, they were seen as separate. Looking at the drawing of Mr. Williams, which gives the appearance of the tail on the same day, the difference of aspect presented to the two observers seems surprising. But if it be true that the tail of the comet pointed directly towards us, the divergence of the rays would be but an effect of perspective, which would necessarily change with great rapidity, considering the extreme relative velocity of the movements of the two bodies.

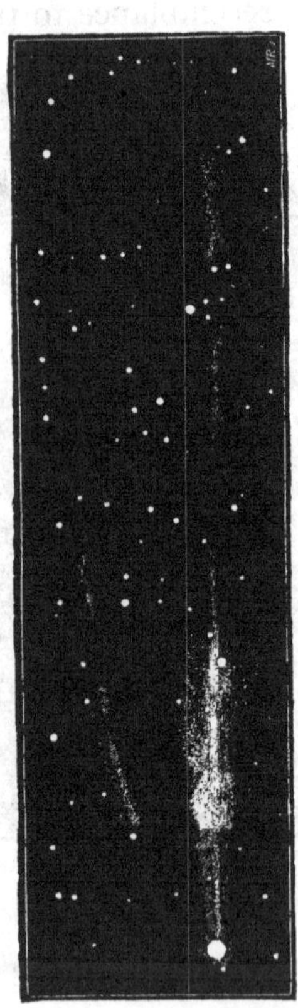

Fig. 30.—The two tails of the comet of 1861, according to Secchi, June 30 and July 2.

The number therefore, as well as the form and dimensions of cometary tails, are variable circumstances, not only as compared one with another, but even for the same comet at different times; and this variation is due to two causes; in the first place, to real changes taking place in the comet itself, frequently with wonderful rapidity; and, in the second place, to the optical effects which the rapid movements of the comet and the earth in their respective orbits necessarily produce in the appearance of the several parts of the head, the nucleus, and the tail.

We have still to mention, amongst comets with multiple tails, the one which was observed in 1825 by Dunlop, in Australia. The tail was formed of five unequal and distinct branches. 'At a distance of $1\frac{1}{2}°$ from the head the rays of the several tails cross each other, and then diverge indefinitely.' Arago, after citing this passage, mentions as a double-tailed comet 1845, III., which 'exhibited a tail of $2\frac{1}{2}°$ long, divided into two branches by a black line.' But, according to this view, a great number of comets might be considered as furnished with double tails, which in point of fact have but one, since it often happens that the outer edges of a tail are more brilliant than the space which separates them; and they are often of unequal length and lustre. Thus, M. Liais considers the tails of the great comets of 1858, 1860, and 1861 as consisting in reality each of two tails, of which the longer and narrower is situated in the prolongation of the radius vector, or line joining the sun to the nucleus; while the other, shorter but more spread out, makes a certain angle with the former. Sometimes, in consequence of the position of the earth with respect to the plane of the comet's orbit, the two tails are projected the one upon the other, and are seen as one alone: as in the case of the comet of 1861. The question is, however, of no great interest. The question of the multiplicity of tails is of no real importance, except as it concerns their origin and the physical causes which occasion their development.

SECTION VI.

DIFFERENT FORMS OF TAILS.

Elementary forms of tails—Rectilinear tails, divergent or convergent, in respect of the head of the comet—Curved tails; comets of 1811 and 1769—Whimsical form of cometary appendages according to ancient observations.

THE tails of comets, under whatever form they may be presented to the observer, are all, whether simple, double, or

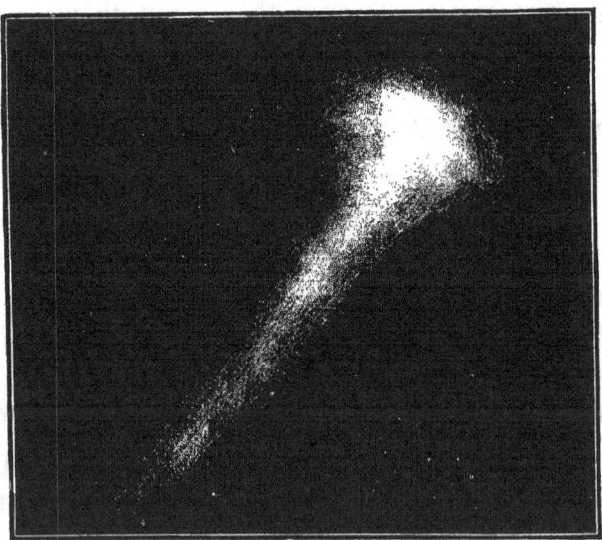

Fig. 31.—Winnecke's comet, June 19, 1868.

multiple, easily reducible to two or three elementary forms.

In the first place, there are comets with rectilinear tails, that is to say, tails whose luminous rays, emerging from the head, are projected in what appear to be right lines against the sky. Sometimes, the tail, as in the comets of 1843 and 1769, and that of Biela, in 1846, resembles a long ribbon of light, nearly of the same width throughout and scarcely varying in intensity. Sometimes it gradually narrows from the head and tapers to a point, like the tail of Halley's comet in 1835 (see fig. 16, page 106), Winnecke's comet in June 1868, and that of P. Henry in August 1873 (figs. 31 and 32). Or it

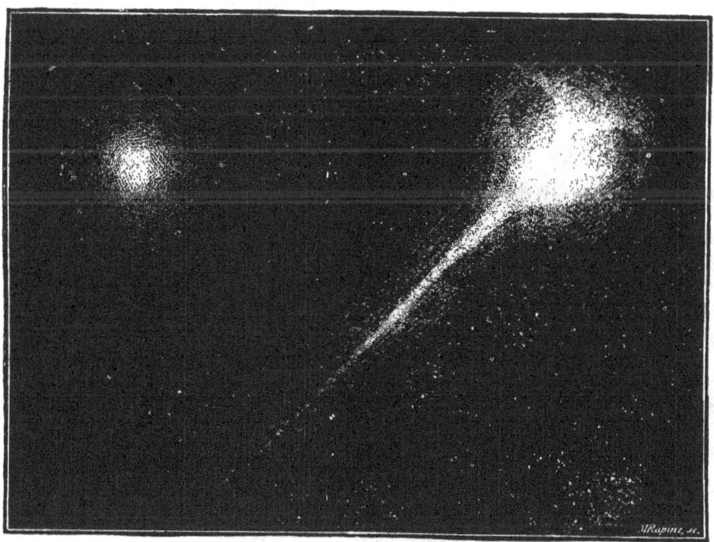

Fig. 32.—Comet of P. Henry, August 26 and 29, 1873.

may happen that the rays of a rectilinear tail may diverge from the head and continue to diverge up to their furthest limit, or so far as their light permits them to be seen; of this kind was the tail of the comet of 1686 (the aspect of which we have given from a contemporary, J. C. Sturm), and also the tail of the great comet of 1264. These are the forms, doubtless, in which the ancients saw the similitude of beams, swords, and lances. But the slightest reflection will serve to convince us that these diverse forms are apparent only, and that the same tail may present itself under any one of these appearances, according to the distance of the earth from the different portions of the cometary appendage. As a simple consequence of the laws of perspective the same tail may appear to be either very short or of great length; or in certain cases it may even disappear, without its real dimensions undergoing any change.

On examining with the aid of a telescope the forms of tails in the vicinity of the nucleus the outline of the tail is frequently observed to sweep round and enclose the head; this curve bears great resemblance to the portion, near the vertex, of a parabola or a very long ellipse, whose focus would be the nucleus. A case in point is supplied by the comet of 1819, whose tail was in the form of a cone with nearly rectilinear boundaries; the great comet of 1811 likewise exhibited a tail whose edges were more luminous than the central portion, and which was curved round the vertex, as if to envelop the nucleus. Besides this curvature near the nucleus the entire tail itself may be curved throughout its length, as was the case with Donati's comet. These are the comets like Turkish sabres, in which our ancestors of the Middle Ages, constantly mindful of the dangers with which the Ottoman empire menaced Christianity, saw threatening

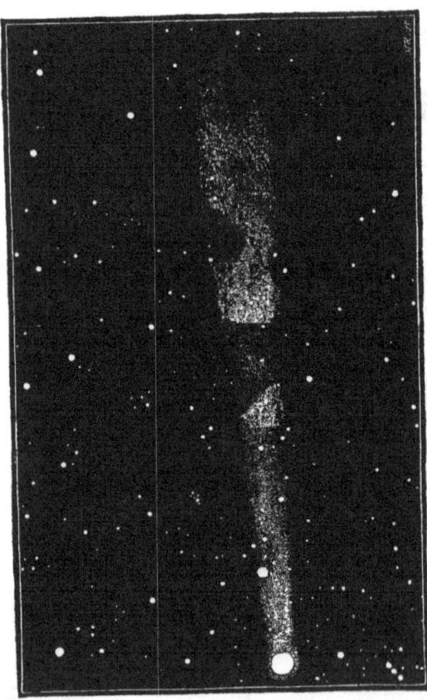

Fig. 33.—The comet of 1264.

presages of war. In all probability they belong to the class called by the ancients the *Horn*, one of the kinds of comets mentioned by Pliny. Examples of it are not unfrequently met with in ancient drawings; but we must not forget that the observers of former times were not always the most exact of draughtsmen, and that they did not hesitate on occasion to improve upon nature according to the dictates of their fancy. A curious instance of this mania for embellishment occurs even in the work of Hevelius. This indefatigable and learned philosopher, wishing to represent in his *Cometographia* the kind of comet which Pliny, under the name of *Xiphias*, has compared to a sword, has not failed to add the handle of the weapon. A fac simile of this remarkable design has been given in Plate III., fig. 9.

Cometary tails are generally curved in the same direction throughout their whole extent; so that one of the boundary-lines of the tail turning its concavity to one region of the heavens, the other boundary will turn its convexity to the region opposite; as, for example, the comet of 1811, Donati's comet, and many others. The two tails of the comet of 1807 were curved in opposite directions; and a drawing of the same comet of 1811, which we find in Chambers's Astronomy and in the Atlas of A. Keith Johnston, represents a similar phenomenon. A more exceptional form, and one of which we know no other example, is mentioned by Pingré in these terms: 'The late M. de la Nux, at the Isle of Bourbon, and ourselves, between Teneriffe and Cadiz, both remarked that the tail of the comet of 1769 was doubly curved towards its extremity; it resembled the figure of an ∾.' But we should bear in mind that Messier has given several drawings of the same comet in which the tail is represented as a rectilinear band, brighter at its edges than either at its axis or in its interior. This last peculiarity is not un-frequent. Nevertheless, the contrary may occur, as was observed in the case of the comet of 1618. 'At Rome,' says Pingré, 'there was seen a kind of nucleus, so called by Hevelius, in the tail of the last comet of 1618; it resembled a line or a dart, which, like the pith of a tree, extended the whole length of the tail, dividing its breadth into two parts. Kepler and Schickard saw the same phenomenon, but it did not then divide the breadth of the tail, it skirted along one of its edges, which is more in conformity with what is generally observed.'

Beyond the forms which we have just described, and which are sufficiently regular to admit of exact definitions, the tails of comets may assume irregular and whimsical appearances. In the accounts extant of great and historic comets, seen with the naked eye by observers who were often themselves astronomers, we find mention made of the most singular appearances; but we can hardly put faith in their descriptions, ingenuous perhaps, but certainly distorted by the superstitious beliefs of the times. It remains for modern astronomers to follow and to depict with scrupulous fidelity all the forms of cometary nuclei, atmospheres, and tails, as exhibited in the field of the telescope. The evolutions of these phenomena are but little known, and they must be studied without preconceived ideas, if we would fabricate a theory which should be exempt from the fallacies of observers. The sole means of discovering truth, in astronomy, as in all the natural sciences, is to begin by collecting facts, and then, relying upon them alone, to deduce reasons.

DONATI'S COMET

as seen at Paris on the 5th of October 1858

SECTION VII.

LENGTH OF TAILS.

Apparent and real dimensions of the largest tails on record—Formation and development of cometary appendages; their disappearance—Variations of length in the tail of Halley's comet at its different apparitions—Great comet of 1858, or comet of Donati.

SINCE we have entered upon the statistics of various cometary elements, let us here give a few particulars respecting the real and apparent dimensions of cometary tails. We will first confine ourselves to the maximum dimensions under which they have been viewed from the earth, dimensions measured in degrees, according to the apparent extent occupied by the train itself in the celestial vault. Passing, then, from the apparent lengths, we will proceed to the actual measures expressed in miles. Under the first head the scale of magnitude will be found to include an enormous range, varying from the tail of $2\frac{1}{2}°$, belonging to the comet of 1851, to the immense tail of 100°, possessed by the comet of 1264, and to the still greater tail of the comet of 1861, which attained a length of 118°, thus exceeding by 28° the apparent distance between the horizon and the zenith. Nor are the differences less considerable when we compare the true dimensions. Whilst, for instance, the second comet of 1811 was provided with a tail about seven millions of miles in length, the great comets of 1811, I., 1847, I., 1687, and 1843 launched into space, in directions opposite to the sun, immense luminous trains measuring from 109 to 199 millions of miles—more than double the distance of the sun from the earth. Some of these elements will be found included in the following table:—

	Perihelion distance.	Length of tail.	
		Apparent, in degrees	Real, in miles
Comet of 1851, I.	1·700	2½	—
" 1860, III.	0·292	15	21,700,000
" 1825, IV.	1·241	17	—
" 1744	0·222	24	18,600,000
" 1811, I.	1·035	25	109,400,000
" 1811, II.	1·582	—	6,800,000
" 1456	0·586	57	—
" 1843, I.	0·005	65	198,800,000
" 1858, VI.	0·578	64	54,600,000
" 1689	0·019	68	—
" 837	0·580	79	—
" 1680	0·006	90	149,000,000
" 1769	0·123	97	39,800,000
" 1264	0·312	100	—
" 1618, II.	0·389	104	49,700,000
" 1847, I.	0·043	—	130,500,000
" 1861, II.	0·822	118	42,200,000

The discordance between the apparent and real lengths is striking. It is hardly necessary to point out the reason of this discordance, as the reader is already aware that it arises from the manner in which the tail of the comet is presented to the observer, and depends upon the visual angle under which a line, more or less inclined, is seen from the earth, according to the relative positions of the earth, the plane of the comet's orbit, and the comet itself. From the apparent length expressed in degrees, and the knowledge of the positions concerned, the true length of the luminous train can be calculated and deduced.

But the observed dimensions of the same tail are far from being always accordant, so that an exact estimation of the real length is often impossible. It is very difficult to distinguish the limits of a light so feeble as is that of most cometary tails, particularly at the extremity further from the nucleus. The clearness of the sky, the power of the instrument employed, even the sight of the observer, are all so many variable elements. On this subject Lalande has said in his *Astronomie*: 'In southern countries, which enjoy a pure and serene sky, the tails of comets are more easily discernible and seem longer. The comet of 1759 at Paris appeared almost destitute of tail, and it was with difficulty that a slight trace of such an appendage was discerned, measuring one or two degrees in length; whilst at Montpellier M. de Ratte estimated its entire length, on April 29, to be 25°, the most luminous portion being about 10°. At the Isle of Bourbon M. de la Nux saw it larger still, owing to the same causes as those which permit the zodiacal light to be seen there constantly.'

SECTION VIII.

FORMATION AND DEVELOPMENT OF TAILS.

Variations of length in the tail of Halley's comet at its different apparitions—Similar phenomena exhibited by Donati's comet in 1858—Does the maximum development of the tail always coincide with the perihelion passage of the comet?

It is now desirable to consider a phenomenon of high importance as regards the physical constitution of comets, viz., the development and variation of their tails according to the position which the comet occupies in its orbit; that is to say, according to its greater or less distance from the sun.

It has been already seen that the tails of comets frequently are formed and developed during the period of the comet's visibility, and generally before the perihelion passage. 'It has been constantly observed,' says Pingré, 'that a comet advancing to its perihelion begins to assume a tail only on its near approach to the sun. The fine comet of 1680 had no tail on the 14th of November, thirty-four days before its perihelion passage. The real length of the tail increases day by day, and the head, or rather the coma surrounding the head, seems, on the contrary, to diminish. The tail attains its greatest length shortly after the comet has passed its perihelion; it then diminishes by degrees, but in such wise that at equal distances from the perihelion the tail is longer after the perihelion passage than before. It has been, moreover, observed that comets whose perihelion distance has much exceeded the mean distance of the sun from the earth have not developed tails, and that the tails of others, all else being the same, have been more magnificent in proportion as the perihelion distances have been less.'

Are we to consider that the laws thus enunciated by the author of the *Cométographie* are general, and apply to all known comets? No, unquestionably, as we are about to see; nevertheless, it is certain that some relation does connect the existence and development of the tails of comets with their greater or less proximity to the sun.

Let us first take, for example, Halley's comet at its apparition in 1835. When it first appeared it had the aspect of a slightly oval nebulosity, and was thus destitute of tail. On October 2, that is to say, six weeks before its perihelion passage, which took place on November 16, the tail was formed, and three days later it attained a length of from four to five degrees. During the following days it continued to increase in length, and on October 15 had attained its maximum of 20°. On the 16th it had become reduced to 10° or 12°, on the 26th to 7°, on the 29th to 3°, and on the 5th of November to 2½°. 'There is every reason to believe,' says Sir John Herschel, 'that before the perihelion the tail had entirely disappeared, as, though it continued to be observed at Pulkowa up to the very day of its perihelion passage, no mention whatever is made of any tail being then seen.' We should add that a drawing made by Sir John Herschel himself, on January 28, leads us to suspect an extension of a part of the comet's atmosphere under the form of a tail; but on May 3, a little more than four months and a half after the perihelion passage, the tail had completely disappeared; the comet had then regained its original form of a round nebulosity.

Q

The same comet, on its apparition in 1759, was at its perihelion on March 12. Now, on April 1, nineteen days after its perihelion passage, the observations of Messier, made at Paris, assign to it a feeble tail of but fifty-three minutes in length. But, taking the observations of La Nux, which were made at the Isle of Bourbon, under much more favourable conditions of visibility, we find the measures of its apparent length to be as follows:—

March 29	3°
April 20	6° to 7°
„ 21	8°
„ 27	19°
„ 28	25°
May 5	47°

It would be necessary to calculate the true lengths in order

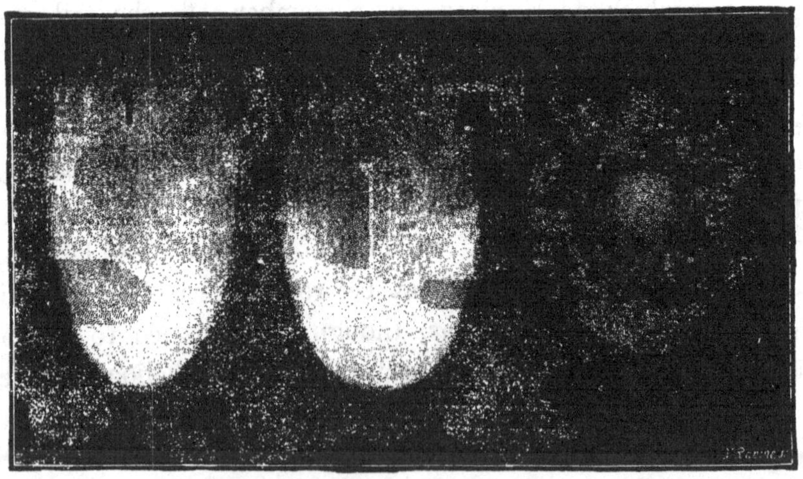

Fig. 34.—Aspect of Donati's comet on December 3, 4, and 6, 1858, according to the observations of M. Liais.

to arrive at positive conclusions respecting the development of the tail during the two apparitions; but it will suffice to remark that in 1759 the tail of the comet did not attain its maximum until long after the perihelion passage, whilst, on the con-

DONATI'S COMET 1858.

Formation and Development of Cometary Appendages, from Drawings by P. G. Bond.

1. September 24, 1858. 2. September 26, 1858.

trary, in 1835 the maximum had been attained before the perihelion.

The great comet of Donati (1858, VI.) likewise furnishes some interesting details on the same point. The first appearance of the tail was observed at Copenhagen and Vienna on August 14, seventy three days after the discovery of the comet, and forty-six days before its perihelion passage; it had

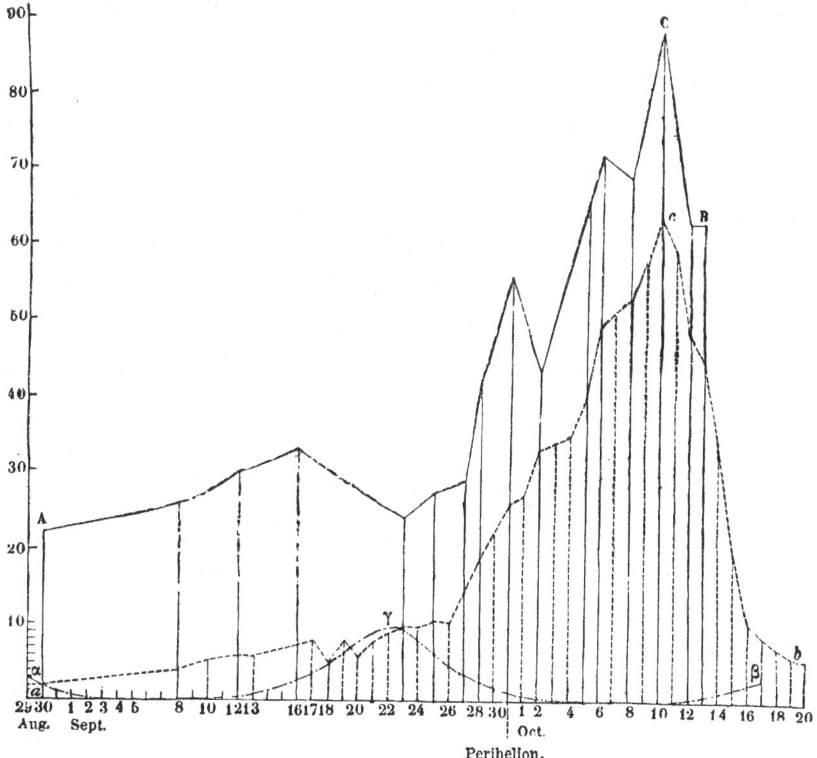

Fig. 35.—Variations of length in the principal tail of Donati's comet.

then an apparent length of but ten minutes. From this date it continued to increase, and at the end of August had attained the length of two degrees. This progressive increase, which underwent but slight fluctuations, the reader may follow either by reference to the table given further on, which is due to

Mr. Bond, or by a glance at the diagram (fig. 35), in which are represented both the apparent and real lengths of the tail. We are here speaking only of the principal tail, which was curved like the edge of a fan, and not of the secondary rectilinear tails, mentioned in Section V. of this chapter. The maximum of apparent length was attained on October 10, eleven days after the perihelion passage; on this day the tail measured sixty-four degrees. From this date it continued to decrease with more rapidity than it had before increased, and on December 3, at Rio de Janeiro, it measured only fifty-five minutes. Three days later it disappeared, 'the comet,' says M. Liais, 'having taken a spherical form, with its nucleus slightly eccentric, and situated in the part nearer to the sun.'

The two curves, *acb*, *ACB*, which in the figure represent the variations in the apparent length and the real length of the tail, exhibit a certain degree of similarity. The differences between the two curves are due, of course, to the changes of distance between the comet and the earth on the successive dates of observation. Figs. 36 and 37, in which the orbits of the comet and the earth are respectively projected, the one upon the other, will enable the reader to determine the real distances between the two bodies on the principal dates of the comet's apparition, and to compare them with the variations observed in the apparent length of the comet's tail. In order to explain completely these variations, however, we must take into consideration all the circumstances that may affect the visibility of the tail, and especially the brightness of the moonlight, which would have the effect of reducing the observed lengths in proportion to its intensity. The curve $\alpha\gamma\beta$, which in fig. 35 marks the varying intensity of the moonlight, reaches its minimum in the nights near October 10. This is the exact date of the maximum apparent length of tail; and the variations, both real and apparent, should for this reason be reduced. But the law of the development of the tail, its formation a certain time before the date of the perihelion passage, its increase in proportion as the comet approached the sun, its diminution, commencing a certain number

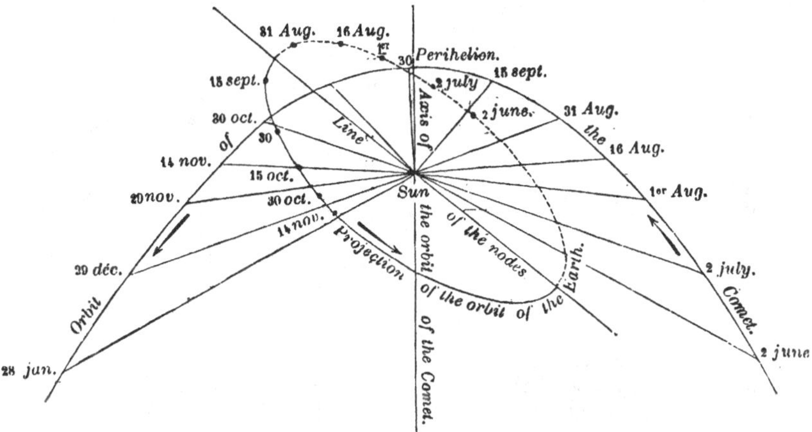

Fig. 36.—Parabolic orbit of Donati's comet. Projection of the earth's orbit upon the comet's orbit. Relative positions of the two bodies.

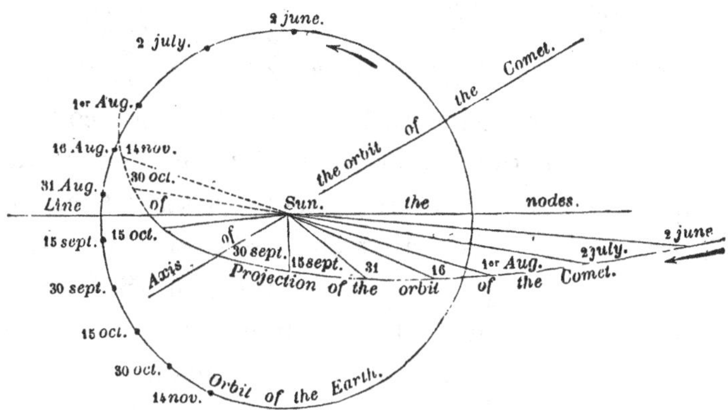

Fig. 37.—Projection of the orbit of Donati's comet upon the plane of the ecliptic. Relative positions of the earth and comet.

of days after the perihelion, its disappearance, effected much more rapidly than its development, are, in our opinion, all incontestable facts which follow from the data which we place before the reader. The following is the table above referred to:—

Length of the Tail of the Great Comet of 1858.

Date	Apparent length in degrees	Real length in miles	Date	Apparent length in degrees	Real length in miles
Aug. 29	2°	14,000,000	Oct. 5	40°	41,000,000
Sept. 8	4°	16,000,000	„ 6	50°	45,000,000
„ 10	5° 24′	—	„ 7	51°	—
„ 12	6°	19,000,000	„ 8	53°	—
„ 13	6°	20,500,000	„ 9	58°	—
„ 16	7°	—	„ 10	64°	54,700,000
„ 17	8°	—	„ 11	60°	—
„ 18	5°	—	„ 12	48°	—
„ 19	8°	—	„ 13	45°	39,000,000
„ 20	6°	—	„ 14	34°	—
„ 21	8°	—	„ 15	20°	—
„ 22	9°	—	„ 16	10°	—
„ 23	10°	14,900,000	„ 17	9°	—
„ 24	10°	—	„ 18	7°	—
„ 25	10° 30′	—	„ 19	6°	—
„ 26	10° 30′	17,000,000	„ 21	12°	—
„ 27	14° 15′	—	„ 22	4°	—
„ 28	19°	26,000,000	„ 24	4° 30′	—
„ 29	22° 30′	—	„ 25	1°	—
„ 30	26°	34,800,000	„ 27	4° 30′	—
Oct. 1	27°	—	„ 30	1° 30′	—
„ 2	33°	37,900,000	„ 31	1° 24′	—
„ 3	34°	—	Dec. 3	0° 55′	—
„ 4	35°	—	„ 6	0°	—

The examples we have just given do not suffice to justify the conclusion that the development of cometary tails depends solely upon the variation of the comet's distance from the sun. At all events, it is clear that comets show very remarkable differences in this respect. For instance, in 1835 the tail of Halley's comet attained its maximum length before the perihelion, and at the date of the perihelion it had entirely disappeared; that of Donati's comet, on the contrary, only attained its maximum after the perihelion; and two whole months then elapsed before the comet again resumed its original form of a round nebulosity.

To conclude our remarks upon this highly interesting but as yet insufficiently studied subject, let us compare the comets

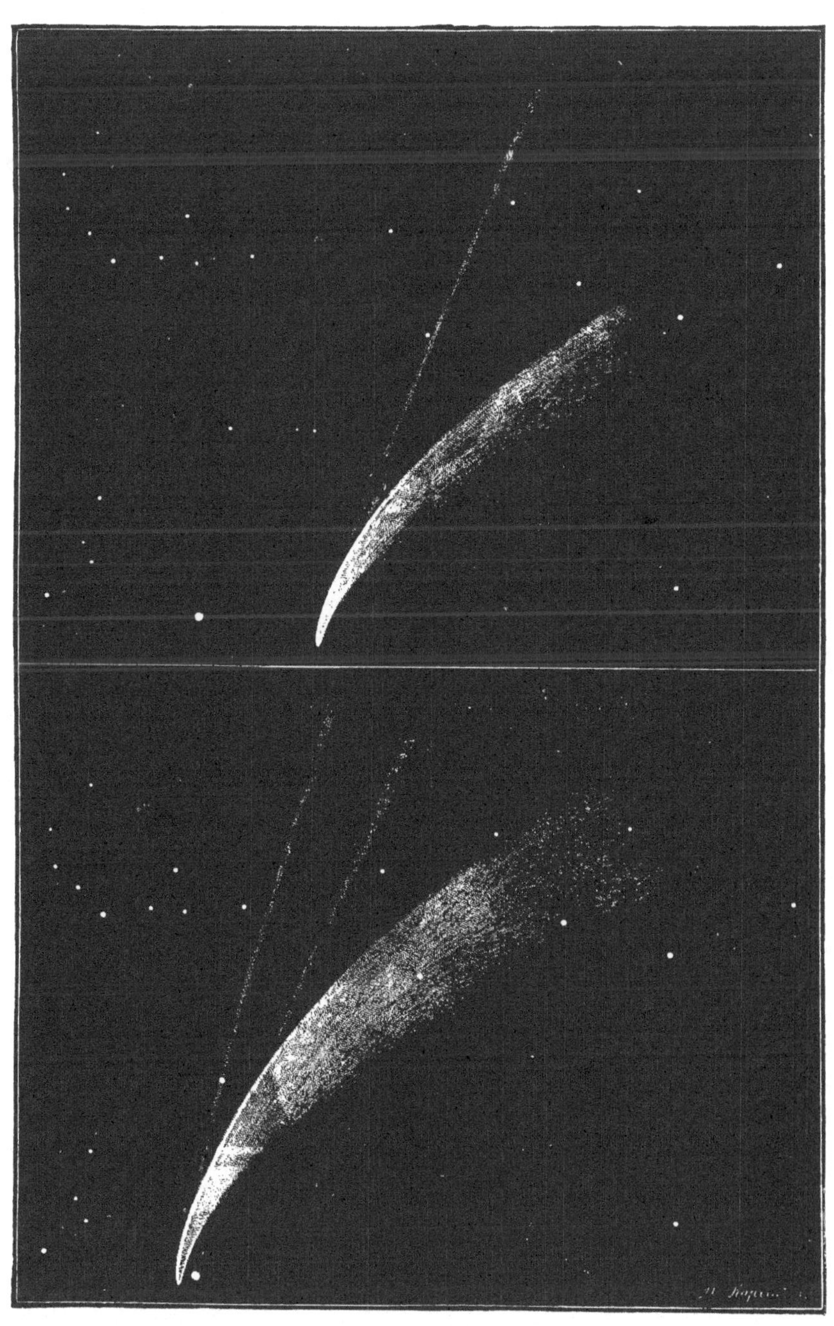

DONATI'S COMET, 1858.

Formation and Development of Cometary Appendages, from drawings by P. G. Bond.

1. October 3, 1858. 2. October 5, 1858.

whose tails we have estimated in miles with reference to their perihelion distances. The table at page 222 will render this comparison easy. In point of fact we see that six only—viz. those of 1843, I., 1680, 1847, I., 1769, 1860, III., and 1811, II. —satisfy the condition that the lengths of their tails are greater in proportion as the perihelion distances are smaller. The comet of 1744 might be substituted for that of 1860, III., without disturbing this relation; but the other four comets do not conform to the rule. So that we are not justified in extending to the comparison of comets, one with another, the law of variation according to distance which we have seen to hold good as regards the development of the same tail.

SECTION IX.

BRILLIANCY OF COMETS.

Estimations of the apparent dimensions or brilliancy of comets—Ancient comets said to be brighter than the sun—Comets visible to the naked eye and comets seen at noonday; great comets of 1744 and 1843.

We will now enter into some particulars respecting the dimensions of comets, their atmospheres, nuclei, and tails. In order to form correct notions concerning this portion of our subject, it is important to distinguish between real and apparent dimensions. This is elementary, but it is here even more necessary than elsewhere, because, from the very nature of cometary orbits, the comet itself, whether periodical or non-periodical, may be situated at the moment of its appearance either very near to or very far from the earth; so that on two successive apparitions the same comet may appear of very different aspect and dimensions, and at one time may present itself as a very conspicuous body in the sky, at another may be hardly visible, or perhaps not visible at all without the aid of a telescope. We have already alluded to this point when speaking of the difficulty of recognising the identity of a new comet with one before observed by its external aspect; and we must here call attention to it again, when we are comparing different comets in respect to their dimensions, either real or apparent.

All comets whose apparitions are anterior to the sixteenth century were visible to the naked eye; their heads, nuclei, or comæ were, therefore, by no means insignificant; the faintest were at least equal in brilliancy to stars of the fifth or sixth magnitude; or if not, the extent of the surrounding nebulosity compensated in point of visibility for the inferiority of the nucleus. This remark applies also to all comets that have been observed with the naked eye since the introduction of the telescope. But, as we are aware, by the aid of instruments comets are detected of so faint a light that they appear as feeble nebulosities, devoid of condensation or nucleus. Many of these last are periodical, and approach to within a moderate distance of the earth; so that it is not their actual remoteness that renders it so difficult for us to see them. There is, therefore, amongst comets the same diversity of dimensions and brilliancy as amongst the stars.

Certain comets have been of enormous dimensions and of great brilliancy. Ancient traditions testify to this intensity, but we must not rely too implicitly upon accounts derived from these sources, for they contain evident exaggerations. As such, for instance, we must regard the comet of B.C. 183, '*which was more brilliant than the sun*, and was seen by daylight in Pisces'; and the comet mentioned by Seneca, which appeared B.C. 146. 'After the death of Demetrius, King of Syria, the father of Demetrius and Antiochus, there appeared shortly before the Achæan war a *comet as large as the sun.* At first it was like a disc of fiery red, and its light dissipated the darkness of the night. Imperceptibly it decreased in size, its light became dim, and it totally disappeared.' Again, the comet which appeared B.C. 136, at the birth of Mithridates, and remained visible for seventy days, seems to have been somewhat magnified by the imagination of observers and historians. 'The heavens appeared on fire; the comet occupied the fourth part of the sky, and its light *exceeded that of the sun.*'

In the *Cométographie* of Pingré we find the following description of comets which were remarkable either for their dimensions or the brilliancy of their light:—

'1006. Haly ben Rodoan being young, a comet was seen in the 15th degree of Scorpio; the head was *three times as large as Venus*; it gave as much light as a quarter of the moon would have given.'

'1106. A great and beautiful comet. On the 4th, or, according to others, on the 5th of February, a star was first seen distant only a foot and a half from the sun; it was there beheld from the third to the ninth hour of the day. Some authors have given to this star the name of comet.

'1208. In this year there appeared a comet. For a fortnight, after sunset, a star was seen of such brilliancy that it produced a great light, not unlike a fire. The Jews regarded it as a sign of the coming of the Messiah.

'1402. A very large and very brilliant comet; no one remembers to have seen such a prodigy. (This is believed to be a prior apparition of the comet of 1532 and 1661.). . . It increased day by day in size and brilliancy as it drew near the sun. On Palm Sunday, the 19th of March, and the two following days, it increased prodigiously; on Sunday, its tail was twenty-five fathoms long;* on Monday, fifty, and even

* In ancient chronicles we frequently find in descriptions of the apparent dimensions of celestial bodies, and more especially the tails of comets, expressions similar to those here mentioned; that is to say, lengths expressed in ordinary measures—*two or three feet*, or *twenty or a hundred fathoms*, &c. It is plainly impossible to give any rational meaning to such statements. Nor have similar expressions for the same kind of estimates entirely passed out of use even at the present day. A person who sees a bolide will say, for instance, that its size was that of an orange, and that it had a train two yards long. He does not understand that this manner of measuring the apparent dimensions of objects, whose real distance is unknown, is altogether indeterminate, and that, although it may

one hundred; on Tuesday, more than two hundred. It then ceased to be visible at night, but during the eight following days it was seen in the daytime close to the sun, which it preceded. Its tail was not more than one or two fathoms long; it was so bright that the light of the sun did not prevent it being seen *at noon-day.*' In 1532, if the identification be correct, the same comet exhibited a degree of brilliancy equal to three times that of Jupiter.

Thus, there have been several comets sufficiently brilliant for their light to have been compared to that of the sun. Three of these were visible during the day. The great comet of 1500, known under the names of *Asta*, and *Il Signor Astone*, was likewise seen in presence of the sun. 'Some voyagers sailing from Brazil to the Cape of Good Hope saw it on the 12th of May; it appeared on the Arabian side of the vessel. Its rays were very long. It was thus continually observed *day and night* for eight or ten days.'

That comets have been brilliant enough for their light to penetrate the sunlit heavens is put beyond a doubt by the observations made by the celebrated Tycho of the comet of

have a precise meaning for him at the moment when he sees the object, it does not follow that at another time his estimate would not be quite different. And, in any case, such estimates, made by different observers, are not comparable one with another. The only proper mode of expressing celestial distances is in degrees and minutes, and the observer, not provided with an instrument, and not accustomed to making such estimations, will find it useful to remember that the diameters of the sun and the moon are pretty nearly equal, and that the diameter of each is about half a degree. By comparison with either of these luminaries it is easy to make a good estimate of a celestial distance. We may also—and this is a good plan, if it be a starlight night—compare the length to be measured with the distance between two well-known stars in any of the constellations, such as the Great or Little Bear, Orion, Pegasus, Cassiopeia, &c. We especially dwell upon this matter, because more than once we have had occasion to deplore the method of measuring celestial dimensions in feet, yards, &c., a method of measuring absolutely without meaning, and which may render valueless an observation which might otherwise be important.

1577, and those of contemporary astronomers of the great comet of 1843. 'On the 13th of November, 1577, whilst the sun was still above the horizon, this new star (the comet) caught the attention of Tycho Brahé. He estimated the diameter of its head at seven minutes.'

With respect to the comet of 1843 the following details, which we borrow from Arago, leave no doubt of its visibility in full sunlight: 'The comet, first perceived by the spectators in broad daylight, and thought to be a meteor, was, at the hour of noon, according to an observation made by M. Amici, 1° 23′ east of the centre of the sun. M. Amici says only of the body that it was *fumous* towards the east. At Parma the observers aver that whilst stationed behind a wall screening the sun from view they distinctly saw a tail of from four to five degrees in length. In Mexico, on the same day (the 28th of February), at eleven o'clock in the morning, according to the *Diario del Gobierno*, "the comet was visible to the naked eye, near to the sun, like a star of the first magnitude, with the first development of a tail directed towards the south." Mr. Bowring, at the mines of Guadaloupe y Calvo (Mexico), saw the comet on the 28th of February, from nine o'clock in the morning until sunset. At Portland (U.S.) the comet was seen with the naked eye in open daylight, to the east of the sun, by Mr. Clarke. Sir John Herschel makes mention of an observation made by one of the passengers on board the *Owen Glendower*, then off the Cape. "The comet was seen as a short dagger-like object, close to the sun, a little before sunset." According to Mr. Clarke, "the nucleus and also certain parts of the tail were as clearly defined as the moon on a clear day." '*

[* Mr. E. C. Otté, in his translation of Humboldt's 'Cosmos' (vol. i., p. 86), states that at New Bedford, Massachusetts, U.S., on February 28, 1843, he distinctly saw the comet between one and two in the afternoon. The sky at the time was intensely blue, and the sun shining with a dazzling brightness unknown in European climates.—ED.]

A century before, in 1743, a comet was observed in Europe —Chéseaux's comet, which we have several times mentioned— which surpassed in brilliancy stars of the first magnitude. On January 9, 1744, the head of the comet was equal to a star of the second magnitude, and its diameter fifteen days after amounted to ten seconds; on January 26 it was equal to a star of the first magnitude; on February 1 it was brighter than Sirius; and finally, during the last few days of this month and the commencement of March, it became so bright that it was visible by daylight in presence of the sun. But a remarkable circumstance related by Chéseaux is this: 'From the 13th of December to the 29th of February (on the following day, the 1st of March, the comet passed its perihelion) the atmosphere of the comet continued to diminish in size,' as if the augmented brilliancy of the head was produced by the disappearance of the nebulosity surrounding the nucleus, or by a condensation of the nebulous atmosphere.

SECTION X.

DIMENSIONS OF NUCLEI AND TAILS.

Real dimensions of the nuclei and atmospheres of various comets—Uncertainty of these elements; variations of the nucleus of Donati's comet—Observations of Hevelius upon the variations of the comet of 1652—Do cometary nebulosities diminish in size when their distance from the sun decreases?—Encke's comet considered in regard to this question at its apparitions in 1828 and 1838.

THE observations that we have just recorded give an idea of the brightness of cometary light, and the intensity to which that brightness may attain; but they afford no certain indication concerning the dimensions of cometary nuclei or atmospheres. Upon this point we are about to give the result of a few measurements; but these measurements, it must be understood, are not so exact as those of the bodies of the solar system, the planets, the moon, and sun. The uncertainty we speak of does not arise from the difficulties experienced in the determination of the measures themselves, although they contribute to it, cometary nuclei being often as deficient in a clear and well-defined outline as the nebulosities; but what more especially prevents us from regarding the numbers we now give as constant, and therefore characteristic elements of the comets to which they belong, is the continual variation to which the different parts of the head are subject during the time of the comet's apparition.

The following two tables contain the values obtained for the dimensions of various cometary nuclei and atmospheres, arranged in order of magnitude:—

Diameters of Cometary Nuclei.

	Miles.
Comet of 1798, I.	28
" 1805	30
" 1799, I.	385
" 1811, I.	429
" 1807	550
" 1811, II.	2,700
" 1819, I.	3,300
" 1847, I.	3,500
" 1780, I.	4,200
" 1843, I.	5,000
" 1815	5,300
" 1858, VI.	5,600
" 1769	28,000

Diameters of Cometary Atmospheres.

	Miles.
Comet of 1799, I.	1,200
" 1807	1,900
" 1847, V.	17,900
" 1847, I.	25,400
" 1849, II.	50,700
" 1843, I.	94,500
" Brorsen, 1846	129,000
" Lexell, 1770	203,000
" 1846, I.	241,000
" Encke, 1828,	264,000
" 1780, I.	267,000
" Halley, 1835	354,000
" 1811, I.	1,120,000

On comparing these tables we find that the six comets, 1799, I., 1811, I., 1807, 1847, I., 1780, I., and 1843, I., whose nuclei and atmospheres have been both measured, do not occupy the same relative positions in each. This is strikingly shown in the case of the great comet of 1811, whose somewhat small nucleus was surrounded by an immense nebulosity, as is evident from the large number in the last line of the second table. The volume of the nucleus was only equal to the 6,300th part of the volume of the earth, whilst that of the coma was 2,800,000 times greater than the volume of the earth; that is to say, more than double the volume of the sun.

In order to justify the remarks at the beginning of this section, let us take for example the beautiful comet of Donati, whose physical elements have been so carefully studied by Bond. The diameter of 5,600 miles, given in the table, has reference to the nucleus on July 19. On August 30 this diameter was reduced by one-sixth, and measured no more than 4,660 miles. It continued to decrease until October 5, on which day it did not exceed 400 miles, less than $\frac{1}{14}$th of its diameter on July 19. The next day it attained 800 miles, having doubled its dimensions between one day and the next, the volume of the nucleus having thus been increased in the proportion of 1 to 8. Finally, on October 8 the diameter attained a new maximum of 1,120 miles, and on the 10th was reduced again by one-half, viz. to 630 miles. We do not now enter into the significance of these rapid variations, of which we shall have to speak hereafter, when treating of the physical constitution of cometary nuclei. It will then be seen that these variations appear to be connected with the changes of distance between the nucleus of the comet and the sun.

Hevelius, in the sixth book of his *Cometographia*, describes the physical aspect of the comet of 1652, the magnitude of the head and tail, together with the brilliancy and the colour of their light. He observes that, the apparent dimensions of the comet having diminished day by day, this diminution was the natural result of the continually increasing distance between the comet and the earth, but that in reality the absolute size of the comet was increasing day by day. This observation, the value of which Pingré denies, because he does not believe that Hevelius could have measured with sufficient accuracy the dimensions of the comet or calculated its distances from the earth, has since been generalized, and several astronomers, including Newton, have remarked that the diameters of cometary nebulosities increase in proportion as the comet becomes more and more distant from the sun. Arago observes that the comets of 1618, II., and 1807, manifestly exhibited this phenomenon. It has, however, been better exemplified by Encke's comet in its two apparitions of 1828 and 1838. The following table shows these remarkable variations:—

Real Diameters of Encke's Comet in 1828.

Days	Distance from the sun	Diameter in miles
October 28	1·46	323,000
November 7	1·32	263,000
„ 30	0·97	122,000
December 7	0·85	82,000
„ 14	0·73	45,000
„ 24	0·54	12,000

The diminution of the diameters is much more rapid than that of the distances from the sun; the six distances decrease, in fact, in the proportion of the numbers 100, 90, 65, 58, 50, and 36, whilst the corresponding diameters are to each other as the numbers 100, 81, 38, 25, 14, and 4; the distances being at length reduced to a third nearly, whilst the diameter is twenty-six times less; and if we pass from the diameter to the volume of the nebulosity, it will be found that between October 28 and December 24 the volume was reduced to the 17,600th part of its original value.

We now proceed to the variations exhibited by the same comet in 1838, the elements of which are as follows:—

Real Diameters of Encke's Comet in 1838.

Days	Distance from the sun	Diameter in miles
October 9	1·42	278,000
" 25	1·19	119,000
November 6	1·00	80,000
" 13	0·88	75,000
" 16	0·83	62,000
" 20	0·76	55,000
" 23	0·71	37,000
" 24	0·69	30,000
December 12	0·39	6,500
" 14	0·36	5,500
" 16	0·35	4,200
" 17	0·34	3,000

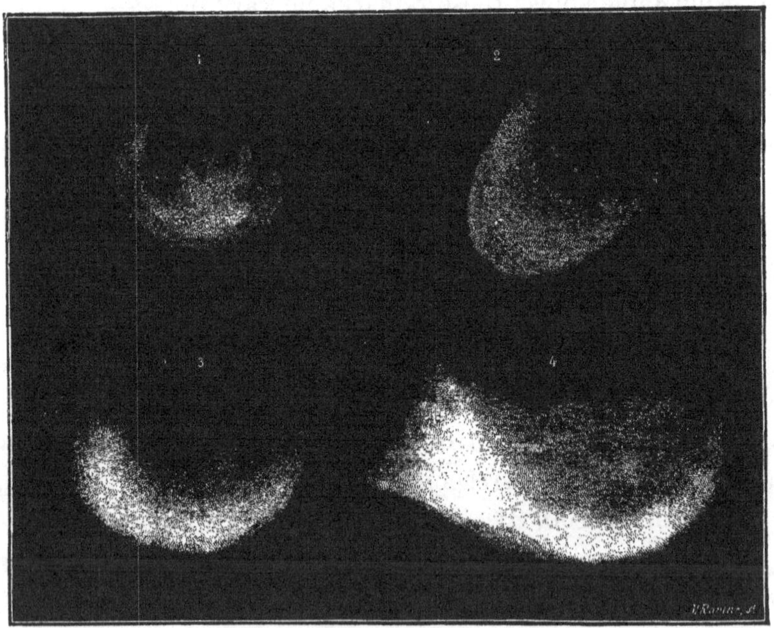

Fig. 38.—Encke's comet, according to the observations of Schwabe. 1. October 19, 1838; 2. November 5; 3. November 10; 4. November 12.

From October 9 to December 17 the distance of the comet from the sun was reduced in the proportion of four to one, while the real diameter of the nebulosity was reduced to the

93rd part of its value; and its volume—supposing the comet to be spherical, or, at all events, only changing its size, not its shape—was reduced to the 813,000th part of the original volume. It appears even certain that on this second apparition the law of decrease of the diameter as compared with the diminution of the distance from the sun followed a still more rapid law of variation. Moreover, it should be remarked that, at equal distances, the comet in 1838 was somewhat less in volume than ten years before. We limit ourselves at present to the statement of the fact, as we shall give further on the explanations offered, and the difficulties which it presents in respect to the physical constitution of comets. As some connexion has been thought to exist between the changes of volume in cometary nebulosities, and the development and formation of tails, we may here remark that Encke's comet is a nebulosity of variable form—sometimes globular, sometimes oval, or more or less irregular (figs. 17, 22, 23, and 38), and that at no time has it ever exhibited a tail.

CHAPTER VIII.

PHYSICAL TRANSFORMATIONS OF COMETS.

SECTION I.

AIGRETTES—LUMINOUS SECTORS—NUCLEAL EMISSIONS.

Predominance of atmosphere in comets—Luminous sectors; emission of vaporous envelopes from the nucleus in the comets of 1835, 1858, 1860, and 1861—Formation of envelopes in Donati's comet; progressive diminution of the velocity of expansion in emissions from the nucleus.

THE planets, as seen through a telescope, are bodies of regular form and definite dimensions, probably invariable, so far as we can judge from observations made in the short period of two centuries and a half that has elapsed since telescopes have been invented. A globular mass, solid or liquid, surrounded on all sides by a light and comparatively thin aëriform envelope, is perhaps, from a physical point of view, the simplest description of a planet. The comparative stability is due, on the one hand, to the preponderance of the central globe, where general phenomena are modified only at long intervals; and on the other to the trifling depth of the atmosphere, the portion of the planet the most subject to variation and internal change.

In comets, we have seen, this relation is reversed, and the atmosphere or nebulous envelope constitutes the entire body, or at all events greatly preponderates. At the utmost we can only conjecture that in some comets the nucleus is solid or liquid. Certainly its volume is generally but a very insignificant portion of the entire nebulosity, even if we except the tail.* A comet which in one part of its orbit seems to be reduced to a simple nebulosity will gradually exhibit a luminous condensation and then a nucleus. This nucleus either increases or decreases in volume and brightness. Nothing appears stable in the constitution of these remarkable bodies; variability of aspect is one of their most distinctive features. We have already seen that the nucleus and the atmosphere of a comet undergo considerable changes in the course of the same apparition; the enormous appendages of certain comets are generated, take form, and develop only to diminish and then disappear. It remains now to study the internal changes, changes that require for their observation instruments of the greatest power, to examine if there is not some connexion between the phenomenon of tails and the movements of the coma and the nucleus, and whether they are not connected with some external influence, such as the solar heat or other natural action. All comets which describe orbits of considerable eccentricity must in the course of a single revolution be exposed to enormous differences of temperature; and the extreme variations of heat and cold to which they are subjected between their perihelia and aphelia cannot fail to create in these masses of vapour, gas, or particles disseminated over enormous volumes—whichever they may be—movements of contraction and dilatation, perhaps even of chemical action, of which upon our globe we can have no idea. The phenomena of solar spots and protuberances are alone comparable with these rapid and singular transformations.

But let us now proceed to the facts which justify these conjectures and invest them with a high degree of probability.

* The tables on page 238 show that the volumes of the nuclei, in the comets of 1799 and 1807, only amount to $\frac{3}{100}$ of the volumes of the nebulosities. This proportion decreases to $\frac{1}{8000}$ in the comet of 1843, and to $\frac{1}{20000000000}$ in the great comet of 1811.

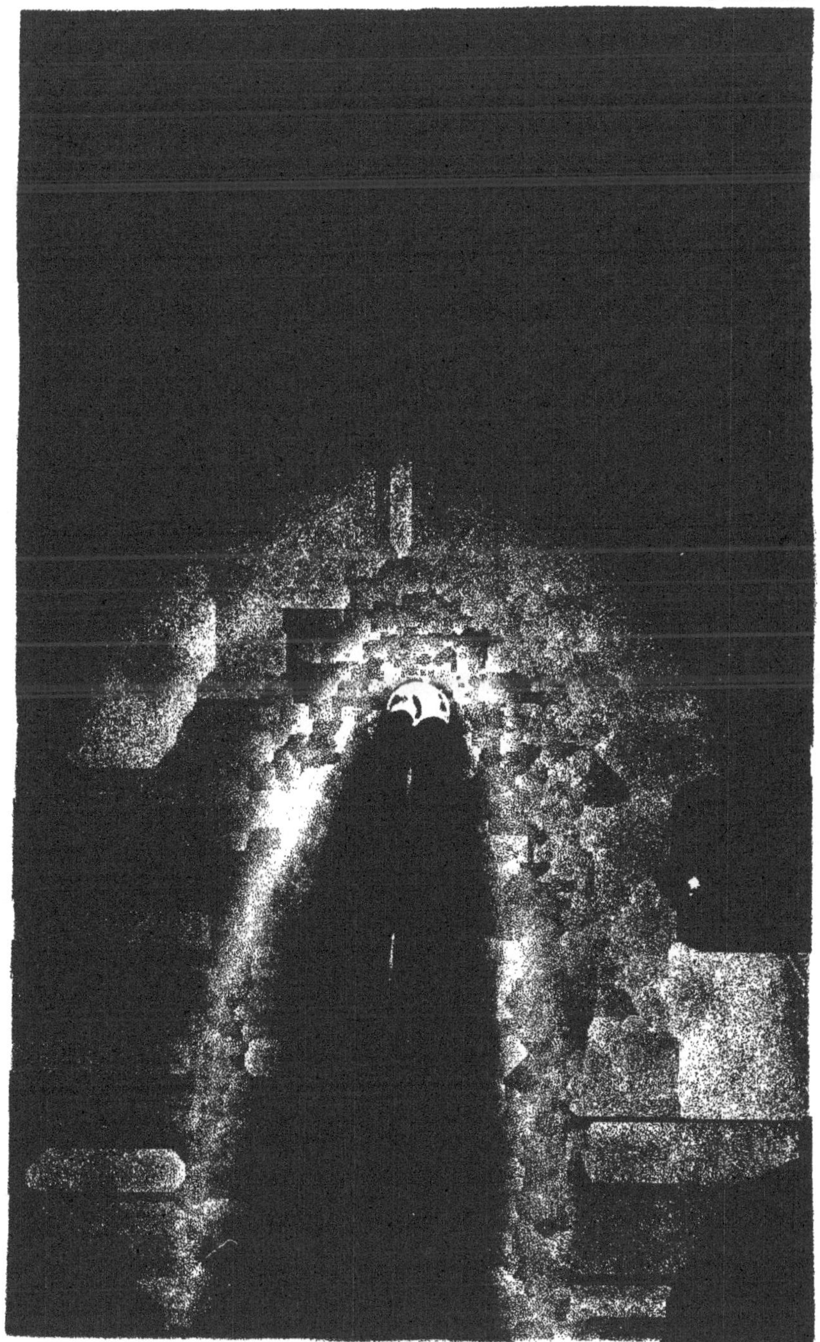

Warren De La Rue. del. July 3 $12^h 40^m$ G.M.T J. Basire sc.

THE GREAT COMET OF 1861

AS SEEN BY WARREN DE LA RUE, D.C.L., F.R.S.

WITH HIS NEWTONIAN EQUATOREAL

OF 13 INCHES APERTURE

The continual changes of which the heads of a certain number of comets are the seat were first put on record by Heinsius, when observing at St. Petersburg the great comet of 1744 (Chéseaux's comet). 'On the 5th day of January,' says Arago, 'Heinsius saw nothing extraordinary about the comet; but on the 25th he discovered a luminous *aigrette*, in the form of a triangle, the apex of which was at the nucleus, whilst the opening was directed towards the sun. The lateral edges of the *aigrette* were curved, as though driven in from outside by the action of the sun. On the 2nd of February these same edges, still more curved, formed the two sides of the commencement of a tail, which became more distinct on the following day.'

No other observations of the same kind were made until the return of Halley's famous comet in 1835, when the formation of luminous sectors, which seemed to spring from the nucleus towards the sun, the variations of their position, number, and brilliancy, and other curious and instructive phenomena, were observed in various parts of Europe: at the Observatory of Paris, by M. F. Arago; at Dessau and Königsberg, by Schwabe and Bessel; at Markree, Ireland, by Mr. Cooper; at Florence, by M. Amici. From October 7 to November 10 the head of the comet presented a succession of appearances of which we subjoin a few examples. (See Plate VI., in which the variations of the comet's atmosphere are represented, according to the observations of Sir John Herschel at the Cape of Good Hope; and fig. 39, in which these appearances are given according to Schwabe.) By the study and interpretation of these phenomena Bessel, the illustrious astronomer of Königsberg, directed the attention of savants and observers to this hitherto much-neglected branch of cometary astronomy; and Arago likewise contributed to the same object by various popular notices in the *Annuaire du Bureau des Longitudes*. Bessel has particularly dwelt upon one fact of high importance: he remarked that the luminous coma, sector, or *aigrette* emanated from the nucleus, and was at first emitted in the direction of the radius vector; it then deviated gradually, and by a marked amount, from its first direction, and finally returned to its original position and deviated again in the opposite direction. He was thus led to infer the existence of a movement of rotation, or rather oscillation of the head and nucleus in the plane of the orbit. It is this oscillation

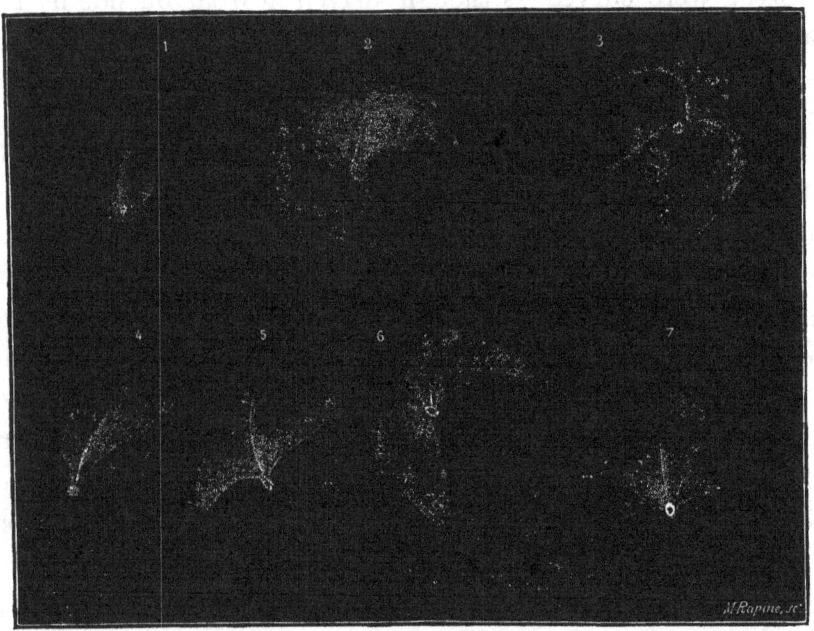

Fig. 39.—Luminous sectors and *aigrettes* of Halley's comet, according to Schwabe. (1) October 7, 1835; (2) October 11; (3) October 15; (4) October 21; (5) October 22; (6) October 23.

which has given rise to the hypothesis, remarkable in all respects, of the existence of a polar force, having its focus of action in the sun, and which causes cometary bodies to oscillate just as a bar magnet causes a magnetic needle to vibrate. Further on we shall devote a section (Chap. XI., Sec. V.) to the exposition of Bessel's theory.

Four other comets have presented analogous phenomena, but with differences that we shall proceed to mention. These are the comet of Donati 1858, those of 1860, III., 1861, II., and lastly the comet of 1862, II., concerning which we shall enter into some details. These details will explain the formation and succession of the luminous *aigrettes*, or sectors, the nebulous envelopes to which they give rise, and lastly the formation of the tail, which the cometary matter that has

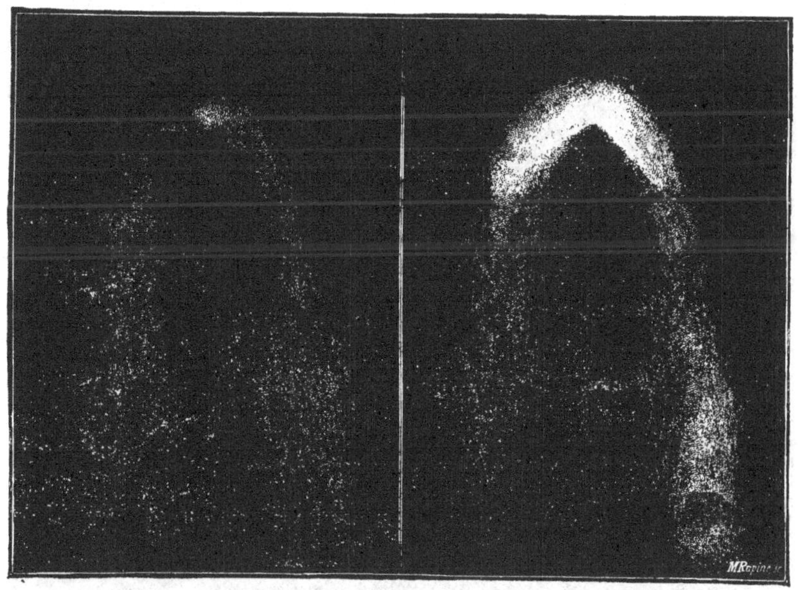

Fig. 40.—Formation of luminous sectors and envelopes. Donati's comet, Sept. 8, 1858.

Fig. 41.—Comet of 1860, III. June 27, according to Bond. *Aigrettes* and envelopes.

thus left the nucleus appears to originate under the influence of a kind of repulsion, the cause of which we shall have later on to consider.

In Donati's comet the jets of luminous matter liberated from the nucleus in the form of luminous sectors, disposed like a fan, produced around the head successive envelopes, which as they receded from the nucleus diminished in brightness and became uniformly blended. This kind of compression was regarded by Bond as the result of progressive diminution in the velocity of expansion of each envelope. Seven successive envelopes, rising above the nucleus, were formed in periods varying from 4 days 16 hours to 7 days 8 hours. Each successive envelope as it arose remained as if retained by the nucleus for a certain time, until, in virtue of some acquired property, it drifted back and contributed to form the two main divisions of the tail. The sectors always appeared in the same direction,

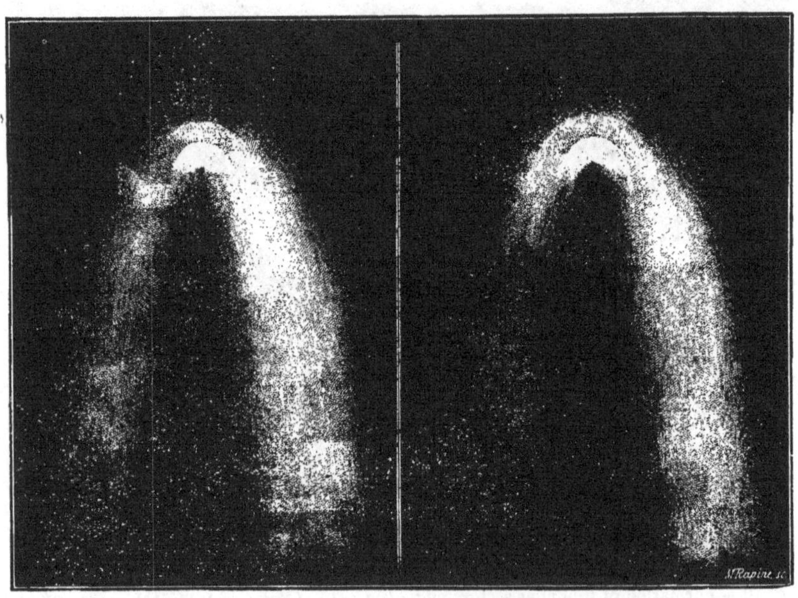

Fig. 42.—Luminous envelopes of Donati's comet. September 30, 1858.

Fig. 43.—The same comet. October 2. From a drawing by Bond.

facing the sun, so that we may conclude that neither the nucleus nor the coma was endowed with a movement of sensible rotation; thus no oscillation of the kind observed by Bessel was manifested in the head of Donati's comet, except the motion necessitated by the constant direction of the luminous sectors towards the sun. Even this absence of rotation, according to Bond, implies the action of a polar force, emanating from the sun, and maintaining the axis of the nucleus in the direction of the focus of movement.

The comets of 1860 and 1861 were also the seats of nucleal emissions in a permanent direction, the first for a fortnight, the second for a month. Eleven successive envelopes

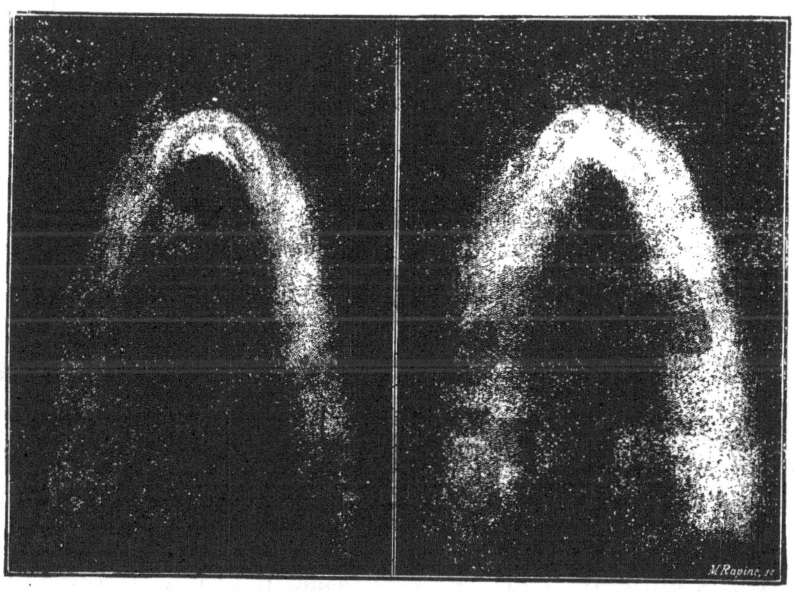

Fig. 44.—Formation of the luminous envelopes in Donati's comet. October 6. Fig. 45.—The same. October 8. Both from drawings by Bond.

were emitted from the nucleus of the comet of 1861, at regular intervals of two days. The development and final dispersion were thus accomplished with much greater rapidity than in the case of Donati's comet.

SECTION II.

OSCILLATIONS OF LUMINOUS SECTORS: COMET OF 1862.

M. Chacornac's observations upon the comet of 1862—Formation of luminous sectors emanating from the nucleus—Oscillation of *aigrettes*, and flowing back of the nucleal matter.

We are now about to give our attention to the evolutions of the luminous sectors of the great comet of 1862, which, on the contrary, presented oscillations analogous to those exhibited by the *aigrettes* of Halley's comet. We shall follow the development of these phenomena by means of the observations of the late M. Chacornac.

On August 10, 1862, M. Chacornac detected in the head of the comet the presence of a luminous *aigrette*, a brilliant sector directed towards the sun. This sector, which at three o'clock in the morning included an angle of 46°, had, by two o'clock on the following day, opened 'like the corolla of a convolvulus, and included 65°. On the 10th the nucleus presented the appearance of a rocket, having a diameter much more extended in the direction of the radius vector than at right angles to it.' It is worthy of remark that the contrary was the case with the nuclei of the comets of 1858 and 1861, which were flattened in the direction of the radius vector. On the 11th the two diameters were nearly equal. New sectors disengaged themselves successively from the nucleus, and on August 26 M. Chacornac determined that between the 10th and the 26th they had succeeded each other to the number of thirteen.

Having carefully followed throughout this interval (with the exception of three nights, when the sky was cloudy) the formation of these successive sectors, M. Chacornac has given in the following terms a brief description of the phenomena which he observed:—

'The nucleus of the comet emits periodically, in the direction of the sun, a gaseous jet from which particles of cometary matter escape like steam escaping from a piston. This jet preserves for a certain time a rectilinear form, as if a force of considerable projecting power, residing in the nucleus, threw off particles in that direction; then it becomes inflected and takes the form of a slightly arched cone. At this same moment the cometary matter, accumulating at the extremity of the jet nearer to the sun, forms a kind of cloud, the rounded outline of which would appear to indicate that at this distance from the nucleus the force of projection has been overcome by some resistance opposed to it; the cloud returns on both sides, like a puff of smoke driven back by the wind; and, opening out into a level sheet, flows away in the direction of the tail.'

'By degrees the vaporous cone, the axis and vertex of which have continued to appear the most luminous portions, assume a diffused and nebulous appearance, as if veiled by an accession of thick atmosphere; the brightness of the centre diminishes, that of the sides increases, and the cone enlarges. The diffused appearance continuing to increase, the gaseous jet loses its form, the light of the axis disappears, and everything seems to indicate that the nucleal emission has ceased in this direction. The nucleus appears round and brilliant. At this time, at an angle with the radius vector of about 30 degrees towards the east, appear the first traces of a new jet destined to succeed the former; and, in proportion as these traces become more apparent, the vaporous jet, originally directed to the sun, continues to enlarge and to curve more and more, until at last, having gradually changed its form and become reduced by imperceptible degrees to a misty haze, it hardly retains a trace of its primitive shape and direction. In this state the hemispherical envelope surrounding the *aigrette* is more brilliant and better defined in the portion corresponding to the jet in process of dispersion than elsewhere.

'During the dispersion of the jet directed to the sun the new jet has been gradually progressing like the first; that is to say, the nucleus has been lengthening by degrees into the form of a cone and disengaging particles from every part of its surface, which, thrown off in the direction of the radius vector, have been actively forming the new jet, which is destined sixteen hours later to pass through the same phases as its predecessor. This new jet exhibits the same changes as the previous one, with this exception, however, that it feeds the eastern portion of the hemispherical envelope, and the other branch of the tail. From the observations which I was enabled to make, it seems that these nucleal emissions have been taking place alternately since August 9, each ray or jet directed towards the sun, or nearly so, being succeeded by another ray or jet inclined to the preceding, so that the number of vaporous jets emitted by the nucleus, from that date up to ten o'clock on the night of August 26, would amount to thirteen. Since the date of the comet's perihelion passage the jet which corresponded very nearly to the direction of the radius vector has gradually inclined to the west, so that the other jet, turned towards the east, is now directed to the sun.'

On comparing these highly interesting phenomena with those previously described it is impossible to avoid being struck, not only by the degree of similarity they present to the phenomena which attracted the attention of Bessel in Halley's comet, but by the differences between these same phenomena and those which were observed in the great comet of 1858. It does not appear that any trace of oscillation was manifested in the luminous sectors of the latter, whilst the development of concentric envelopes was, on the contrary, the distinctive feature. On the other hand, in the comets of 1835 and 1862 the oscillatory motion, more or less rapid, chiefly characterised the jets of vapour and luminous matter emitted by the nucleus. These differences and analogies it will be necessary to bear in mind, when we endeavour to trace to a unique physical cause the transformations that are continually occurring within the atmospheres of comets, especially in the vicinity of the perihelia.

SECTION III.

DUPLICATION OF BIELA'S COMET.

First signs of the doubling of Biela's comet, in the month of January 1846—Observations of the twin comets in America and Europe—Gradual separation and approach of the fragments—The two comets return and are observed in 1852; their distances found to have increased—Elements of the orbits of the two comets.

WE now come to transformations still more singular in the outward appearance of cometary nebulosities, and more radical in their nature.

The second return of Biela's comet (period $6\frac{3}{4}$ years) since the epoch of its discovery as a periodical comet in 1826, or the eleventh of its returns since it was first observed in 1772, was marked by a memorable event, viz. its duplication and division into two distinct and separate comets. We here subjoin a few details on the subject of this event.

On December 21, 1845, the comet was observed by Encke at Berlin; on the 25th of the same month it was seen by M. Valz at Marseilles. Neither of these two astronomers perceived the slightest trace of separation. On the 19th, however, Mr. Hind remarked towards the north of the nucleus what appeared to be a kind of protuberance: was this a premonitory sign of the doubling of the comet? However this may be, it appears certain that the comet was first seen to be double on January 13, 1846, at Washington. In Europe the existence of two separate nuclei was observed by Professor Challis,* at Cambridge, on January 15; and by M. Valz at Marseilles, and Encke at Berlin, on the 27th, only fifteen days before the perihelion passage of the comet.

[* The doubling of Biela's comet was so remarkable an event in the history of cometary astronomy, that I think Professor Challis's own account of what he saw will be found interesting to the reader. He announced the extraordinary appearance of the comet to the President of the Royal Astronomical Society as follows:—

'On the evening of January 15, when I first sat down to observe it, I said to my assistant, "I see *two* comets." However, on altering the focus of the eye-glass and letting in a little illumination the smaller of the two comets appeared to resolve itself into a minute star, with some haze about it. I observed the comet that evening but a short time, being in a hurry to proceed to observations of the new planet. On first catching sight of it on this evening (Jan. 23) I again saw two comets. Clouds immediately after obscured the comet for half an hour. On resuming my observations I suspected at first sight that *both* comets had moved. This suspicion was afterwards confirmed: the two comets have moved in equal degree, retaining their relative positions. I compared both with Piazzi, 0^h 120, and the motion of each in 50^m was about 7^s in R.A. and $10''$ in N.P.D. What can be the meaning of this? Are they two independent comets? or is it a binary comet? or does my glass tell a false story? I incline to the opinion that this is a binary or double comet, on account of my suspicion on Jan. 15. But I never heard of such a thing. Kepler supposed that a certain comet separated into two, and for this Pingré said of him, "*Quandoque bonus dormitat Homerus.*" I am anxious to know whether other observers have seen the same thing. In the meanwhile, I thought, with the evidence I have, I had better not delay giving you this information.'

In a subsequent letter Professor Challis says: 'There are certainly two comets. The north preceding is less bright and of less apparent diameter than the other, and, as seen in the Northumberland telescope, has a minute stellar nucleus. I compared the two comets independently with A.S.C. [Astronomical Society's Catalogue] 51 on the evenings of January 23 and 24, and obtained the following places . . .

'The greater apparent distance between the comets on January 24 is partly accounted for by their approaching the earth. I saw the comets on January 25, but took no observation. The relative positions were apparently unchanged.

'I think it can scarcely be doubted, from the above observations, that the two comets are not only apparently but really near each other, and that they are physically connected. When I first saw the smaller, on January 15, it was faint, and might easily have been overlooked. *Now* it is a very conspicuous object, and a telescope of moderate power will readily exhibit the most singular celestial phenomenon that has occurred for many years—a double comet.'

'On the 18th and 20th of January,' says M. Valz, 'there was nothing remarkable in the appearance of the comet; but the central luminous condensation seemed to be more intense than on preceding apparitions. Cloudy weather did not permit me to see the comet again until the 27th. I was then struck with amazement to find two nebulosities, separated by an interval of 2′, instead of one nebulosity alone ... Yesterday, on the 29th, in spite of clouds, I again observed the double head; the secondary head is much fainter than the other.' Each head was followed by a short tail, whose direction was perpendicular to the line joining the centres of the nuclei. The two nuclei were moving with the same velocity and in the same direction. On January 31 Mr. Hind verified the rapid separation of the nuclei. Less than a month later the distance between the twin comets had tripled, and the aspect of each varied from day to day. Sometimes the one nucleus would excel in brightness, sometimes the other, so that it was difficult to say which was the original comet and which the secondary.

Fig. 46 shows the aspects and relative positions of the nuclei and their tails on February 21, according to a drawing of the Russian astronomer, Otto Struve. At this time there was no apparent connexion, no material communication between the two bodies. 'The part of the heavens separating them was,' as Humboldt observes, 'remarkably free from all nebulosity, as seen at Pulkowa. Now, some days later, Lieutenant Maury observed at Washington, with a telescope furnished with a Munich object-glass of 9 inches diameter, rays sent out by the old comet towards the new, so that for some time a kind of bridge extended from the one to the other. On the 24th

At Königsberg, M. Wichmann observed the comet on the 14th, and saw nothing of the companion. There was, however, some vapour in the air. On January 15, the air being purer and the moon not risen, he saw the companion comet immediately with a power of 45.—*Monthly Notices of the Royal Astronomical Society*, vol. vii. pp. 73–75 (March 13, 1846).—Ed.]

of March the little comet, insensibly diminishing in brightness, was hardly recognisable. The larger one, however, continued visible until about the 16th or 20th of April, when it also disappeared.' ('Cosmos,' vol. iii.)

The increase of the apparent interval between the two nuclei proved no actual increase of distance between the two fragments of the comet, since they were approaching the earth during the time of observation; but the calculation of the true distances was performed by M. Laugier and subsequently by M. Plantamour and M. D'Arrest; and it results from

Fig. 46.—Biela's comet after the duplication on February 21, 1846. According to Struve.

the following table, due to the last-named astronomer, that the two comets continued to separate till February 13, and after that date gradually approached one another:—

Distance between the two nuclei, 1846—	January	14,	177,000 miles.
,, ,,	,,	24,	186,000 ,,
,, ,,	February	3,	191,000 ,,
,, ,,	,,	12,	193,000 ,,
,, ,,	,,	23,	191,000 ,,
,, ,,	March	5,	190,000 ,,
,, ,,	,,	15,	180,000 ,,
,, ,,	,,	25,	172,000 ,,

The variations of brilliancy and size presented by the two

comets were not less remarkable than the variations of the distance between them. Both had nuclei, both had short tails, parallel in direction and nearly perpendicular to the line of junction. 'At its first observation, on January 13, the new comet was extremely small and faint in comparison with the old, but the difference both in point of light and apparent magnitude continually diminished. On the 10th of February they were nearly equal, although the day before the moonlight had effaced the new one, leaving the other bright enough to be well observed. On the 14th and 16th, however, the new comet had gained a decided superiority over the old, presenting at the same time a sharp and starlike nucleus, compared by Lieut. Maury to a diamond-spark. But this state of things was not to continue. Already, on the 18th, the old comet had regained its superiority, being nearly twice as bright as its companion, and offering an unusually bright and starlike nucleus. From this period the new companion began to fade away, but continued visible up to the 15th of March. On the 24th the comet appeared again single, and on the 22nd of April both had disappeared.' (Herschel, 'Outlines of Astronomy.')

The luminous communication, mentioned above, which Maury observed between the two bodies is also worthy of attention. Besides the tails of the comets Maury saw a fine luminous arc, which extended from one nucleus to the other like the arch of a bridge. This was when the new comet was at its maximum brilliancy: and when the old comet had regained its superiority, it threw out new rays, which gave it the appearance of a comet with three tails, making angles of 120° with one another, and one of which joined the two comets.

These curious phenomena raise questions of the highest interest. What cause determined the separation of Biela's comet? Did it arise from a disturbing force foreign to itself, or was it due to some intestine convulsion? Whence proceeded the variations of brilliancy, too striking to be attributed to optical illusion? When the separation was first observed, had it already been accomplished some time? We think with M. Liais that this is likely, and that in four-and-twenty hours the original comet could not have projected to a distance of 177 thousand miles a fragment which subsequently only receded very slowly from the parent body.

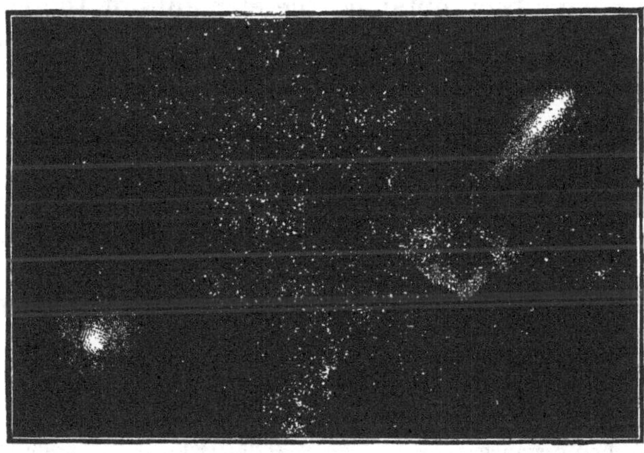

Fig. 47.—The twin comets of Biela at their return in 1852. According to Secchi.

It may be readily imagined that astronomers were on the look-out to again observe Biela's comet when it should return in 1852. Accordingly, in August and September of that year the two comets, which had performed their revolution in company, were seen by Professor Challis at Cambridge, by Father Secchi at Rome, and by M. Struve. This time the mean distance of the two comets from each other was eight times greater than on the occasion of the former passage, in 1846. The following table gives the distances between them during the time they were observed:—

August	27, 1852	1,502,000 miles.
September	4 „	1,560,000 „
„	12 „	1,603,000 „
„	20 „	1,624,000 „
„	28 „	1,615,000 „

It is to be remarked that the maximum distance of the two nuclei corresponds, both in 1846 and 1852, to within a few days of the date of the perihelion passage of the comets, which took place in 1846 on February 12, and in 1852 on September 23 and 24.

The two comets, therefore, may be considered henceforth as two distinct bodies; and in fact, from their respective movements have been obtained the elements of the two orbits, which, however much they may resemble each other and betray the community of their origin, exhibit none the less marked differences. These elements, according to D'Arrest, for the passage of 1852, were as follows:—

Elements of the two Comets of Biela.

Perihelion passage,	1852—Sept. $24^{d}\ 5^{h}\ 14^{m}$ p.m.	1852—Sept. $23^{d}\ 10^{h}\ 50^{m}$ p.m.
Longitude of perihelion	109° 20′ 24″	109° 13′ 21″
Longitude of node	246 5 16	246 9 11
Inclination	12 33 25	12 33 47
Perihelion distance	0 860161	0·860592
Eccentricity	0·7552007	0·7561187
Period	6 yrs. 214 days.	6 yrs. 229½ days.

Movement direct.

These elements agree very closely, as may be seen, but show a difference of 15½ days in the periods of the revolutions. Have other perturbations been since experienced? This was a possibility sufficiently obvious to make the return of Biela's comet a matter of doubt, if not in 1859, at least in 1866. In 1872 one of the nuclei of the comet was situated, at its node, so near the earth that a rencontre between the two may be conjectured to have actually taken place. To this rencontre some astronomers have attributed the splendid phenomenon witnessed by European observers on the night of November 27—the thousands of shooting stars which then fell like a rain of sparks were, according to them, an integral part of one of these two comets. Other astronomers believe that the phenomenon in question was due to the meeting of the earth, not with one of the fragments of Biela's comet, but with a swarm of meteors which had once formed part of the same nebulosity. If the first conjecture be well-founded it is not improbable that, under the powerful influence of the earth's mass, the nebulosity of the comet has been completely shattered. Future observations will perhaps furnish the elements requisite for a solution of this problem.

[The circumstances connected with the passage of Biela's comet in 1872 were of so extraordinary a character that it seems desirable to give an account of them here.

Professor Klinkerfues, of Göttingen, on comparing the brilliant meteor-shower of November 27, 1872, with those of other years, was led to the assumption that in this instance we were in the closest proximity to Biela's comet. Under these circumstances the comet would remain almost stationary in the neighbourhood of the radiant of convergence for a few days after the meteor-shower, and Professor Klinkerfues concluded that there was even a hope of finding the comet itself, provided the intelligence could be at once transmitted to an observatory sufficiently far south. Accordingly, having determined the radiant point from the tracks of eighty meteors, he sent the following telegram to Madras: 'Biela touched earth on 27th, search near *theta Centauri*.' This telegram reached Madras by way of Russia in one hour and thirty-five minutes. The consequences of it are best told in the words of Mr. Pogson, the Government astronomer at Madras, who writes, under date December 5, 1872: 'A startling telegram from Professor Klinkerfues on the night of November 30 ran thus: "Biela touched earth on 27th, search near *theta Centauri*." I was on the look-out from comet-rise (16^h) to sunrise the next two mornings, but clouds and rain disappointed me. On the third attempt, however, I had better luck. Just about $17\frac{1}{4}^h$ mean time, a brief blue space enabled me to find Biela, and though I could only get four comparisons with an anonymous star, it had moved forward $2^s.5$ in four minutes, and that settled its being the right object. I recorded it as "Circular; bright, with a decided nucleus, but no tail, and about 45″ in diameter." This was in strong twilight. Next morning, December 3, I got a much better observation of it: seven comparisons with another anonymous star, two with one of our current Madras catalogue stars, and two with 7734 Taylor. This time my notes were "Circular; diameter 75″; bright nucleus, a faint but distinct tail, 8′ in length and spreading, a position angle from nucleus about 280°." I had no time to spare to look for the other comet, and the next morning clouds and rain had returned.' For three mornings the sky was quite overcast, and afterwards the comet would rise in daylight, and could not therefore be observed.

The positions of the comet observed by Mr. Pogson do not well accord with the calculated places of either part of Biela's comet, or of the meteor-stream through which the earth passed on November 27. Capt. Tupman (R.A.S. *Notices*, xxxiii. p. 318, March 1873) gives reason for his opinion, that the body seen by Mr. Pogson was neither Biela's comet nor a meteoric aggregation travelling in the same orbit, nor a body that had passed near the earth on November 27. Dr. Oppolzer (*Ast. Nach.*, Nos. 1920 and 1938, January 31 and May 13, 1873), although he originally held the same view, was led by the investigations he undertook to consider it highly probable that Pogson's comet is closely connected with the shower of shooting-stars on November 27, and that it is even possible that it was a head of Biela's comet; but Dr. Bruhns (*Ast. Nach.*, No. 2054, September 10, 1875) arrives at the conclusion 'that it is very probable that Pogson's comet was unconnected with Biela's comet or with the meteor-swarm, and that it was a new comet.' If it were Biela's comet, the latter must have been about twelve weeks behind its time. It has been suggested that the observations made on December 2 and 3 referred to different heads of the comet, but there seems no doubt that the body observed was the same on both occasions. In any case, the fact that Professor Klinkerfues should have felt sufficient confidence in the truth of his hypothesis to send the telegram to Mr. Pogson, and that the latter should have actually detected a comet in the neighbourhood of the position indicated, forms a very striking, I might almost say, romantic episode in astronomical history, whether the body thus found was a portion of Biela's comet, or a meteor-swarm on its track, or even an independent body.

In consequence of the interest excited by the above observations, Mr. Hind communicated to the Royal Astronomical Society (*Notices*, xxxiii. p. 320) an account of the actual state of the calculations with regard to Biela's comet, from which it appears that 'both nuclei of the comet were last observed in the autumn of 1852, having been found much further from their calculated places than was expected, a circumference which undoubtedly affected the number of observations, and which was occasioned by the unfortunate substitution by Professor Santini of a semi-axis major depending wholly upon the observations of the previous appearance in 1846, in place of that which he had deduced from observation in 1832, and carried forward by perturbation to 1846. This source of error in the prediction for 1852 is indicated by Professor Santini in a communication made to the Venetian Institute in November 1854. There is no reason to suppose that any perturbations beyond those resulting from known causes operated between the appearances of the comet in 1846 and 1852; indeed, the observations of these years have been connected without difficulty by the application of planetary perturbations during the interval.' The effect of the perturbations was calculated by Professors Santini, Clausen, Hubbard, and Michez, for the period from 1852 to 1866, and the perihelion passages were fixed for May 24, 1859, and January 26, 1866. 'In 1859,' Mr. Hind proceeds, 'the position of the comet in the heavens rendered its discovery almost hopeless, and its having passed by us unobserved is thus accounted for; but it is not so as regards the return in 1866. I believe it is certain that the comet did not pass its perihelion in that year within seven weeks of the time predicted... So far as I know at present the calculation of perturbations from 1866 to 1872 has not been undertaken by anyone... it has probably been felt to be a useless labour to carry forward the elements from the predicted time of perihelion in 1866, considering the want of success attending the endeavours to find the comet in the corresponding track.... If we suppose that the comet did really encounter the earth [on November 27, 1872] in descending to perihelion on December 27, there will be found since 1852 three mean revolutions of 6·754 years, and the perturbations being small from 1866 to 1872, the comet might have been in perihelion about March 28, 1866, instead of January 26. It is clear, therefore, that if the perihelion passage of Biela's comet took place in 1866, six or eight weeks later than anticipated, its having passed unobserved need occasion no surprise.'—Ed.]

SECTION IV.

DOUBLE COMETS MENTIONED IN HISTORY.

Is there any example in history of the division of a comet into several parts?—The comet of B.C. 371—Ephorus, Seneca and Pingré—Similar observations in Europe and China—The Olinda double comet, observed in Brazil, in 1860, by M. Liais.

THE doubling of Biela's comet did not fail to direct attention to the several instances on record of analogous phenomena which had hitherto been looked upon as little worthy of belief. It was then remembered that Democritus had, according to Aristotle, related the fact of a comet having suddenly divided into a great number of little stars. It was this, perhaps, that gave rise to the opinion of certain philosophers of antiquity that comets were composed of two or more wandering stars. Seneca, in endeavouring to refute this opinion, mentions the account given by Ephorus, the Greek historian, of the division of the comet of the year B.C. 371 into two stars. He thus expresses himself:—

'Ephorus, who is far from being an historian of unimpeachable veracity, is often deceived—often a deceiver. This comet, for example, upon which all eyes were so intently fixed on account of the immense catastrophe produced by its apparition—the submersion of the towns of Helice and Bura—Ephorus pretends *divided into two stars.* No one but himself has related this fact. Who could possibly have observed at what moment the comet dissolved and divided into two? And besides, if this division was actually seen to take place, how is it that no one saw the comet form itself into two stars? Why has not Ephorus given the names of these two stars?'

These two last arguments appear of little value, whilst the fact itself mentioned by Ephorus, since the observations of January 1846, appears no longer impossible.

In 1618 several comets were observed, some in Europe, others in Persia, which could not be identified or distinguished from one another with certainty. Two of these comets were seen at the same time and in the same region of the heavens, and this led Kepler to suspect they were parts of one and the same comet, which had divided into two. When recording this opinion of Kepler's, Pingré, who took part with Seneca against Ephorus, now considers the great astronomer at fault, and exclaims, '*Quandoque bonus dormitat Homerus!*' At the present time, although unable to affirm that this division did actually take place, we are forced to consider the conjecture of Kepler as at least probable. There are, besides, other facts on record not altogether dissimilar, and which are narrated by Pingré himself:—

'In the year B.C. 14, Hantching-Ti ascended the throne of China, in the twenty-sixth year of the fourth cycle; in the eighteenth year of his reign a *star was seen to resolve itself into fine rain and entirely disappear.*'

'Under the consulate of M. Valerius Massala Barbatus, and P. Sulpicius Quirinus, before the death of Agrippa, a comet was seen for several days suspended over the City of Rome; it then appeared to resolve into a number of little torches.' This is related by Dion Cassius.

According to observations recorded by the Chinese analists, and collected by Edward Biot, three comets joined together appeared in the year 896, and described their orbits in company. Here again is a passage from Nangis, quoted by Pingré, from which it appears that the comet of 1348 separated into several fragments: 'In the early part of the night, in our presence and to our great astonishment, this very large star divided into several beams, which spread eastward over Paris and entirely disappeared. Was this phenomenon a comet, or other star, or was it formed of exhalations, and again resolved into vapour? These are questions I must leave to the judgment of astronomers.'

It is very probable that some of the phenomena of sudden

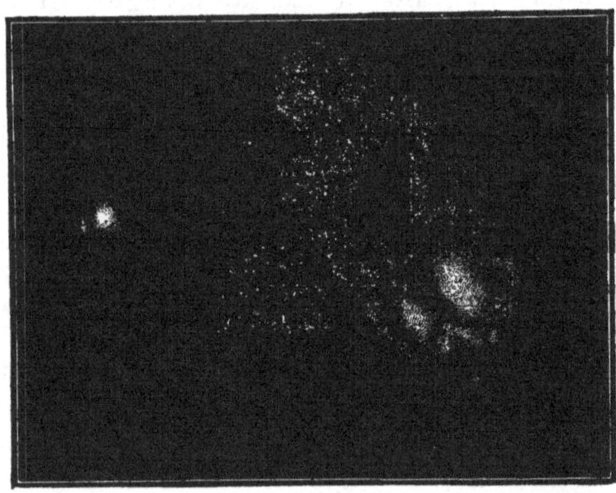

Fig. 48.—The Olinda double comet on February 27, 1860, according to M. Liais.

division mentioned by the ancients may have reference to bolides or, as Pingré says, to 'meteors'; but the facts are none the less curious, since between comets, bolides, and shooting stars a real relation and community of origin have been proved to exist.

But, before closing this section, we must not forget to mention a phenomenon analogous, if not to the doubling of Biela's comet, at least to the fact of two or more nuclei existing in the same comet. The observations were made at Olinda in Brazil by M. Liais, a contemporary French astronomer, in the course of the months of February and March 1860. Fig. 48 gives the aspect of the Olinda double comet on February 27, 1860, the day after its discovery; and the two following figures 49 and 50—which, like the first, reproduce the drawings of M. Liais—suffice to define the positions and forms of the two nuclei on the two latter dates. On February 27 the principal comet, which was greatly superior both in size and brilliancy to the secondary nebulosity, exhibited a nucleus from which sprung two luminous sectors directed towards the sun. The nebulosity which formed the head enveloped these

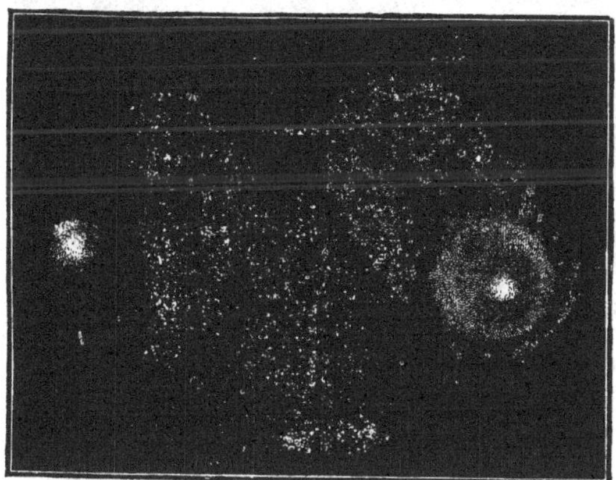

Fig. 49.—The Olinda double comet on March 10, 1860, according to M. Liais.

sectors, and, extending in the opposite direction, contributed to form the tail, whilst the secondary comet consisted only of a nebulosity preserving a marked condensation at its centre. Twelve days after, on March 10, it had hardly undergone any change, whilst the sectors had disappeared in the nucleus of the principal comet, and were seemingly replaced by a nearly circular envelope around the nucleus. On March 11 the principal comet, instead of a condensation or single nucleus, exhibited 'two other and smaller centres situated almost upon the greater axis. The second nebulosity appeared of uniform intensity upon its circumference. It was much fainter than the day before, and hardly visible. On March 11 there was an evident tendency of the great nebulosity to divide, in which case there would have been a triple comet.' On the next day, March 12, the aspect had once more changed; a single centre of condensation, situated like that of the 10th, was visible, and the fainter of the two comets was with difficulty distinguished. On March 13, the last day of observation, it had completely disappeared.

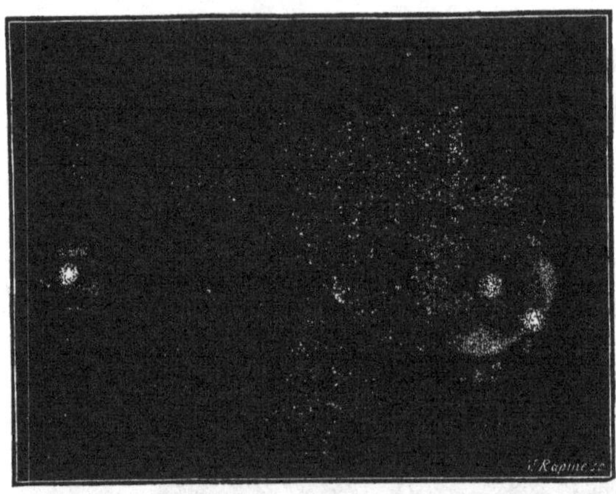

Fig. 50.—The Olinda double comet on March 11, 1860, according to M. Liais.

The Olinda comet (the first comet of 1860), not being periodical, or, what comes to the same thing, having a very long period, it will not be possible to study the changes it will in all likelihood have sustained at the very remote epoch of its probable return. But astronomers, warned by the disruption of Biela's comet, are now watching for the transformations of these celestial embryos, these nebulous masses without consistence, which the planets unsettle in their orbits, and which, sometimes divided under the influence of a more powerful perturbation, or scattered in different fragments, as the ancient observations lead us to believe, shed throughout the regions of space the matter of which they were originally formed.*

* Since writing the above lines a new comet [Coggia's comet, 1874] of considerable interest, visible to the naked eye, has made its appearance in the skies of Europe. In the last days of its visibility—too brief, unfortunately—the head appeared to undergo singular transformations, and to evince a certain tendency to become double. Further on (Chap. X. sec. vii.) will be found the particulars of these phenomena and drawings of the various appearances presented.

CHAPTER IX.

MASS AND DENSITY OF COMETS.

SECTION I.

FIRST DETERMINATION OF THE MASSES OF COMETS.

Lexell's comet, and the calculations of Laplace—The smallness of cometary masses deduced from the fact that comets exercise no disturbing influence upon the earth, the planets, or their satellites.

THE educated have long since ceased to believe in the mysterious influence of comets upon human events; such a belief, in fact, would imply a degree of superstition very little in accordance with the spirit of modern times, and would denote complete ignorance of astronomical phenomena. But if comets, by their unexpected apparitions, no longer announce to the world some great event or terrible catastrophe, are they not capable of acting yet more directly for the overthrow of our planet, either by disturbing it in its movement or by striking against it in a rencontre which might prove fatal to its inhabitants? We will further on consider the probability of such a rencontre, and the effect it would produce upon our globe and its inhabitants. But it is easy to understand that these effects would very greatly depend upon two elements of which we have not yet spoken, viz. the mass and density of the comet.

I have elsewhere * endeavoured to give an elementary idea of the methods which astronomers have recourse to in order to calculate the mass of any celestial body; that is to say, the

[* *Le Ciel* ('The Heavens'), part iv. An English translation, edited by Mr. Lockyer, has been published by Messrs. Bentley & Co.—ED.]

quantity of matter it contains as compared with the mass of the earth or of the sun; in short, to weigh it. To this work, therefore, I may be permitted to refer those of my readers who are unfamiliar with astronomical determinations of the kind in question. The methods employed differ from each other, but all are based upon the principle of universal gravitation.

Let us now proceed to the results, and see what is known of cometary masses.

We have seen that certain comets, in describing their orbits, have approached sufficiently near to several of the planets, Jupiter and Saturn, Mars and the Earth, to be sensibly disturbed in their movements by the perturbations so produced. These perturbations, the effect of which is to alter the form and dimensions of the comet's orbits, have been predicted and calculated beforehand; and the result has proved that the accelerations and retardations assigned by theory were due, as had been anticipated, to the disturbing action of the planetary masses. If the masses of comets were of the same order of magnitude as the planets themselves, they would reciprocally cause an appreciable degree of change in the movements of Jupiter or the other planets. Nothing of the kind has been detected.

Let us take, for example, the comet of 1770 (Lexell's comet), that famous comet which was compelled, in the first instance, by the powerful attraction of Jupiter to describe an elliptic orbit of short period, and, by a subsequent action of the same planet, was consigned for ever to the depths of space. Not only did this comet fail to exercise an appreciable influence upon the mass of Jupiter at the two epochs of its passage in the vicinity of the planet in 1767 and 1779, but it in no respect disturbed any of its four satellites.* The same comet in 1770

* According to the calculations of Burckhardt, undertaken at the instigation of Laplace, the comet in 1779 traversed the system of Jupiter's satellites, since its

passed very near the earth; its least distance from our globe was but a sixtieth part of the distance of the earth from the sun, viz., about 1,500,000 miles, or six times the distance of the moon. Of all observed comets Lexell's comet has most nearly approached the earth, which would have been sensibly disturbed in its movement if the mass of the comet had been at all comparable to that of our globe. 'Had the two masses been equal,' says Laplace, 'the action of the comet would have caused an increase of 11,612 seconds (centesimal) in the length of the sidereal year.' We are certain, from the numerous comparisons that Delambre and Burckhardt made, in order to construct their tables of the sun, that since the year 1770 the length of the sidereal year has not increased by 3″ (2″·6 sexagesimal); the mass of the comet, therefore, was not $\frac{1}{5000}$th part of that of the earth.*

distance from the planet was less than the mean radius of the orbit of the furthest satellite. But it follows from the researches of M. Le Verrier, published in 1844, that the distance was in reality equal to at least three and a half times this radius. The conclusions, therefore, that were obtained relative to the small mass of the comet are not justified.

* 'Not only,' says Laplace elsewhere, 'do comets fail by their attraction to disturb the movements of the planets and their satellites, but if, as is very probable, in the course of past ages any comets have come in contact with these bodies, the shock of the rencontre would not appear to have exercised much influence upon their movements. It is difficult not to believe that the orbits of the planets and their satellites were nearly circular in the beginning, and that their small eccentricities, as well as the common direction of their movements from west to east, depend upon the initial circumstances of the solar system. Neither the action of comets nor collisions with them have changed these phenomena; and yet, if any comet meeting with the moon, or with one of Jupiter's satellites had had a mass equal to that of the moon it would in all probability have rendered their orbits very eccentric. Astronomy presents two other very remarkable phenomena, dating apparently from the origin of the planetary system, and which a very moderate shock would have destroyed entirely: I mean the equality of the movements of rotation and revolution of the moon and the librations of the first three satellites of Jupiter. It is very evident that the blow of a comet whose mass did not exceed the thousandth part of that of the moon would suffice to give a very sensible value to the real librations of the moon and the satellites. We may, therefore, be reassured as

to the influence of comets, and astronomers have no reason to fear that they can in any respect interfere with the accuracy of astronomical tables.'—*Mécanique Céleste*, t. iv. p. 256.

There is, however, in the planetary system an anomaly which might be considered as arising from the perturbations due to a rencontre with a comet. We mean the great inclination and the retrograde movement of the satellites of Uranus. Such an hypothesis appears to us not improbable; and it would invalidate the conclusions drawn by Laplace from the uniformity and the constancy of the motions of the planets and their satellites.

[No doubt the anomalous motion of the satellites of the distant planet Uranus weakens somewhat the force of Laplace's argument, but not, it seems to me, to any serious extent. Laplace's arguments, in regard to the moon's rotation and the librations of Jupiter's satellites, remain of course unaffected.—Ed.]

SECTION II.

METHOD OF ESTIMATING THE MASSES OF COMETS BY OPTICAL CONSIDERATIONS.

The masses of Encke's comet and the comet of Taurus determined by M. Babinet—Objections to this method of determination.

We have thus a determination of cometary masses deduced from the reciprocal disturbances exercised by comets and the planets on one another. It shows that comets have extremely small masses, since, greatly disturbed themselves in their course when they approach a planet, they appear never to have exercised any disturbing influence upon the movements of the planet itself. But, from the value found for the mass of Lexell's comet—a value which, however, is only a maximum limit—it may be seen how far a comet is from being considered a visible nonentity (*rien visible*), to make use of the forcible expression of M. Babinet. The 5,000th part of the mass of the terrestrial globe is equivalent to the sixtieth part of the mass of the moon, a quantity, it will be agreed, far from negligible.

For the justification of his expression M. Babinet has relied upon the following optical considerations. He has called attention to the known fact that stars of exceedingly faint light may be seen through cometary nebulosities without their light losing any of its intensity. Considering, for example, the comet of Encke, which in 1828 had the appearance of a globe-shaped nebulous mass of 311,000 miles in diameter, and through which Struve saw, without any apparent diminution of lustre, a star of the eleventh magnitude, M. Babinet reasons as follows:—The cometary nebulosity having in no respect altered the luminous intensity of the star, we may conclude that its intensity could not have been the sixtieth part of that of the star. Now, the atmosphere illuminated by the full moon obliterates all stars of less than the fourth magnitude, and yet the lunar light has, according to Wollaston, an illuminating power 800,000 times less than that of the sun. Lastly, taking into consideration the relative thicknesses of our atmosphere and of the comet, M. Babinet has arrived at this conclusion: that the substance of a comet is of no greater density than that of our atmosphere divided by the enormous number *forty-five thousand billions*. According to this reckoning Encke's comet would hardly weigh twelve hundred tons.

The same method of estimating cometary masses by optical considerations has also been applied by M. Babinet to the comet of 1825, the so-called comet of Taurus. We have seen that the comet, when interposed before a star of the fifth magnitude, altered its brightness in no perceptible degree. The star in question had, therefore, not lost more than half a magnitude, or about a fifth of its light. It had consequently preserved at least four-fifths of its normal brightness. Now, its light was then traversing a stratum of about 5,000 miles in thickness; that is to say, a thousand times the thickness of the atmosphere, supposing it to be throughout of the density of the air at the surface of the earth. And as it is known that light in traversing perpendicularly our atmosphere loses more than a quarter of its intensity, it follows that the brightness of the star must have been reduced to the fraction $(\frac{3}{4})^{1,000}$ of its real brightness, if the density of the cometary nebulosity were the same as that of the air. This density is, therefore, enormously less, and it is expressed by a fraction having unity for its numerator, and for its denominator a number consisting of 126 figures. ‘When,’ he says, in conclusion, ‘Sir John Herschel, in his last work upon astronomy, spoke of a few ounces as the mass of a comet's tail, his statement was received with almost general incredulity. Nevertheless this estimate is quite exaggerated in comparison with the preceding.’

We will not seek to enquire if the calculations to which these ingenious methods lead are based upon data beyond all dispute, if the density is proportional to the absorption of light, and if the substance of which cometary nebulosities are formed is comparable to that of known gases, both in respect to their molecular composition and respective optical properties. But, granting the conclusion to be legitimate, it must be noticed that it is one which applies only to the comet of 1825 and to that of Encke, or at most to comets of no higher luminous intensity. The whole argument of M. Babinet depends upon the extremely feeble intensity of cometary light as compared with the illumination of the atmosphere by the sun, and the great extent of the nebulosity traversed by the stellar light. This reasoning, therefore, does not hold good in respect to very luminous comets—those, for example, which have been seen at noonday and in sunshine with the naked eye—such as that of the year B.C. 43, and those of the years 1006, 1402, 1532, 1577, 1618, 1744, and especially the great comet of 1843, which was observed at Florence at noonday, at 1° 23′ distance from the sun. The first comet of 1847 was visible at London in the vicinity of the sun. Even if we set aside these as exceptionally brilliant comets, we have seen that the observations of the 5th comet of 1857, on September 8, were in no respect obstructed by the light of the moon. Such comets are not to be compared with Encke's comet, a feeble nebulosity, with hardly any central condensation.* Besides, it is not certain that the stars which have been seen through cometary nebulosities would not have been changed in their intensity, perhaps even eclipsed, if the occultation, instead of taking place behind some portion of the nebulosity, had occurred strictly behind the nucleus, the most luminous portion of the head of the comet. No occultation of this kind has yet, to our knowledge, been proved with certainty to have taken place.† It would therefore be wrong to generalise upon the foregoing conclusion, for, whilst everything leads us to believe that cometary masses are in general greatly inferior to the planetary masses, there is nothing to prove that certain amongst them may not attain a value sufficiently great to produce, in the event of a rencontre with the earth, or with any other planet, a shock or some other kind of sensible perturbation.

* This condensation, however, has been sometimes much less feeble. M. Faye remarks: 'The relative density of Encke's comet must be pretty considerable, since it can appear to the naked eye as a star of the fourth magnitude.'

† [See Chapter X. sec. ii. p. 294.—Ed.]

SECTION III.

THIRD METHOD OF DETERMINING THE MASSES OF COMETS.

Theory of the formation and development of cometary atmospheres under the influence of gravitation and a repulsive force—Calculations of M. Edouard Roche—Masses of the comets of Donati and Encke as determined by this method.

WE are now about to see the same question, when investigated by another method, lead to results quite different to those of M. Babinet.* Between the opinions—entirely conjectural, be it observed—of the savants of the eighteenth century who held that comets were bodies dense and massive as the planets, and those of some contemporary astronomers who regard them as visible nonentities, there is room for a determination which is removed from both extremes, and is moreover better justified.

For this method of determination we are indebted to M. Edouard Roche, professor in the Faculty of Sciences at Montpellier. In a series of very remarkable researches into the theory of cometary phenomena, which we shall analyse further on, M. Roche shows that there is a determinate relation

* The following is the passage from the *Outlines of Astronomy*, to which Babinet alludes (*ante*, p. 283): 'Newton has calculated (*Princ.*, iii. p. 512) that a globe of air of ordinary density at the earth's surface of one inch in diameter, if reduced to the density due to the altitude above the surface of one radius of the earth, would occupy a sphere exceeding in radius the orbit of Saturn. The tail of a great comet, then, for aught we can tell, may consist of only a very few pounds or even ounces of matter.' But Herschel, it will be noted, speaks only of tails, not of atmospheres and nuclei.

between the distance of the comet from the sun, its mass, and the diameter of the portion of its nebulosity subject to the attraction of the nucleus, otherwise called the diameter of its true atmosphere. This relation holds at distances so remote from the sun that the repulsive force, either apparent or real which engenders the tail may be neglected. Another element —the repulsive force—comes into operation when the comet approaches the perihelion; or rather when the comet is near the sun, it is necessary to take account of this force.

Relying, then, upon micrometric observations, which furnish an approximate estimate of the diameters of the nebulosities of the comets of Encke and Donati, M. Roche has arrived at the following result.

As compared with the mass of the earth the mass of Donati's comet would be equal to 0·000047; that is to say, to about the twenty-thousandth part of the former, or about fifty-three times the mass of the terrestrial atmosphere. It would be equal in weight to a sphere of water of 250 miles radius, or to about 268,000 billions of tons—a very different estimate indeed to the pounds of M. Babinet! As regards the density, M. Roche deduces it from the diameters of the nucleus and the nebulosity, which in October 1858 were nearly 4″ and 50″, or 990 miles and 12,400 miles respectively. Assuming that the mass remained unchanged from June, the date of the first determination, to the month of October, and that the mass of the nebulosity was the 1,000th part of the total mass, the density of the nucleus would be almost an eighth of that of water, and the density of the nebulosity about the 154,000th part of that of atmospheric air.

The mass of Encke's comet, estimated by the same method, is found to be about the 1,000th part of the mass of the earth. 'Although we have found,' says M. Roche, 'for the comet of Encke a mass superior to what might have been supposed, *à priori*, these numbers are not inadmissible, and it seems to us that no serious objection can be made to our theory.'

Of the three methods for the determination of cometary masses which we have just passed in review the first is the most certain; but it has furnished as yet only negative solutions of the problem, and these few in number. It leads us to believe that the masses of comets are very small in comparison with the planetary masses; and it is from the absence of all perturbation caused by comets that we have been able to deduce a superior limit to their masses. The second method, founded upon optical considerations, is the most conjectural, because it assumes that the transparency is inversely as the density, an hypothesis entirely gratuitous, considering how completely ignorant we are of the true physical condition of the substance of which comets are composed. To the third method, therefore, it seems to us the preference should be given, and it is this, in fact, which has furnished the most positive results.* But the subject, it cannot be denied, continues to be involved in much obscurity.

The foregoing remarks, it must be borne in mind, have reference only to the mass of comets. In speaking of their density it would be evidently necessary to distinguish between the nucleus, either solid or liquid, when sufficiently distinct, and the nebulosity of the comet. To the density of this nebulosity, or that of the tail, the calculations and results of M. Babinet might reasonably be applied. The density of the nucleus could, it is true, be easily deduced from the mass of the comet, if we neglect the mass of the matter that envelopes it;

[* I cannot refrain from expressing my own opinion that the first method (viz. by means of the perturbations produced by comets) is the only satisfactory one. The other two both involve hypotheses and assumptions which render the results obtained by means of them most uncertain.—ED.]

but for this we should require very exact measurements of the nucleus, which would be difficult to obtain. It is known, moreover, that the dimensions of the nucleus vary in the same comet with the distance of the comet from the sun. The density, therefore, must itself vary with the distance and these dimensions.

CHAPTER X.

THE LIGHT OF COMETS.

SECTION I.

INTEREST ATTACHING TO THE PHYSICAL STUDY OF COMETARY LIGHT.

WE have seen what the telescope has taught us of the structure of comets, so complex and wonderfully mobile, so different in this respect from that of the planets or the sun. On the one hand we see solid or liquid bodies, bearing the most striking analogy to the terrestrial globe, surrounded like it by atmospheres of comparatively small extent, stable in every portion; these are the planets, the moon, and the satellites of the planets. As regards the sun and the stars—which shine, like the sun, by their own light, and are, like him, as everything leads us to suppose, foci of light and heat to other planetary groups—if these bodies are incandescent gaseous masses, their condensation is so enormous and their physical constitution is comparatively so stable, that the changes of which they are perpetually the theatre have no appreciable effect upon their equilibrium. In comparison with comets they are permanent stars; while comets seem to be nothing more than clouds—wandering nebulæ, to employ the expression of Laplace, who has but reproduced in a more happy form the term so happily applied by Xenophanes and Theon of Alexandria.

But it is not merely by its concentration in the field of a telescope that the light of a comet may be made subservient to the study of its physical constitution; the undulations of which it is composed, after passing through the depths of space and arriving at the confines of the atmosphere, after passing through the atmosphere and penetrating the crystal lens of the instrument, retain certain distinctive qualities by which the savant who subjects it to analysis may distinguish whether this light has emanated directly from the body itself or, on the contrary, has undergone reflexion within the cometary mass, and is consequently only light reflected from the sun. Other methods of analysis will permit us to penetrate yet more deeply into the inmost constitution of a comet and its different parts, and the light with which it shines is again the agent to which we have recourse, and which will reveal to us the chemical nature of the cometary matter. So that only one more difficulty remains to be surmounted in order to unveil completely the composition of these once mysterious bodies; that is, to penetrate actually and materially into the interior of a cometary mass. And, as we shall shortly see, there is great probability of such an event being realised, if it has not already partially happened. In any case we have already said enough to show the interest attaching to the study of cometary light, the subject of the following sections of this chapter.

SECTION II.

TRANSPARENCY OF NUCLEI, ATMOSPHERES, AND TAILS.

Visibility of stars through the atmospheres and tails of comets; ancient and modern observations upon this point—Are the nuclei of comets opaque, or transparent like the atmospheres and tails?—Reported eclipses of the sun and moon produced by comets.

THE visibility of stars, even of very small ones, through the comæ and tails of comets is a fact which had been observed by the ancients. Aristotle in his Meteorology mentions the stars seen by Democritus notwithstanding the interposition of a comet. Seneca says likewise, in his *Quæstiones Naturales*, 'that we may see stars through a comet as through a cloud;' and further on, 'the stars are not transparent, and we can see them through comets—not through the body of the comet where the flame is dense and solid, but through the thin and scattered rays which form the hair; it is through the intervals of the fire, not through the fire itself, that you see.' Humboldt, in quoting this last passage, 'per intervalla ignium, non per ipsos vides,' adds: 'This last remark was unnecessary, for it is possible to see through a flame the thickness of which is not too great.' This is true; but Seneca has merely recorded the fact that up to his time stars had been seen behind the tail or coma, not behind the nucleus itself. The want of the telescope did not, in fact, permit the ancients to distinguish the body or nucleus of a comet, even when the comet had a nucleus.

Modern astronomers themselves are not in a better position, since all observed occultations of stars by comets, one alone excepted, refer to the interposition of the nebulosity forming the coma, not to that of the nucleus properly so called. We here mention the principal instances observed, and more especially those which have been invoked to prove the transparency of cometary light, beginning with the single exception above referred to, of which Arago gives the following account: 'On the 27th of October, 1774, Montaigne saw at Limoges a star of the sixth magnitude in Aquarius through the nucleus of a small comet.' * Let us now proceed to the others.

On November 9, 1795, Sir William Herschel distinctly perceived a double star of the eleventh or twelfth magnitude through the central part of the nebulosity of a comet. The two component stars, one of which was much fainter than the other, were both clearly visible. The comet was Encke's, which is generally destitute of nucleus, and very rarely exhibits more than a faint condensation of light in the centre of its nebulosity. On November 7, 1828, Struve saw in the centre of the same comet a star of the eleventh magnitude, which for a moment he mistook for a cometary nucleus, and whose brightness appeared in no respect diminished. Now, the thickness of the nebulosity interposed was not less than 310,000 miles. This is the observation upon which, as we have seen, M. Babinet has founded his calculation of the mass and density of the nebulosity itself. According to an observa-

* [I myself saw the nucleus of Halley's comet at its apparition in 1835 pass over a star, when I was at the Cambridge Observatory. I remember the circumstance distinctly, and my impression is that there was *no diminution at all* in the brightness of the star. The printed record of my observation runs as follows: 'Sept. 25, $9^h\ 45^m$ to 12^h. During the whole time the comet (seen with the equatorial $3\frac{3}{4}$ inch aperture) appeared to continue changing its figure. It passed over three stars (the nucleus covering one), which were distinctly visible during the whole time.'—*Cambridge Observations*, vol. viii., 1835, p. 216.—Ed.]

tion made at Geneva twenty-one days later by M. Wartmann a star of the eighth magnitude was, on the contrary, completely eclipsed by the comet. It is interesting to compare these two observations, which show the comet's condensation between the dates mentioned; in this interval the volume of the nebulosity had become reduced to one-eighth, and there must have been a corresponding luminous condensation and increased brilliancy, which would explain the occultation seen by Wartmann. In April 1796 Olbers remarked a similar fact in respect to a star of the sixth or seventh magnitude, which, hardly weakened in intensity, appeared a little to the north of the centre of the nebulosity; the star, therefore, was not occulted by the nucleus, but its light was sufficiently bright to render the nucleus for some time invisible in its vicinity.

Cacciatore observed, at Palermo, the occultation of a star by the comet of 1819. 'On the 5th of August,' he states, 'I observed through the nebulosity, very close to the nucleus, a star which at the most was of the tenth magnitude.'

When we add to these observations that of Struve, who, on October 29, 1824, saw a star of the tenth magnitude at 2″ from the centre of the comet, without the light of the star being at all diminished; those of Pons and Valz, in 1825, who saw, the former a star of the fifth magnitude, and the latter one of the seventh magnitude occulted by the famous comet of Taurus, it will be seen that the light of comets, not only that of their tails, but also that of their nebulosities in close proximity to the nucleus, is transparent in the highest degree. But is the nucleus properly so called equally transparent? This we have not yet data to determine, since we have no observation of the occultation of a star by a comet, which indicates with certainty the interposition of the nucleus, excepting that mentioned by Montaigne. Pons, in 1825, mentions the passage of the star to the centre of the nebulosity, but not its passage behind the luminous nucleus.

From the foregoing facts we are forced to conclude that the matter of cometary tails and nebulosities, if gaseous, is of extreme tenuity; but it is perhaps so discrete—*i.e.* the particles are so far apart—as not to occasion any perceptible occultation of a light seen through them. This was the opinion of M. Babinet, who, from the calculations above quoted respecting this extreme tenuity of cometary matter, has come to the conclusion that 'the substance of comets, therefore, is a kind of very divided matter, consisting of isolated particles, without mutual elastic reaction.' An observation made by Bessel helps to confirm this view, as Humboldt, when recording it, justly remarks. On September 29, 1835, Bessel saw, about 8″ from the centre of the head of Halley's comet, a star of the tenth magnitude. At this moment its light was traversing a considerable portion of the nebulosity; but the luminous ray was not deflected out of its rectilinear course, as the illustrious astronomer satisfied himself, by measuring the distance between the occulted star and a star visible on the verge of, but outside, the nebulosity. 'So complete an absence of refracting power,' says Humboldt, 'scarcely allows us to suppose that the matter of comets is a gaseous fluid. Must we have recourse to the hypothesis of a gas infinitely rarefied, or are we to believe that comets consist of independent molecules, the union of which constitutes cosmical clouds, devoid of power to act upon the luminous rays that pass through them, just as the clouds of our own atmosphere do not alter the zenith distances of the stars?'

It, therefore, still remains an open question whether the cometary nucleus, the luminous and brilliant central portion of a comet—that part of it, in short, which gives to the comet the appearance of a star—is opaque or transparent. In any case, let us repeat, it is clear that we must refrain from generalising, for it would be absurd to identify, from this point of view, the faint nuclei, hardly visible in the telescope, of many of the smaller comets with those of comets which have shone like stars of the first magnitude, and have been luminous enough to appear in broad daylight and shine in the most brilliant regions of the heavens in the vicinity of the sun.

In support of the opacity of cometary nuclei various anciently recorded facts have been adduced; but these facts are either apocryphal or at least very doubtful. Thus the eclipse of the year B.C. 480, mentioned by Herodotus, and the eclipse mentioned by Dion Cassius, which took place in the year in which Augustus died, not admitting of explanation from the movement and interposition of the moon, were supposed to be due to the intervention of comets, a supposition altogether without foundation and very improbable. In the *Cométographie* of Pingré, under the date of 1184, we find the following: 'On the 1st of May, about the sixth hour of the day, a sign was visible in the sun: its lower portion was totally obscured. In the centre it was traversed by what appeared to be a beam! The rest of its disc was so pale that it impressed the same pallor upon the faces of those who looked upon it. Was this phenomenon the effect of a comet situated between the sun and ourselves? I do not know, but I consider it possible.' The total obscuration of the lower part of the sun would be, on this hypothesis, the partial eclipse produced by the opaque nucleus, and the beam traversing the disc the densest portion of the tail. Lastly, the pallor of the sun could be explained by the interposition of the vapours composing the nebulosity. But this is mere supposition.

Not an eclipse of the sun but an eclipse of the moon would appear to have been caused by the comet of 1454, according to Phranza, the *protovestiare*, or master of the wardrobe, of the

Turkish emperors. But it has been proved that the Latin version of the text is corrupt, and that Phranza has simply chronicled the fact of the simultaneous presence in the heavens of the comet and the full moon at the time when the latter was eclipsed.

SECTION III.

COLOUR OF COMETARY LIGHT.

Different colours of the heads and tails of comets—Examples of colour taken from the observations of the ancients: red, blood-red, and yellow comets—Difference of colour between the nucleus and the nebulosity—Blue comets—The diversity of colour exhibited by comets is doubtless connected with cometary physics, and with the temperature and chemical nature of cometary matter.

THE light of many comets has been sensibly coloured. The comet of B.C. 146 exhibited a reddish tinge, according to Seneca: 'A comet as large as the sun appeared. Its disc was at first red and like fire.'...

A little further on Seneca again observes: 'Comets are in great number, and of more than one kind; their dimensions are unequal, their colours are different; some are red, without lustre; others are white and shine with a pure liquid light. ...Some are blood-red, sinister presage of that which will soon be shed.' The ancients had, therefore, observed the difference of colour in the light of comets. And we shall mention a number of similar examples taken from the chronicles of the Middle Ages and from modern observers.

The comets of 662 and 1526 are cited by Arago as having been 'of a beautiful red;' and we have seen that Pliny in his classification speaks of comets whose 'mane is the colour of blood.' Such was the comet which appeared in November 1457; according to an ancient chronicle 'its coma or tail resembled the colour of flame.' The horrible comet which, according to Comiers, appeared in 1508 was very red, representing human heads, dissevered limbs, instruments of war, &c.' The first comet of 1471 'was very large and of a reddish colour; it rose before the dawn.' In 1545 'a comet whose coma was the colour of blood burned for several days; it then became pale and soon disappeared.' Gemma, when speaking of the comet of 1556, thus expresses himself: 'Although Paul Fabricius has stated that the comet appeared small to him, I can affirm that, from the commencement of its apparition, I found it not less than Jupiter in size; the colour of the comet resembled that of Mars; its ruddy colour, however, degenerated to paleness.' This remark refers more especially to the nucleus, for, according to another eyewitness, 'the colour of the tail towards its extremity continued pale, livid, and similar to that of lead.' The opposite was the case with the comets of 1577 and 1618. Tycho relates of the first that its head was round, brilliant, and remarkable for a certain leaden whiteness, whilst the tail, turned towards the east, darted in a direction opposite the sun rays of a more ruddy colour. As regards the second, its tail appeared, says Arago, of a very bright colour.

The comet of 1769 had a slightly reddish nucleus, as also had that of 1811, observed by Sir William Herschel; but the nebulosity of the latter was of a bluish green, which caused Arago to conjecture that this appearance of colour might be due to the simple effect of contrast. It is clear, however, that the colour of the nucleus and that of the nebulosity were very sensibly different. A brilliant zone, narrow and semicircular, surrounding the head of the comet of 1811 on the side nearest to the sun, was of a decided yellow colour.

Amongst observations of earlier date we find mention made of comets which have shone with a golden yellow light.

Such was the comet of 1555, whose rays shone like gold; and that of 1533, whose tail was of a beautiful yellow. Of Halley's comet, at its apparition in 1456, it is said 'the colour of the comet resembled that of gold.' It is true that 'at other times and perhaps in other places it appeared pale and whitish; it sometimes resembled a glistening flame.'

This last remark suggests a very natural reflection, and leads us to consider how far the state of the atmosphere, its more or less purity, and the greater or less height of the comet above the horizon, may have contributed to invest these bodies and their tails with the tints above described. It appears certain, however, that the light of comets is far from being always of the same colour. The great comet of 1106 was of a remarkable whiteness. 'Situated towards that quarter of the heavens where the sun sets in winter, it extended a whitish beam, resembling a linen cloth. From the commencement of its apparition both the comet and its beam, which was as white as snow, diminished day by day.' According to other chronicles 'its rays were whiter than milk.' This, as may be seen, forms a complete contrast to the red and yellow colours of the preceding comets; nor is the contrast less with the comets of which we are about to speak.

Under the date of 1217 Pingré has the following: 'Several prodigies were observed; blue comets were seen.' The comet of 1356, observed in China, was of a whitish colour bordering upon blue. The comet of 1457, the tail of which resembled an upright spear, was of 'a livid dusky colour, very like that of lead.' The second comet of 1468 also 'was blue, but somewhat pale.' The one which appeared at the end of 1476 was of pale blue bordering upon black. And we must not forget the two comets, 'terrible and of a blackish hue,' whose apparition in 1456, before that of Halley's comet, has been mentioned by some authors.

Modern observers appear to have paid but little attention to the study of cometary light. Nevertheless, we find in Arago the following comparison between the tail of the great comet of 1843 and the zodiacal light: 'On the 19th of March the tail of the comet, situated close to the zodiacal light, was evidently tinged with red, inclining to yellow.' He says nothing about the colour of the nucleus. Amongst the numerous observations of the comet of Donati (1858) we have only met with the following mention of its colour, made by an observer at Neufchâtel, M. Jacquet: 'On Sunday, the 3rd of October, after a cloudless day, splendid twilight. The irregular line of mountains near where the sun has disappeared is traced against a sky glowing with gold and fire. It is six o'clock. We endeavour to see if the comet, in consequence of the purity of the air, may not be already visible. After a few moments' search we discover it, extremely small and pale, and of the silvery brightness of a planet seen by daylight.' Two days after, on October 5, at the same hour, the comet was visible in the neighbourhood of Arcturus. 'The clouds,' says M. Jacquet, 'pass from the region of Arcturus far too slowly for my patience; they disperse at last; I see a yellow star, and a little underneath and to the right a small white plume. My attention is caught by these two colours; one would say the comet is of gold and the plume of silver' (*Souvenirs de la Comète de* 1858). This evidently refers to the colour of the tail and the envelope surrounding the nucleus, for the next day the same observer speaks of the nucleus as 'small, bright, and of a reddish yellow.'

Coggia's comet (1874, III.), observed this summer, was distinguished by very appreciable phenomena of colour. Father Secchi remarks: 'The comet, when observed with an ordinary eyepiece, was magnificent. On the 9th of July it formed a fan, of a reddish tint (by contrast), of about 180 degrees of opening, composed of curvilinear rays, springing from a nucleus of yellowish green.' The Roman astronomer thus attributes the colour of the tail to the effect of contrast. M. Tacchini is of a different opinion. After having described the continuous spectrum upon which was projected the discontinuous spectrum of the nucleus, he proceeds to add: 'This beautiful coloured band, which presented itself only at the passage of the nucleus, when seen through a simple eyepiece appeared of a greenish white, whilst the fan itself was sensibly reddish, even when occulting the nucleus.' *

The question of the colour of the light in the several portions of a comet, its nucleus, atmosphere, and tail, is an interesting one, for it is intimately connected with the physical nature of the light itself. In conjunction with results afforded by spectroscopes and polariscopes it will doubtless help to determine if comets shine by their own light, or only reflect

* 'Questo bel nastro colorato presentavasi al solo passagio del nucleo, il quale guardato coll' oculare semplice appariva bianco-verdastro, mentre il ventaglio era sensibilmente roseo anche occultando il nucleo.' (*Memorie della Societa degli Spettroscopisti Italiani.* Luglio, 1874.)

[Mr. Huggins, describing the appearance of the comet in the telescope, writes, 'The nucleus [of Coggia's comet] appeared of an orange colour. This may be due in part to the effect of contrast with the greenish light of the coma. Sir John Herschel described the head of the comet of 1811 to be of a greenish or bluish-green colour, while the central point appeared to be of a pale ruddy tint. The elder Strube's representations of Halley's comet, at its appearance in 1835, are coloured green, and the nucleus is coloured reddish yellow. He describes the nucleus on October 9 thus, "Der Kern zeigte sich wie eine kleine, etwas in gelbliche spielende, glühende Kohle von länglicher Form." Dr. Winnecke describes similar colours in the great comet of 1862' (Proc. Roy. Soc., vol. xxiii., p. 157). According to Mr. Lockyer, the colour, both of the nucleus and of the head, as observed in Mr. Newall's telescope, was a distinct orange yellow. Mr. Newall says the colour of the comet was greenish yellow. Messrs. Wilson and Seabroke, observing the comet on July 14, at Rugby, considered that it was reddish in colour (R.A.S. *Notices*, vol. xxxv., p. 84).—Ed.]

that which they receive from the sun. Perhaps both hypotheses are true; but if so, to what extent do the atmosphere and the nucleus participate in this double cause of visibility? This is a question we are not yet in a position to answer, although, as we shall see, several steps have already been taken in this direction.

SECTION IV.

SUDDEN CHANGES OF BRILLIANCY IN THE LIGHT OF COMETARY TAILS.

Rapid undulations occasionally observed in the light of cometary tails; observations of Kepler, Hevelius, Cysatus, and Pingré; comets of 1607, 1618, 1652, 1661, and 1769—Undulations in the tails of the comets of 1843 and 1860; do these undulations arise from a cause peculiar to the comet itself, or do they depend upon the state of the atmosphere?—Objection made by Olbers to the first of these hypotheses; refutation by M. Liais.

THE tails of certain comets have exhibited variations of brilliancy, sudden changes of intensity, analogous to the phenomena of the same kind which are observed in the aurora borealis, and which, it is believed, have been remarked in the zodiacal light. This fact was unknown to the ancients; and when Seneca speaks of the augmented or diminished brilliancy of comets, it is evident that he alludes to the changes produced, in the course of their apparition, by the variations of their distance from the earth. He compares them 'to other stars which throw out more light and appear larger and more luminous in proportion as they descend and come nearer to us, and are smaller and less luminous as they are returning and increasing their distance from us.' (*Quæstiones Naturales*, vii. 17.)

Kepler is the first observer* who has made mention of

* There are, however, some earlier observations of the same fact. The tail of the comet of 582 appeared, according to Gregory of Tours, like the smoke

these singular changes. 'Those,' he says, 'who have observed with some degree of attention the comet of 1607 (an apparition of Halley's comet), will bear witness that the tail, short at first, became long in *the twinkling of an eye.*' Several astronomers, Kepler, Wendelinus, and Snell, saw in the comet of 1618 jets of light, coruscations and marked undulations. According to Father Cysatus the tail appeared as if agitated by the wind; the rays of the coma seemed to dart forth from the head and instantly return again. Similar movements were observed by Hevelius in the tails of the comets of 1652 and 1661; and Pingré, describing the observations of the comet of 1769, made at sea, between August 27 and September 16, by La Nux, Fleurien, and himself, thus describes the phenomenon of which he was a witness: 'I believe that I very distinctly saw, especially on September 4, undulations in the tail similar to those which may be seen in the aurora borealis. The stars which I had seen decidedly included within the tail were shortly after sensibly distant from it.'

M. Liais has given the following account of the observations made by him of the great comet of 1860: 'On the evening of the 5th of July, whilst I was observing the comet at sea, I saw a rather intense light from time to time arise in those portions of the tail that were furthest from the nucleus. Sometimes instantaneous, and appearing upon a small extension of the extremity of the tail, which then became more visible, these fugitive gleams reminded me of the pulsations of the aurora borealis. At other times they were less fleeting, and their propagation in rapid succession could be followed for

of a great fire burning in the distance. The comet of 615, observed by the Chinese, had what appeared to be a movement of libration in its point. But the analogy of these phenomena with those that we shall describe does not seem very evident.

some seconds in the direction of the nucleus near the extremity of the tail. These appearances then resembled the progressive undulations of the aurora borealis, but even in this case they were only visible in the last third of the length of the tail. The gleams in question were similar to those that I remember to have seen in the tail of the great comet of 1843, and which were observed by very many astronomers.'

Are these variations incidental to the comet itself? It has been doubted: it has been supposed that they are produced by sudden changes in the transparency of our atmosphere. Olbers has made the objection that, if a real and instantaneous change had taken place in the brightness of the tail, it could not have been seen from the earth in so short a time as a few seconds, as, the different parts of a tail several millions of miles in length being situated at very unequal distances from the earth, the times of transmission of the light from each extremity to the observer would not be identical, and hence an interval of some minutes would be required to produce the appearance of the propagation of a luminous change from one end of the tail to the other. Now observers speak of variations much more rapid—of some seconds, in fact. M. Liais reduces this objection to its just value by pointing out that long cometary tails generally 'front' us and are not seen as it were sideways, so that the difference of distance between the earth and each extremity of the tail is not so great as Olbers had supposed. For example, 'the difference of time occupied by the light in coming from the two extremities of the tail of the comet of 1860 to the earth did not amount to four seconds on the 5th of July.' The same observer likewise remarks that the undulations seen by him took place only in a portion of the tail, and that on the same evening he made comparative observations of the Milky Way and the zodiacal light, but without being able to detect in either luminous movements similar to those exhibited by the cometary light. It appears clear, therefore, that the phenomenon was not occasioned by variations in the transparency of our atmosphere. It will be necessary, therefore, to seek the true explanation in the comet itself, in the actual variations of its light either in the nucleus or in the tail.

SECTION V.

DO COMETS SHINE BY THEIR OWN OR BY REFLECTED LIGHT?

Do the nuclei of comets exhibit phases?—Polarisation of cometary light—Experiments of Arago and of several contemporary astronomers—The light of nebulosities and atmospheres is partly light reflected from the sun.

In the last century astronomers were almost entirely preoccupied with the study of cometary movements, the nature of cometary orbits, the periodicity of comets, and with every question, in fact, that tended to prove that, like the planets, these bodies are subjected to the universal law of gravitation. Astronomical physics was then hardly recognised, and conjecture filled the place of modern analytical research. It was doubtless owing to this preoccupation that comets were at that time looked upon as bodies of kindred nature to the planets. There was a kind of reaction against the ancient hypothesis of terrestrial meteors and transient fires. 'Planets are opaque bodies,' says Pingré; 'they only send back the light which they receive from the sun. We ought not, perhaps, to conclude definitively that comets are also opaque bodies; it is not absolutely proved that a luminous body may not circulate around some other body. But the light of comets is feeble and dull; its intensity varies; we can perceive in it sensible inequalities and even gaps. It does not appear that these phenomena can be explained otherwise than by supposing comets to be opaque bodies, possessed of no other light than that which they receive from the sun, and surrounded by an atmosphere similar to that of the earth. Clouds are formed within this atmosphere, just as in our own atmosphere; these clouds weaken or totally intercept the rays of the sun, and successively deprive us of the sight of a portion of the comet. This hypothesis would explain everything. . .' The same author says elsewhere: 'The nucleus or the head of a comet is the most brilliant and at the same time the smallest part of it, and is supposed, with reason, to be a solid body of no great size, and probably of small density.' This, the reader will see, is mere conjecture.

Other savants have assumed that comets are planets of a particular kind, and do not receive their light from the sun, but shine by their own brilliancy; but no observations or proofs have been given in support of this opinion.

We must, however, remark that certain comets have been thought to exhibit phases. Cassini, when observing the comet of Chéseaux, in 1744, 'noticed the phase of that comet, the illuminated portion of which was only half-visible.' These last words are Lalande's; Cassini himself only mentions the irregularity of the nucleus of the comet. In the year 813 'a comet appeared which resembled two moons joined together; they separated, and after taking different forms resembled at last a man without a head.' Pingré explains this singular appearance by the phases of the nucleus and tail, the comet being then near its conjunction with the sun. More precise testimony is afforded by an observer who saw, first as a crescent, and then in first quarter, the nucleus of the comet of 1769 when it was approaching the sun. Arago has discussed the observations made at Palermo in 1819 by Cacciatore, and which had induced that astronomer to believe that the comet in question had exhibited phases. Arago bases his refutation upon a drawing made by Cacciatore on July 5, 1819 (fig. 51), in which the cusps of the crescent are situated in a line directed towards the sun, instead of at right angles to it, as they were ten days later, on July 15.

The absence of phases in cometary nuclei is not an argument against their opacity. Pingré remarked: 'If comets are true planets, either their heads or their nuclei must of necessity be opaque bodies, illuminated by the rays of the sun; but these rays also penetrate the atmospheres, which often send on to us even more light than the body of the comet.

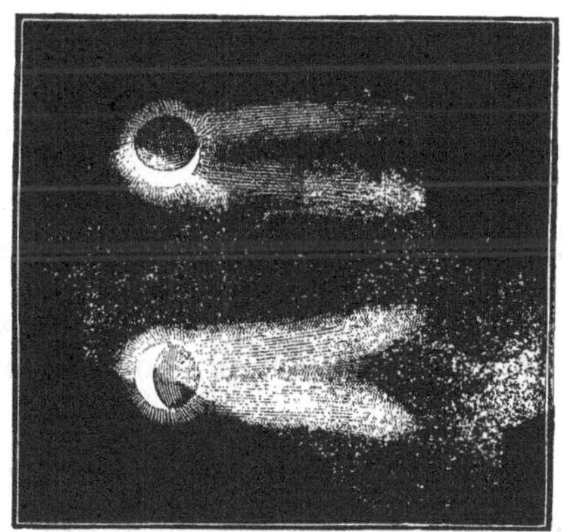

Fig. 51.—Supposed phases of the Comet of 1819, according to Cacciatore: observations of July 5 and 15.

It is doubtless for this reason that comets are not seen crescent-shaped or in quadrature, as is the case with the moon, Mercury, and Venus.' The same reason has been given by Arago in other terms. 'I confess,' he observes, 'that the absence of phases in a nucleus, perhaps diaphanous, and surrounded, as is that of a comet, by a thick atmosphere, which, by reflexion, is able to distribute light in all directions, cannot lead us to any certain conclusions.'

Lastly, let me here add a remark which it is hardly possible anyone could fail to make, on comparing together the telescopic views of certain comets, as, for example, those of the head of

Donati's comet. The luminous sectors issuing from the head of the comet might easily be mistaken for phases in instruments of insufficient power, whilst it is evident that these variable phenomena, the succession and oscillations of which are so remarkable, are of quite a different nature, and are not simple optical appearances.

The problem involved in the nature and constitution of cometary light was at length attacked by Arago in a new direction, and by a method which enables the observer to determine whether the light of an observed comet is that of matter luminous in itself, or whether it is, wholly or in part, solar reflected light. This method has been applied to the nuclei, we shall see further on, as well as to the atmospheres and to the light of cometary tails. The first researches of Arago on this subject date from 1819; eleven years after Malus had discovered polarisation by reflexion, and eight years after Arago himself had remarked the phenomena of the colours of polarised light.

This is not the place to explain how the nature of any source of light may be studied by the aid of an optical apparatus called a *polariscope*. We shall only state that when a luminous object is examined by the aid of a Nicol's prism or a thin plate of tourmaline, two images are formed, which vary in intensity and colour as the apparatus is turned completely round through four right angles, if the light emitted from the object is polarised. But if, on the contrary, the light is natural, the images manifest neither difference in intensity nor difference in shade of colour. And further, when the light is polarised we can determine whether it has been polarised by reflexion, and if so, in what plane, so that we can thus obtain information about the source from which it was emitted.

Applying these principles to the study of comets, Arago

subjected to examination the light of the comet of 1819, and afterwards that of Halley's comet, in 1835. 'I directed,' he remarks, 'upon the comet (that of 1819) a small telescope furnished with a double refracting prism : the two images of the tail of the comet presented a slight difference of intensity, which was verified by the concordant observations of Humboldt, Bouvard and Mathieu. On the 23rd of October, 1835, having applied my new apparatus (the telescope-polariscope) to the observation of Halley's comet, I saw immediately two images exhibiting complementary tints, the one red, the other green. On turning the telescope through 180° the red image became green, and the green became red. The light of the comet, therefore, was not composed of rays having the properties of direct light; it was reflected or polarised; that is to say, definitely, it was light that had proceeded from the sun.' But might not the results be due to the terrestrial atmosphere? To be assured on this point, Arago directed the same telescope at the time of his first observation upon Capella, and found the two images of the star perfectly equal in intensity. The light of the atmosphere not being polarised, it became evident that the polarisation was effected at the surface of the cometary matter.

These observations have since been confirmed by numerous savants. Chacornac, at Paris; Ronzoni and Govi, in Italy; Poey, at Havana; and Liais, in Brazil, have found that the light of Donati's comet was polarised either in the nucleus or in the part of the tail adjacent to the nucleus. The condition which Brewster insisted upon as essential to the removal of all doubt in regard to the possibility of polarisation by refraction in the terrestrial atmosphere has been fulfilled, for M. Poey found that the plane of polarisation passed through the sun, the comet, and the eye of the observer; so that some portion, at least, of the light of the comet was reflected solar light.

But to what extent is this the case? In addition to the light reflected from the sun have comets no light of their own? It is for spectral analysis to reply. We are about to see if this method of observation, which as yet has been applied only to a few not remarkable comets, may not afford some information on the subject. Let us beforehand, however, call attention to two interesting observations made in 1861 and 1868 by Father Secchi. The first relates to the great double-tailed comet of 1861. At first the nucleus presented no trace of polarisation, whilst the light of the tail was strongly polarised. On July 3 the nucleus gave traces of polarisation. Father Secchi concluded that, during the first few days, the nucleus shone by its own light—'perhaps,' he observes, 'on account of the incandescent state to which the comet had been brought by its close proximity to the sun.' The second observation was made in 1868, and refers to Winnecke's comet. Having examined its light by the aid of a telescopic polariscope, the same observer found no appreciable difference of colour in the images of the nucleus; whilst the light of the aureola about the comet exhibited an evident trace of complementary colour. 'Thus,' he concludes, 'the light of the nucleus is principally its own.' We shall presently see that further observations of the same kind have been made recently on Coggia's comet of 1874.

SECTION VI.

SPECTRAL ANALYSIS OF THE LIGHT OF COMETS.

Researches of Huggins, Secchi, Wolf, and Rayet—Spectra of different comets: bright bands upon a continuous luminous ground—Analysis of the light of Coggia's comet in 1874—Chemical composition of different nuclei and nebulosities.

PHYSICISTS, it is well known, recognise three orders of spectra as produced by sources of light when a luminous beam emanating from these sources has been decomposed in its passage through a prism or a system of prisms.

A spectrum of the first order consists of a continuous coloured strip, exhibiting neither dark lines, nor bright bands separated by dark intervals; it is, in fact, the solar spectrum, more or less brilliant in colour, and of more or less extent, but destitute of the fine black lines which belong to the spectrum of the sun. Incandescent solids or liquids produce these continuous spectra. Spectra of the second order are those which arise from sources of light composed of vapours or incandescent gas; they consist of a greater or less number of lines or brilliantly coloured bands, separated by dark intervals; the number, the position, and consequently the colours of these lines or luminous bands are characteristic of the gaseous substance under ignition. Every chemically simple body, every compound body which has become luminous without decomposition, has a spectrum peculiar to itself. By the inspection of the brilliant lines furnished by a gas or incandescent vapour we can discover the chemical elements of which it is composed. Lastly, a spectrum of the third order is one which, like the spectrum of solar light, may be regarded as formed of a continuous spectrum intersected by black lines, more or less fine, but generally much narrower than the luminous intervals between them. These dark lines indicate the existence of absorbent vapours in front of the source of light which produces the continuous spectrum. Wherever a black line is found to exist, the luminous wave, the refrangibility of which is determined by the position of the line, has become extinct. Experiment has shown that the substances of which these vapours are formed have the property of intercepting luminous rays of the same refrangibility as those which they themselves emit in an incandescent state. Incandescent sodium, for example, gives a spectrum of one luminous line, situated in the yellow portion of the spectrum; on the other hand, solid incandescent carbon gives a continuous spectrum; but if the vapour of sodium surround the carbon, the spectrum will show a black line in place of the yellow sodium line. A spectrum of the third order, therefore, indicates a light emanating from a solid or liquid body, itself surrounded by an atmosphere of absorbent vapours.

Having stated these elementary facts, let us now see what results have been obtained by the application of the prism to the analysis of cometary light.

The spectrum of the comet of 1864 was found by Donati to consist of three brilliant lines. It is the first observation of the kind with which we are acquainted. 'The spectrum of this comet,' he observes, 'resembles the spectra of metals: the dark portions are broader than those that are more luminous; it may, therefore, be considered as a spectrum formed of three brilliant lines.' This simple observation contains nearly all that spectral analysis has made known in its application to the light of comets. The spectrum consisting of three bright lines or luminous bands has been found, up to the present time, in every comet that has been analysed ; only the refrangibility of these bands appears to vary with different comets, indicating either a difference of physical condition or a difference otherwise but little apparent, in their chemical constitution. There are other peculiarities, however, which deserve mention, and these we will successively call attention to.

The comet of 1866, I., discovered by Tempel, was analysed both by Mr. Huggins and by Father Secchi. 'The light which emanated from the nucleus,' says the first observer, 'was that of a broad continuous spectrum fading away gradually at both edges. These fainter parts of the spectrum corresponded to the more diffused marginal portion of the comet. Nearly in the middle of this broad and faint spectrum, and in a position in the spectrum about midway between *b* and *F* of the solar spectrum, a bright point was seen. The absence of breadth of this bright point in a direction at right angles to that of the dispersion showed that this monochromatic light was emitted from an object possessing no sensible magnitude in the telescope. This observation gives us the information that the light of the coma of this comet is different from that of the minute nucleus. The nucleus is self-luminous, and the matter of which it consists is in the state of ignited gas. As we cannot suppose the coma to consist of incandescent solid matter, the continuous spectrum of its light indicates that it shines by reflected solar light.

'Since the spectrum of the light of the coma is unlike that which characterises the light emitted by the nucleus, it is evident that the nucleus is not the source of the light by which the coma is rendered visible to us. It does not seem probable that the matter in the state of extreme tenuity and diffusion in what we know the material of the comæ and tails of comets to be could retain the degree of heat necessary for the incandescence of solid or liquid matter within them. We must conclude, therefore, that the coma of this comet reflects light received from without; and the only available foreign source of light is the sun.'

In the light of the same comet the Roman astronomer distinguished three lines, one of which—the middle of the three, of moderate brightness, and possibly that which was seen by Mr. Huggins—was situated in the green portion of the continuous spectrum, between the lines *b* and *F* of Fraunhofer; the two others, which were very faint, were situated, the one in the red, the other towards the violet. Beyond these lines appeared matter slightly diffused.

The comet of 1867, II., gave a spectrum probably analogous, but less distinct. 'In the spectroscope,' says Mr. Huggins, 'the light of the coma formed a continuous spectrum. I was unable, on account of the faintness of the nucleus, to distinguish with certainty the spectrum of its light which was projected upon the large spectrum of the coma. Once or twice I suspected the presence of two or three bright lines, but I could not be certain on this point.'

The comet of 1868, I. (Brorsen's), exhibited to the same observer a spectrum of three brilliant bands projected upon a faint continuous spectrum. 'The middle band,' says Mr. Huggins, 'is so much brighter than the others that it may be considered to represent three-fourths, or nearly so, of the whole of the light which we receive from the comet... In this nebulous band, however, I detected occasionally two bright lines, which appeared to be shorter than the band, and may be due to the nucleus itself.' Father Secchi, at Rome, analysed the light of the same comet, the spectrum of which likewise appeared to him discontinuous, and formed of luminous bands, upon a ground slightly luminous. The brightest of these bands was situated in the green, near to the magnesium line *b*. Another was visible in the blue, beyond the line *E*, but was less vivid and more vaporous. Two other lines were seen as well, the one in the yellow, the other, which was hardly perceptible, in the red. This makes in all four bands, instead of the three seen by Mr. Huggins; but the faintness of one of them perfectly explains this difference in the results of the two observations.

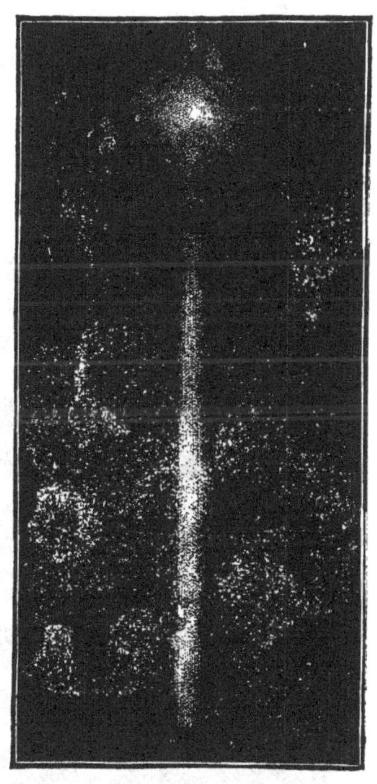

Fig. 52.—Comet of 1868, II. (Winnecke's). From a drawing made by Mr. Huggins.

Three luminous bands likewise formed the spectrum of the light of the comet of 1868, II. (Winnecke's). 'The middle one,' says Father Secchi, 'which is the brightest, is in the green; another, moderately brilliant, is situated in the yellow; and the last and faintest in the blue. The field of the telescope is full of a faint diffused light.' The positions of the luminous bands were measured by M. Wolf, who found that the most brilliant was situated between *b* and *F* of Fraunhofer, nearly in contact with *b*. Of the two others, the one was situated between *D* and *E*, a little nearer *E* than *D*; the third beyond *F*, but close to it. In fig. 53 are shown the two spectra of the comets of Brorsen and Winnecke, compared with the spectra obtained from an induction spark in olive-oil and in a current of olefiant gas. This interesting comparison is due to Mr. Huggins. There is a very close accordance between the spectrum of olefiant gas (C^2H^4) and that of Winnecke's comet, whilst the spectrum of Brorsen's comet is notably different, if not in its composition, at least in the fact of its bands being situated nearer together.

Comet I. of 1870, observed by Messrs. Wolf and Rayet, gave three brilliant bands similar to the preceding, projected upon a faint continuous spectrum. According to Mr. Huggins the same result was afforded by the comet of 1871, I. and

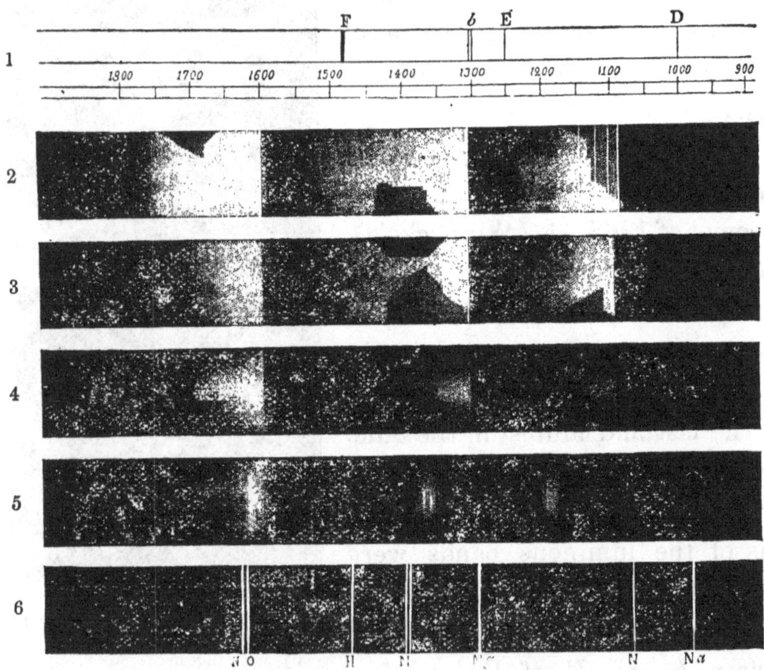

Fig. 53.—Spectra of the light of the comets of 1868, I. (Brorsen), and 1868, II. (Winnecke) from the observations of Mr. Huggins : (1) Solar spectrum; (2) spectrum of carbon spark taken in olive-oil ; (3) spectrum of carbon spark taken in olefiant gas ; (4) spectrum of comet of 1868, II. ; (5) spectrum of Brorsen's comet, 1868, I. ; (6) spectrum of an induction spark.

Encke's comet, with the difference, that the spectrum of the latter was not continuous ; this Mr. Huggins attributes to the small size and slight brilliancy of the nucleus.

Two other comets were in like manner analysed by Messrs. Wolf and Rayet with the following result:—

'The comet discovered at Marseilles by M. Borrelly,' they observe, 'on the night of the 20th-21st of August (1873, III.), presents the form of a circular nebulosity, about two minutes in diameter, provided with a tolerably brilliant nucleus in the centre. The spectrum is composed of a continuous spectrum extending from the yellow nearly to the violet, due in part to the solar reflected light, and of two luminous bands, the one in the green, the other in the blue. The green band is intense, clearly defined towards the red, but diffused towards the violet. The continuous spectrum is much brighter than that which we have observed in preceding comets, and much narrower. Perhaps this is due to a solid nucleus.'

In the same year was discovered at the Observatory at Paris, by MM. Paul and Prosper Henry, two young astronomers, a comet (1873, IV.), the light of which, when analysed on two occasions, gave the spectrum represented in fig. 54. On the

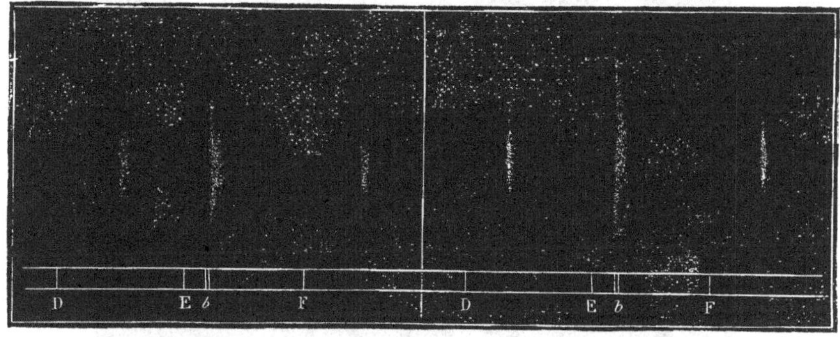

Fig. 54.—Spectrum of the Comet 1873, IV. (Henry's) (1) August 26; (2) August 29.

nights of the 26th–27th of August the comet exhibited the form of a circular nebulosity with a very bright condensation of light at its centre; its appearance was that of the stellar mass in Hercules when seen through an instrument of insufficient optical power to resolve it into stars.* The spectrum was

* Fig. 32 (p. 217) represents Henry's comet, as seen in the telescope at the date of these observations.

composed of the three usual luminous bands, with this peculiarity, that the most brilliant line, that in the green, was twice as long as either of the other two. There was no trace of a continuous spectrum. On the night of the 29th–30th the comet had a tail 25′ long, and its central nucleus had increased in

Fig. 55.—Coggia's comet, June 10, 1874, according to the drawing of M. G. Rayet.

brightness from the seventh to the sixth magnitude. The head of the comet always gave a spectrum composed of three luminous bands, but traversed this time by a very faint continuous spectrum. The brightness of the comet having increased, we were enabled to make the spectral observations with a comparatively narrow slit, and the band in the green then became more distinctly visible. In one portion of its length it was bounded on both sides by straight lines, but was throughout more brilliant on the side of the red. The brilliancy of the red and blue lines had also increased a little.

Since writing the above lines five new comets have been discovered and observed in the first six months of the year 1874: the first on February 20, by M. Winnecke; the second on April 11, by MM. Winnecke and Tempel; the third, and the most brilliant, which was visible in July to the naked eye, was discovered by M. Coggia, at Marseilles, on April 17. The other two were discovered, the one by M. Borrelly, the other by M. Coggia. But it is to the spectra of the second and third that the following results refer:—

'On the morning of the 20th of April,' says Father Secchi, 'the light of the comet (the second) was moderately bright; it exhibited a nucleus surrounded by an irregular fan-shaped nebulosity. The simple spectroscope applied to the great telescope of Merz showed traces of bands, but the diffusion of the object did not permit the use of this instrument. The compound spectroscope was applied, the telescope being also used, for so faint was the object that nothing could be made out with certainty. Then, on removing the telescope and looking with the unassisted eye, the spectrum appeared very clearly formed of three distinct bands, well separated: one in the blue-green, another in the green, and the third in the yellow-green. The first was the most brilliant and most extended. My impression is that these bands occupied the same places as the bands of the other comets, but I was unable to make exact measurements.'

The same astronomer thus describes his two spectroscopic observations of Coggia's comet: 'On the 17th of May I was able to ascertain that the spectrum consisted of bands; two especially were very bright in the green and the yellow-green. Having illuminated the tube of the telescope in front of the slit with the diffused light of various gases, the two bright bands were found to correspond with the bands of carbonic oxide and carbonic acid. The faintness of the light did not allow of the recognition of the other bands.'

The light of the same comet was subjected to analysis by MM. Wolf and Rayet. 'On the 19th of May,' observes the latter, 'I was able, in conjunction with M. Wolf, to make a first spectroscopic observation with some completeness. The diameter of the comet was nearly three minutes, and a tail was beginning to develop itself. The light, when analysed by the prism, gave a continuous spectrum from the orange to the blue (the spectrum of the solid nucleus), crossed by three bright bands (the spectrum of the gaseous nebulosity). It was the well-known cometary spectrum, but it differed from the ordinary spectrum in the dimensions and relative brilliancy of its different parts. Thus, whilst the continuous spectrum of the nucleus is in general wide and diffused, the spectrum given by Coggia's comet was very narrow. And again, the luminous transverse bands, instead of being ill-defined towards the most refrangible side, were terminated both towards the red and violet by straight and tolerably sharp lines. The remarkable fact of the central band being the longest and the most luminous struck me forcibly, as I had never witnessed it before.'

This last-mentioned fact was confirmed by a second observation, made on the night of June 4th–5th. 'The continuous spectrum,' says M. Rayet, 'corresponding to the nucleus, is remarkably narrow—nearly as narrow as that of a star seen through the same instrument; it is not unlike the spectrum of a star of the sixth magnitude, but it is colourless towards the extremities. The spectrum extends on both sides beyond the luminous bands. The spectrum of bands is composed of three lines, which by their refrangibility correspond to the yellow, the green, and the blue. The central band is long and very luminous; and when the aperture of the slit is suitably diminished it is terminated, towards the red and violet, by sharply-defined lines; it shows, therefore, none of that fading-off appearance towards the violet which is found in the spectra of ordinary telescopic comets. . . The bands in the yellow and blue are about half as bright as the middle one; they are slightly diffused towards the edges, and approximate to the ordinary type.

'If, instead of directing the slit of the spectroscope upon the focal image of the nucleus, so as to obtain at once the spectrum of the nucleus and that of the nebulosity, the slit is so turned as to cut the image of the tail, a spectrum is then obtained which presents the three bright bands above described, without a trace of the continuous spectrum, and separated from each other by dark intervals. In the tail, therefore, there is no solid incandescent matter of sensible amount.'

In the next section of this chapter other details will be found regarding the analysis of the light of the comet of 1874. They were received too late to be inserted in this section to which they naturally belong. These details, we may remark, confirm the results of MM. Wolf and Rayet.

Such are the results that have been afforded up to the present time by the spectral analysis of light. They are important on account of the conclusions we may even now permit ourselves to draw from them respecting the physical and chemical constitution of several cometary bodies.

In the first place, there is one fact common to all comets whose light has been analysed—the fact that their spectrum principally consists of a certain number of light bands separated by dark intervals of some extent. The continuous and very faint spectrum upon which these bands are projected existed, or at least was visible, only in some cases. Comets whose nuclei are very faint, like that of Encke's comet, or not sufficiently luminous (comet 1873, IV.), have failed to give a continuous spectrum. We may consider, therefore, that the bright bands are not produced by the light of cometary atmospheres or comæ. From his first observations Mr. Huggins came to an opposite conclusion, but this was doubtless owing to the impossibility of comparing the results then obtained with those afforded by the comets which have been analysed since.

We may thus regard the comets with nuclei which have been analysed by the spectroscope as constituted as follows:—

In the centre of the nebulosity a nucleus giving a continuous spectrum. Does this necessarily imply a liquid or solid incandescent matter? We might answer in the affirmative if the continuity of the spectrum could be regarded as complete; but it is so faint that it is difficult to say with certainty whether the light with which it shines really belongs to the incandescent matter of which it is composed, or if it is light reflected from the sun. It is not improbable that this light is of both kinds, especially when the comet is drawing near the sun and is subjected to a continually increasing temperature. The observations of polarisation by reflection prove that in any case a part of the light is reflected from the sun.

As regards the light of the atmospheres and tails, the spectrum of bright bands denotes alike the gaseous and the incandescent state of the matter of which they are composed. The identity in this respect of the tail and coma of Coggia's comet clearly shows that it is the matter of the atmosphere which, under the influence of a repulsive action, helps to form the cometary appendage opposite the sun. As, on the other hand, the phenomena of sectors emanating from the nucleus prove that the atmospheric envelopes are formed at the expense of the nucleus, it is very difficult to admit the incandescent state of the cometary atmosphere and tail without admitting that the nucleus, the seat of their continual formation, is likewise in an incandescent state. It is, then, probable that the nucleus, at all events in the vicinity of the perihelion, emits, besides light reflected from the sun, direct light that has emanated from its own substance.

In a chemical point of view the comets—few in number, it is true—which have as yet been subjected to examination are of very simple constitution. They consist of simple carbon, or of a compound of carbon and hydrogen, according to the comparisons made by Mr. Huggins; carbonic oxide or carbonic acid, according to the researches of Father Secchi. The Italian astronomer was, therefore, justified in saying: 'It is very remarkable that all the comets observed up to the present time have the bands of carbon.'

SECTION VII.

THE COMET OF 1874, OR COGGIA'S COMET.

Of the five comets of 1874 the third, or comet of Coggia, was alone visible to the naked eye—Telescopic aspect and spectrum of the comet during the early part of its apparition, according to Messrs. Wolf and Rayet—Observations of Secchi, Bredichin, Tacchini, and Wright; polarisation of the light of the nucleus and tail—Transformations in the head of the comet between the 10th of June and the 14th of July, according to Messrs. Rayet and Wolf.

THE comets, and not the comet, of 1874 should form the title, strictly speaking, of the present section of our work. Indeed, at the time of adding these lines to this chapter—that is to say, in the last few days of the month of August of this year [1874]—five new comets have been discovered and observed. But one only, the third in order of date, has attracted the attention of the public, for the simple reason that it alone became bright enough during the time of its apparition to be visible to the naked eye. The other four continued to remain telescopic comets, accessible only to professional astronomers. Although its visibility in Europe was of brief duration, the comet of 1874, III., or comet of Coggia, presented in its physical aspect, and in the changes of form in its head and tail, sufficiently curious phenomena to merit special mention and some detailed description.

At the Observatory of Marseilles, on the night of April 17, the new comet was discovered by an astronomer of that establishment, M. Coggia, already known in the scien-

COGGIA'S COMET, 1874.

Seen from the Pont-Neuf, Pàris.

tific world by his discovery of the planet Ægle, the second comet of 1870, and last year by his discovery of the seventh comet of that year, of which we have already made mention in the chapter on Periodical Comets. The new comet on its first appearance was a very faint nebulosity, but as it advanced nearer to the sun and the earth it grew rapidly brighter, and became visible to the naked eye about the beginning of July. From this date the comet continued to increase in brilliancy up to the night of July 14, when its own and the diurnal movement combined caused it to subside into the mists of the horizon and finally disappear from our latitude. It is to be hoped that it will have been observed in regions nearer the equator and in the southern hemisphere. This is greatly to be desired, for it disappeared from us at the very moment when its telescopic study had become of the very highest interest. We will, therefore, limit ourselves to the facts observed, letting each observer speak for himself.

'At the date of its discovery,' says M. Rayet, 'the comet was faint and of a circular form, with a very marked central condensation, resembling a luminous point. The nebulosity was about two minutes in diameter. The light was of such small intensity that it was hardly possible to verify the existence of a spectrum. The comet continued to approach the sun and the earth, and its brilliancy steadily increased.'

The spectral analysis of its light made jointly by Messrs. Rayet and Wolf, on May 19, has been given in the preceding section as well as that made on the night of the 4th–5th of June.

On the night of the 4th–5th of June the comet exhibited a round and very brilliant nucleus, about equal in brilliancy to a star of the eighth magnitude. The surrounding nebulosity, from which the nucleus stood out very distinctly, measured four minutes in diameter, and was prolonged opposite the sun into a tail eight minutes in length. Its light, which had quadrupled in intensity since April 17, gave a clearly visible spectrum, which we have already described on page 324 according to M. Rayet.

The light of the comet was likewise analysed at Rome by Father Secchi, whose observations confirm in their principal results those of the French savants. The three bright bands and the continuous spectrum which crossed them transversely presented, on the dates of June 18 and July 9, the appearance shown in fig. 56. Father Secchi directs attention to a peculiarity which is readily to be perceived, viz. the discontinuity in the continuous spectrum in the neighbourhood of each band, and which is more especially apparent in the second observation. 'On examining,' he observes, 'the spectrum thus composed with a Nicol's prism, the continuous portions were diminished in intensity, whilst the bands themselves lost none of their brilliancy. This observation would lead us to believe that the continuous spectrum was derived from reflected light.' We see that in this respect Father Secchi differs in opinion from the French astronomers quoted above, who consider the continuous spectrum as produced by a solid nucleus in a certain state of incandescence. It appears certain that the light of the comet was polarised; this was proved by the observations made at Rome in the course of July; but may not the nucleus at the same time both emit its own light and reflect that of the sun? This is a question not yet solved, and doubt still exists concerning the nature of cometary light.

We add further details, due to the same astronomer, who was enabled to observe the comet for a longer time than was possible in France, as is shown by the date of July 17, mentioned below.

'The comet,' he remarks, 'when observed with an ordinary

eyepiece was magnificent. On the 9th of July it formed a fan of a reddish tint (by contrast with the nucleus) of about 180 degrees of opening, composed of curvilinear rays springing from a nucleus of yellowish green. On increasing the magnifying power to 100 the nucleus was seen surmounted only by very faint plumes and reduced in size to a small diffused sphere hardly two seconds in diameter. The absence of all defined limit, an effect produced by the high magnifying power employed, proves that no solid body was contained in the nucleus. The same power shows, in fact, the satellites of Jupiter with clearly-defined discs.

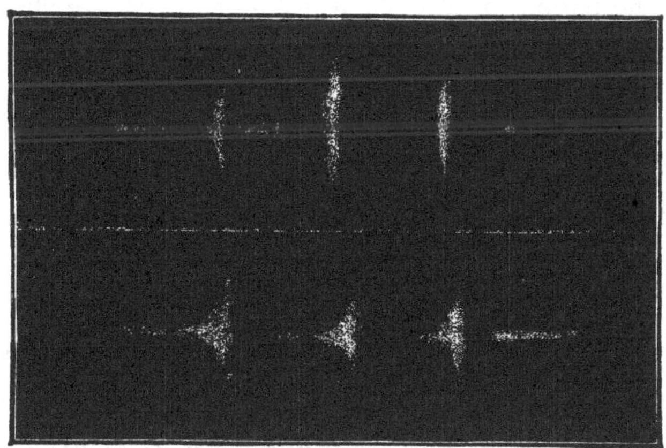

Fig. 56.—Spectra of the Comet of 1874, III., (Coggia's), according to Father Secchi.

'At the request of Mr. Hind we have looked for the comet during the daytime, but without success. There appeared little probability of seeing it under these conditions, for Jupiter, a much more brilliant object, was not visible. On the 17th of July the tail was enormous; it extended to the star υ of the Great Bear, the head being hidden below the horizon. It must have been at least 45 degrees in length. On the 13th it was very expanded near the head.'

On comparing the position of the bright bands of the

cometary spectrum with the spectra given by carbon and carbonic acid, Father Secchi found them to correspond; but, on employing hydrocarbons, no hydrogen line appeared to coincide with those of the comet. These results show that astronomers are not yet well agreed in their interpretation of the facts afforded by spectral analysis, for we read in a letter addressed from Moscow to the Italian Spectroscopical Society by Professor Bredichin, that this savant compared the positions of the bands of the comet with those of a hydrocarbon in a Geissler's tube; and he adds: 'Within the limits of the errors of the observations (I made ten) the bands of the comet coincide with the bands of the hydrocarbon whose wave-lengths are 5633, 5164, 4742 of Ångström's scale'

At Palermo, M. Tacchini has also made the following observations upon the spectrum of the comet and the polarisation of its light:—

'The bright lines observed in the spectrum of the comet were four in number, corresponding, when referred to the solar spectrum, to the following positions of Angström's scale: 6770, 5620, 5110, and 4800. The position of these lines cannot be looked upon as strictly accurate, on account of the manner in which they were obtained; but it is evident that the three last correspond to the spectrum of carbon. The red line was less distinct than the others, because in this part the red was bright and diffused. This line was only well seen in the last days of June and the first days of July. The three other lines were not of equal length, and the longest was the 5620 line; the 5110 line was the brightest of all, and appeared almost as white as the magnesium line after solar eruptions. The continuous spectrum of the comet's nucleus was projected seemingly upon a ground formed of a more intense solar spectrum, in which the red was, as has just been said, the most extended. This beautifully coloured band or ribbon was seen only at the passage of the nucleus, which, observed through a simple eyepiece, appeared of a greenish white, whilst the fan was sensibly rose-coloured, even when occulting the nucleus. In the bright solar light reflected by the nucleus, traces of polarised light were to be expected; and to test their existence we invited Signor Pisati, professor of physics, to make experiments by the aid of the polariscopes at his disposal. On applying a bi-quartz to the telescope, traces of polarisation were observed, but they were very feeble. By the aid of a Nicol's prism the light appeared strongly polarised, and the greatest diminution of light took place when the principal section of the prism was coincident with the direction of the tail; from which it follows that the light was polarised in a plane passing through the sun. The experiment was again repeated upon the brightest portion of the tail, and with the same result. Towards the middle of the tail the light was so feeble that nothing certain could be determined; but it seems probable that reflexion took place throughout the entire length of the tail.' (*Memorie della Societa degli Spettroscopisti Italiani.* Luglio, 1874.)

These very decisive conclusions respecting the polarisation of the comet's light derive further confirmation from the observations of Mr. Wright, at Yale College (U.S.), which led him to infer that a considerable portion of the light of the comet was derived from the sun by reflexion.

Let us now return to the aspect of the nucleus and nebulosity, as shown in the telescope during the most interesting period of the comet's apparition. In order to follow the various appearances presented we shall avail ourselves of the detailed and very careful descriptions placed at our disposal by MM. Wolf and Rayet. Thanks to the courtesy of these gentlemen, we shall be able to study the different changes exhibited by the comet from original drawings hitherto unpublished, and which we have received permission from MM. Wolf and Rayet to engrave.

'On the 10th of June,' they remark, 'the comet preserved unchanged the same general aspect as on the preceding days; it was still a circular nebulosity about four minutes in diameter, and provided with a central nucleus very brilliant and remarkably distinct, which gave to the comet a remarkable appearance. In a direction opposite to the sun the nebulosity was lengthened out, and thus formed a tail, which, narrow at its base, expanded into a fan about twenty-four minutes in length. The coma was more brilliant in the centre than towards the edges. (See fig. 55, p. 322.)

'The comet preserved the same appearance, whilst increasing rapidly in size, till about June 22, at least so far as it was possible to judge from observations much impeded by the light of the moon.

'The spectrum remained as above described, viz. it consisted of a very narrow continuous spectrum, and of three bright transverse bands.

'On the 22nd of June a series of changes in the head of the comet began. On this day the comet, when examined with the Foucault's telescope, 15¾ inches aperture, appeared to be enclosed in the interior of a very elongated parabola. Starting from the nucleus, situated where the focus of the parabola would be, the light diminished regularly towards the vertex; but towards the interior of the parabola the diminution of the light was abrupt, and its line of separation was another parabola, slightly more open than the first, and having for its vertex the brilliant nucleus itself. (See fig. 57.) The parabola passing through the nucleus formed, when prolonged, the sides of the tail, the edges of which were clearly defined and were much more brilliant than the inner portion. The tail had, therefore, the appearance of a luminous envelope, hollow in the interior. The nucleus continued sharp and bright.

'On the 1st of July the general form of the comet remained unchanged: it still appeared bounded on the outside by an arc of a parabola. The luminous point, however, had shifted

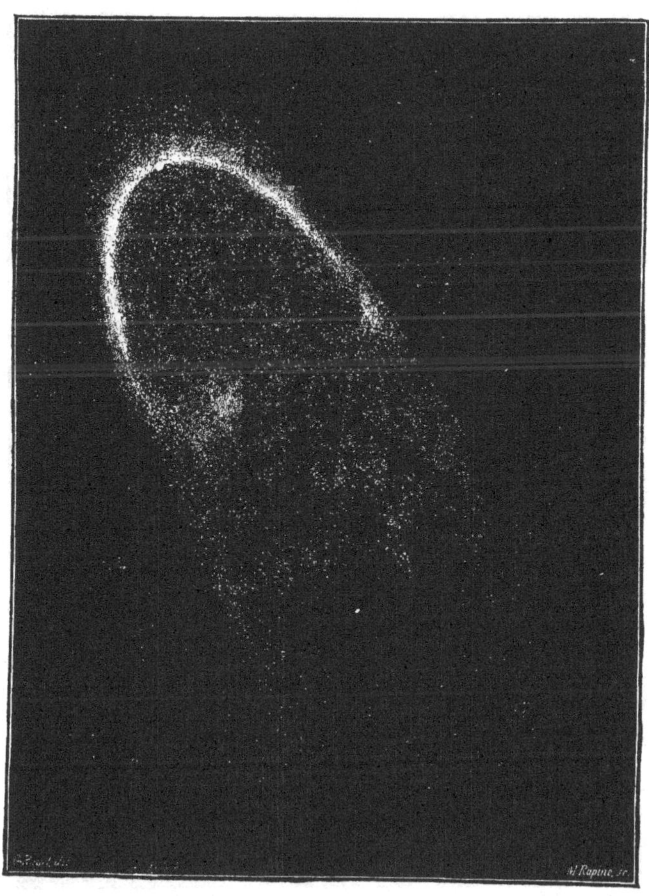

Fig. 57.—Coggia's Comet seen in the telescope on June 22, 1874, according to a drawing by M. G. Rayet.

forward into the interior of the second parabola, and the two sides of the tail were not symmetrical. (Fig. 58.) The west side (the side on which the right ascension is the greater) was sensibly more luminous than the other. The spectrum of bright bands given by the nebulosity was moderately luminous, and colours were distinguishable in the narrow spectrum of the nucleus; the red at one extremity, and a tint of blue or violet at the other.

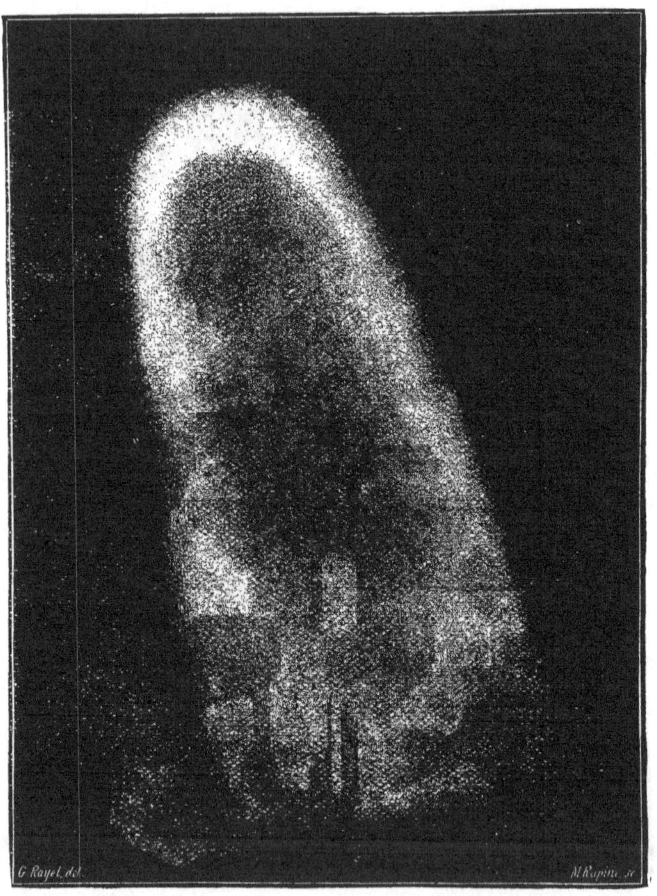

Fig. 58.—Coggia's Comet on July 1, 1874, according to a drawing by M. G. Rayet.

'Since the 5th of July the comet's want of symmetry has continued to increase in a marked degree, and towards the head the diminution of the light has become less regular.

'On the 7th the want of symmetry was striking, the west

portion of the tail being about twice as brilliant as the east portion. At the same time the nucleus appeared to have become diffused and "fuzzy" towards the head, whilst it was still clear and distinct towards the tail. It suggested the idea of an open fan.

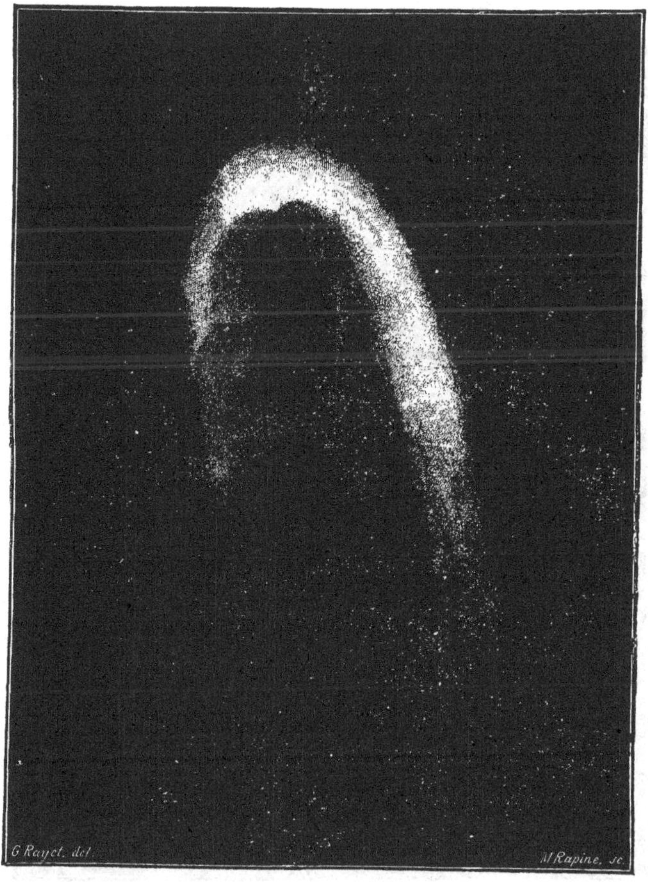

Fig. 59.—Coggia's Comet on July 13, 1874, according to M. G. Rayet.

'From the 7th to the 13th of July the weather was unfavourable for observations; but in the interval the comet had undergone no material change, for on the 13th it was visible again, having the same form, but somewhat more pronounced. The fan of light, however, formed at the expense of the nucleus, had assumed greater importance, and was inclined in a very marked manner towards the western portion of the coma. At the moment of observation (fig. 59), about 10 P.M., the northern portion of the sky was slightly foggy, and the

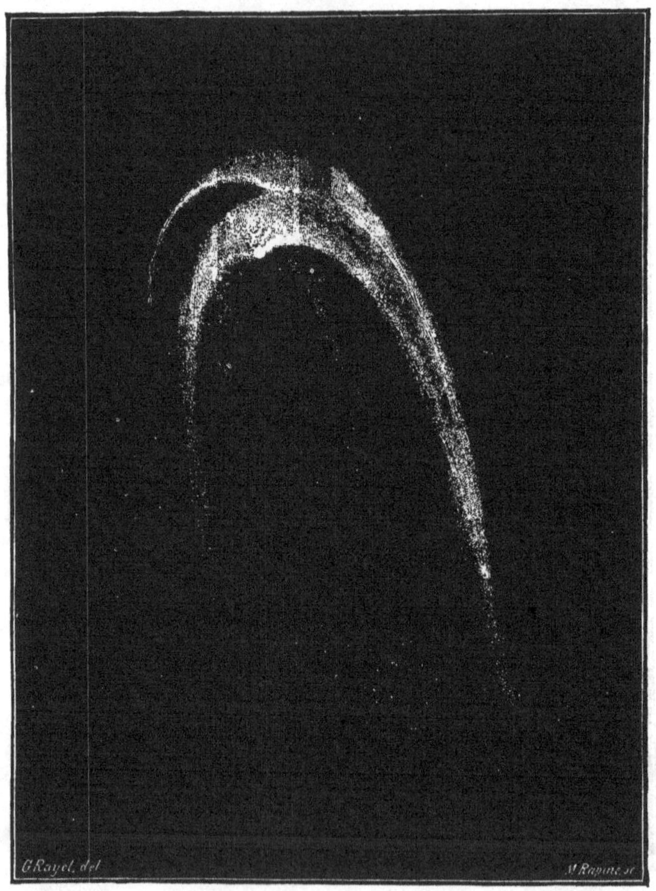

Fig. 60.—Coggia's Comet on July 14, according to M. G. Rayet.

comet already close to the horizon. As for the tail, it extended nearly to *o* of the Great Bear, and thus had an apparent length of about 15 degrees.

'Our last observation of the comet was on the 14th, at

$9^h\ 30^m$ P.M. Important changes had taken place in the aspect of the head. (Fig. 60.) The fan of light was altogether thrown towards the west, and on this side was prolonged into a long train, losing itself far into the coma. Towards the east the fan terminated abruptly, and the line of termination made only a small angle with the axis of the comet. Two plumes or jets were visible, projecting forward, one on the right, the other on the left. These plumes seemed to rise from the edge of the tail, of which they formed as it were the prolongation. The eastern plume soon curved back towards the tail; it was faint, and was soon lost in the nebulosity. The plume directed towards the west was much more brilliant and curved back immediately towards the tail, the bright outer edge of which it helped to define.'

MM. Wolf and Rayet call attention to the fact that the comets of 1858 and 1861 exhibited transformations similar to those of the comet of Coggia. The analogy is evident, but at the same time there are marked differences. The aspect of the comet of 1874, on the night of July 14, was especially remarkable for the phenomena indicated in the drawing of M. Rayet, which we think are unprecedented. The plumes which have just been described indicate the commencement of a radical transformation in the form of the head and tail—one would have said that two different comets were in juxtaposition, the one projected upon the other. Was this, as has been suggested, a premonitory sign of duplication? This is what we shall learn, if the series of observations unfortunately interrupted in Europe has been continued in the southern hemisphere. There is, also, a peculiarity which calls for remark on comparing the drawings of the French astronomers with those of Mr. Newall, taken nearly at the same time. The two plumes in the English sketch form two very regular plumes, symmetrically placed with regard to the axis of the tail, the nucleus and the head of the comet; they irresistibly remind us of the antennæ of certain moths. We must confess that the very carefully-studied drawing of M. Rayet appears to us to merit entire confidence.

But let us return to the observations of MM. Wolf and Rayet. Their concluding remarks relate to the spectral analysis of the comet's light during the month of July:—

'Whilst the comet was changing form, its spectrum preserved the same character and appearance, and continued to increase in brightness. It was not until July 13 that it became modified by the exaggerated importance of one of its parts. At this time the nucleus had become diffused, and the solid matter of which it had been composed appeared to be distributed throughout the head of the comet, so that the spectrum consisted of a luminous and vividly coloured streak, continuous from the red to the violet, standing out from a continuous and broader spectrum. The three luminous bands had nearly disappeared, probably drowned in the light of the continuous spectrum. The comet moreover was situated low down in the mists near to the horizon. In the continuous spectrum we looked in vain for the presence of bright lines or black bands.

'On the 1st and 6th of July, whilst the luminous bands were yet visible, we referred micrometrically the position of the most brilliant of them—the middle one—to the lines E and b. The wave-lengths, on the least refrangible side of the line, were thus found to be: 1st of July, 5161; 6th of July, 5165. The wave-length of the three lines b being 5174, this band is slightly more refrangible.

'We believe that this measure is accurate; but the difficulty of such determinations is so great, that we think it useless to identify this band with the bright lines of any gas.'

Such are the facts that have been as yet collected concerning the physical and chemical constitution of the comet discovered by M. Coggia. We shall refrain from drawing any conclusions from them, as all discussion at the present moment would be incomplete and consequently premature. The new comet has certainly been observed by many astronomers both in Europe and America; and we must wait for the observations made subsequently to the disappearance of the comet from our latitudes.

We shall say nothing of the comets I., II., and V. of 1874, except that they were discovered, the first two by M. Winnecke,

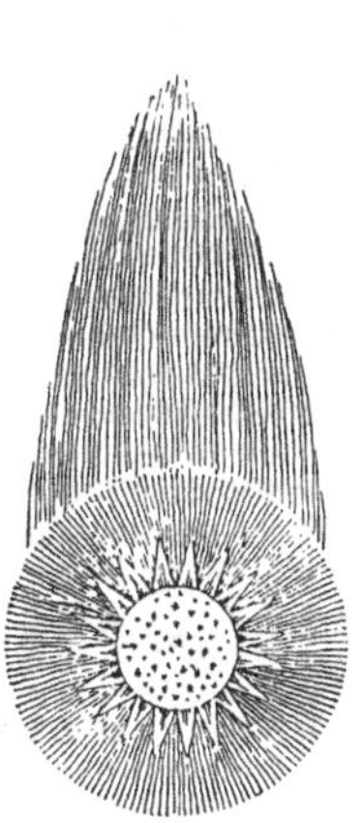

Fig. 61.—Comet of 1618, according to Hevelius. Multiple nuclei.

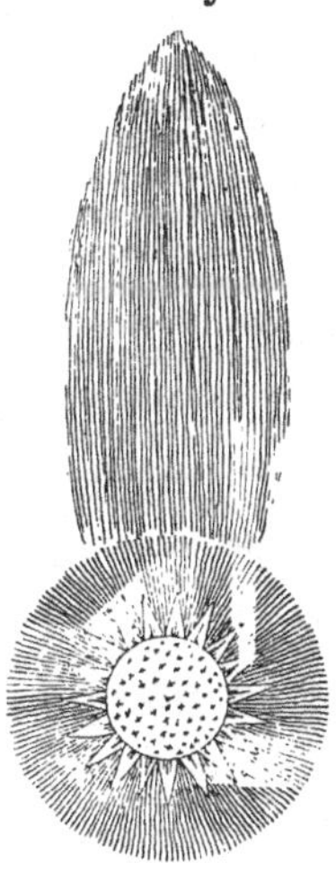

Fig. 62.—Comet of 1661, according to Hevelius. Multiple nuclei.

on February 20 and April 11 respectively; the third by M. Coggia, on August 20. But comet IV., 1874, discovered by M. Borrelly (Marseilles), exhibited a very interesting structure, which gives some reason to believe that the observations of the comet of 1618 and 1661 by Hevelius are worthy of more consideration than they have hitherto received by astronomers. The heads of these comets (figs. 61 and 62) were furnished with multiple nuclei, so that they were formed to all appearance of an assemblage of little stars. We give the fac-simile of the drawings by which Hevelius has endeavoured to represent their nucleal structure.

The comets 1869, I., 1869, III., and 1871, I., were in like manner formed of nebulosities studded with a great number of luminous points, which gave them the appearance of certain resolvable nebulæ. The comet discovered in 1874 by M. Borrelly belongs evidently to the same class. M. Wolf thus describes the appearance presented by the fourth comet of the year: 'The new comet,' he observes, 'discovered at Marseilles by M. Borrelly presented from the first the appearance of a somewhat faint but nearly resolvable nebula. Upon the whitish ground of the nebulosity appear a number of little brilliant points, of which the most conspicuous is excentric and situated behind and to the north of the centre of figure. This comet seems, therefore, to belong to a class whose representatives are few in number, and to which Schiaparelli has called attention, viz. to the class of comets composed of a mass of little nuclei. On August 3 the aspect of Borrelly's comet was not unlike that of the stellar mass in Hercules, but of less extent and brilliancy. On August 8 the principal excentric nucleus had become more brilliant, while the nebulosity had increased in extent.'

ON COGGIA'S COMET (III., 1874).

[ADDITION BY THE EDITOR.]

M. Guillemin's book was published at the end of 1874, before it was possible to compare together and discuss all the observations and drawings that had been made of the comet. I therefore propose in this addition to the chapter to give a brief account of some of the other observations of this comet, which must be in the remembrance of every one of our readers.

The following remarkable and interesting letter from Mr. J. Norman Lockyer appeared in the *Times* of July 16, 1874. It was reproduced in *Nature* for July 23, 1874:—

'Mr. Newall's Observatory, Ferndene, Gateshead.

'I was enabled on Sunday night (12th inst.), by Mr. Newall's kindness, to spend several hours in examining the beautiful comet which is now visiting us, by means of his monster telescope—a refractor of 25-in. aperture, which may safely be pronounced the finest telescope in the world, or, at all events, in the Old World.

'The view of the comet which I obtained utterly exceeded my expectations, although I confess they were by no means moderate; and as some of the points suggested by the observations are, I think, new, and throw light upon many recorded facts, I beg a small portion of space in the *Times* to refer to them, as it is important that observers have their attention called to them before the comet leaves us.

'I will first deal with the telescopic view of the comet. Perhaps I can give the best idea of the appearance of the bright head in Mr. Newall's telescope, with a low power, by asking the reader to imagine a lady's fan opened out (160°) until each side is almost a prolongation of the other. An object resembling this is the first thing that strikes the eye, and the nucleus, marvellously small and definite, is situated a little to the left of the pin of the fan—not exactly, that is, at the point held in the hand. The nucleus is, of course, brighter than the fan.

'Now, if this comet, outside the circular outline of the fan, offered indications of other similar concentric circular outlines, astronomers would have recognised in it a great similarity to

Donati's beautiful comet of 1858, with its "concentric envelopes." But it does not do so. The envelopes are there undoubtedly, but, instead of being concentric, they are excentric, and this is the point to which I am anxious to draw attention, and, at the risk of being tedious, I must endeavour to give an idea of the appearance presented by these excentric envelopes. Still referring to the fan, imagine a circle to be struck from the left-hand corner with the right-hand corner as a centre, and make the arc a little longer than the arc of the fan. Do the same with the right-hand corner.

'Then with a gentle curve connect the end of each arc with a point in the arc of the fan half-way between the centre and the nearest corner. If these complicated operations have been properly performed, the reader will have superadded to the fan two ear-like things, one on each side. Such "ears," as we may for convenience call them, are to be observed in the comet, and they at times are but little dimmer than the fan.

'At first it looked as if these ears were the parts of the head furthest from the nucleus along the comet's axis, but careful scrutiny revealed, still in advance, a cloudy mass, the outer surface of which was regularly curved, convex side outwards, while the contour by the inner surface exactly fitted the outer outline of the ears and the intervening depression. This mass is at times so faint as to be invisible, but at other times it is brighter than all the other details of the comet which remain to be described, now that I have sketched the groundwork. These details consist of prolongations of all the curves I have referred to backwards in the tail.

'Thus, behind the bright nucleus is a region of darkness (a black fan, with its pin near the pin of the other pendant from it, and opened out 45° or 60° only will represent this), the left-hand boundary of which is a continuation of the lower curve of the right ear. The right-hand boundary is similarly a continuation of the lower curve of the left ear. Indeed, I may say generally—not to enter into too minute description in this place—that all the boundaries of the several different shells which show themselves, not in the head in front of the fan, but in the root of the tail behind the nucleus, are continuous in this way—the boundary of an interior shell on one side of the axis bends over in the head to form the boundary of an exterior shell on the other side of the axis.

'At last, then, I have finished my poor, and, I fear, tiresome description of the magnificent and truly wonderful sight presented to me as it was observed, on the whole, during some hours' close scrutiny under exceptional atmospheric conditions.

'I next draw attention to the kind of change observed. To speak in the most general terms, any great change in one "ear" was counterbalanced by a change of an opposite character in the other, so that when one ear thinned or elongated, the other widened; when one was dim, the other was bright; when one was more "pricked" than usual, the other at times appeared to lie more along the curve of the fan and to form part of it. Another kind of change was in the fan itself, especially in the regularity of its curved outline and in the manner in which the straight sides of it were obliterated altogether by light, as it were, streaming down into the tail.

'The only constant feature in the comet was the exquisitely soft darkness of the region extending for some little distance behind the nucleus. Further behind, where the envelopes of the tail were less marked, the delicate veil which was over even the darkest portion became less delicate, and all the features were merged into a mere luminous haze. Here all structure, if it existed, was non-recognisable, in striking contrast with the region round and immediately behind the fan.

'Next it has to be borne in mind that the telescopic object is after all only a section, from which the true figure has to be built up, and it is when this is attempted that the unique character of this comet becomes apparent. There are no jets, there are no concentric envelopes; but, as I have said, in place of the latter, excentric envelopes indicated by the ears and their strange backward curvings, and possibly also by the fan itself.

'I prefer rather to lay the facts before observers than to state the conclusions to be derived from them, but I cannot help remarking that, supposing the comet to be a meteor-whirl, the greatest brilliancy is observable where the whirls cut or appear to cut each other ; where we should have the greatest number of particles, of whatever nature they may be, in the line of sight; and not only so, but regions of greatest possible number of collisions associated with greatest luminosity.

'It would be a comfort if the comet, to partly untie a hard knot for us, would divide itself as Biela's did. Then, I think, the whirl idea would be considerably strengthened. I could not help contemplating the possibility of this when the meaning of the "ears" first forced itself upon my attention.

'The spectroscopic observations which I attempted, after the telescopic scrutiny, brought into strong relief the littleness of the planet on which we dwell, for a seven hours' rail journey from London had sufficed to bring me to a latitude in which the twilight at midnight was strong enough to show the middle part of the spectrum of the sky, while to the naked eye the tail of the comet was not so long as I saw it in London a week ago.

'I had already, in observations in my own observatory, with my $6\frac{1}{4}$-in. refractor (an instrument smaller than one of Mr. Newall's four finders!), obtained indications that the blue rays were singularly deficient in the continuous spectrum of the nucleus of the comet, and in a communication to *Nature* I had suggested that this fact would appear to indicate a low temperature.

'This conclusion had been strengthened by Sunday night's

observations, and it was the chief point to which I directed my attention. The reasoning on which such a conclusion is based is very simple. If a poker be heated, the hotter it gets the more do the more refrangible—*i.e.*, the blue—rays make their appearance if its spectrum be examined. The red colour of a merely red-hot poker and the yellow colour of a candle-flame are due, the former to an entire, the latter to a partial, absence of the blue rays. The colour, both of the nucleus and of the head of the comet, as observed in the telescope, was a distinct orange-yellow, and this, of course, lends confirmation to the view expressed above.

'The fan also gave a continuous spectrum but little inferior in brilliancy to that of the nucleus itself; while over these, and even the dark space behind the nucleus, were to be seen the spectrum of bands which indicates the presence of a rare vapour of some kind, while the continuous spectrum of the nucleus and fan, less precise in its indications, may be referred either to the presence of denser vapour, or even of solid particles.

'I found that the mixture of continuous band spectrum in different parts was very unequal, and further that the continuous spectrum changed its character and position. Over some regions it was limited almost to the region between the less refrangible bands.

'It is more than possible, I think, that the cometary spectrum, therefore, is not so simple as it has been supposed to be, and that the evidence in favour of mixed vapours is not to be neglected. This, fortunately, is a question on which I think much light can be thrown by laboratory experiments.

'P.S.—(By Telegraph.)—Wednesday night.—Sunday's observations are confirmed. The cometary nucleus is now throwing off an ear-like fan. Ten minutes' exposure of a photographic plate gave no impression of the comet, while two minutes gave results for the faintest of seven stars in the Great Bear.'

A rough outline sketch of the head and envelopes of Coggia's comet, as seen in Mr. Newall's 25-inch refractor on the night of July 12, appeared in *Nature* for July 16, 1874.

On July 8, Mr. G. H. With, at Hereford, observing the comet with an 8½-inch aperture Newtonian reflector, noticed a remarkable oscillatory motion of the fan-shaped jet, upon the nucleus as a centre, and which occurred at intervals of from three to eight seconds. 'The fan seemed,' says Mr. With, 'to tilt over from the preceding towards the following side, and then, for an instant, appeared sharply defined and fibrous in structure. Suddenly it became nebulous, all appearance of structure vanished, and the outline became merged in the surrounding matter. At the moment of this change a pulsation was transmitted from the head through the coma, as though luminous vapour had been projected from the former into the latter. These phenomena were observed many times during the evening, both by myself and a well-trained optical assistant.' *

Both Mr. With and Mr. Newall also speak of a faint luminous cloud that preceded the head of the comet, *i.e.* in front of it, on the opposite side to the tail, and apparently separate from the comet. The latter also states that the effect of motion was conveyed in a remarkable manner by the flickering of the tail.

Mr. Huggins's paper on the spectrum of Coggia's comet was read before the Royal Society on January 7, 1875, and is printed in *Proc. Roy. Soc.*, vol. xxiii., pp. 154–159. The following are some extracts from his account:—

* *R.A.S. Notices*, May 1876.

'The comet now visible, which was detected by M. Coggia, April 17, 1874, is the first bright comet to which the spectroscope has been applied. The following spectroscopic observations of this comet were made from July 1 to July 15 :—

'When the slit of the spectroscope was placed across the nucleus and coma, there was seen in the instrument a broad spectrum, consisting of the three bright bands which were exhibited by Comet II., 1868, crossed by a linear continuous spectrum from the light of the nucleus.

'In the continuous spectrum of the nucleus I was not able to distinguish with certainty any dark lines of absorption, or any bright lines other than the three bright bands.

'Besides these spectra, there was also present a faint broad continuous spectrum between and beyond the bright bands.

'When the slit was moved on to different parts of the coma, the bright bands and the faint continuous spectrum were observed to vary in relative intensity.

'When the slit was brought back past the nucleus on to the commencement of the tail, the gaseous spectrum became rapidly fainter, until, at a short distance from the nucleus, the continuous spectrum predominated so strongly that the middle band only, which is the brightest, could be detected on it.

'We have presented to us, therefore, by the light of the comet three spectra :—

'1. The spectrum of bright bands.

'2. The continuous spectrum of the nucleus.

'3. The continuous spectrum which accompanies the gaseous spectrum in the coma, and which represents almost entirely the light of the tail.'

Mr. Huggins then describes in detail the spectrum of bright bands and the continuous spectrum of the nucleus, and proceeds :—

'When the nucleus was examined in the telescope, it

appeared as a well-defined minute point of light, of great brilliancy. I suspected at times a sort of intermittent flashing in the bright point. The nucleus suggested to me an object on fire, of which the substance was not uniform in composition, so that at intervals it burned with a more vivid light. On July 6 the diameter of the nucleus, when measured with a power of 800, was $1''\cdot8$; on July 13 the measure was nearly double, viz., $3''$; but at this time the point of light was less defined. On July 15 the nucleus appeared elongated towards the following side of the comet, at an angle of about 40° to the comet's axis. The nucleus appeared of an orange colour. This may be due in part to the effect of contrast with the greenish light of the coma.'

'The continuous spectrum which accompanies the gaseous spectrum was observed in every part of the coma; near its boundary, and in the dark space behind the nucleus, the continuous spectrum became so faint as to be detected with difficulty, at the same time that the bright bands were distinctly visible. The more distant parts of the tail gave probably a continuous spectrum only.'

Mr. Huggins thus concludes his remarks:—

'On several evenings I satisfied myself that polarised light was present in every part of the comet. I do not think that the proportion of polarised light exceeded one-fifth of the total light. The polarisation, as exhibited by the partial extinction of one of the images formed by a double-image prism, appeared to be more marked in the tail. It must be remembered that such would appear to be the case to some extent even if the proportion were not really greater, because the same proportional diminution in a faint object is more appreciated by the eye. Still there was probably a relatively large proportion of polarised light in the tail.

'The reflected solar light would account for a large part of the continuous spectrum. To what source are we to ascribe the remaining light which the prism resolves into a continuous spectrum? Is it due to reflexion from discrete particles, too large relatively to the wave-lengths of the light for polarisation to take place? or is it due to incandescent solid particles? From the co-existence of the band-spectrum, we can scarcely think of distinct masses of gas dense enough to give a continuous spectrum.

'The difficulty which presents itself in accounting for sufficient heat to maintain this matter and the nucleus in a state of incandescence has also to be encountered in respect of the gaseous matter which emits the light which is resolved into the bright bands.

'The solar radiation to which the comet was subjected would be inadequate to account for this state of things directly. Is there chemical action set up within the comet by the sun's heat? Is the comet's light due to electricity in any form excited by the effect of the solar radiation upon the matter of the comet? Are we to look for the source of the light to the friction of the particles of the cometary matter which has been thrown into violent agitation by the comet's approach to the sun?' *

Mr. Christie found that on July 3, 6, 7, and 13 the tail and coma were partially polarised in a plane through the axis of the tail.

On p. 340 M. Guillemin calls attention to the difference between M. Rayet's drawing of Coggia's comet on July 14 and Mr. Newall's drawing made at the same time, and adds that the former appears to merit entire confidence. I do not wish for a moment to say otherwise, but it is only fair to state that

* Mr. Huggins's other papers on cometary spectra are to be found in *Proc. Royal Soc.*, vol. xvi., p. 386 (1868); vol. xix., p. 488 (1871); vol. xx., p. 45 (1872); and *Phil. Trans.*, vol. clviii., p. 555 (1868).

Mr. Newall's drawing is confirmed by nearly all those of the same date that I have seen. It will be found engraved in the *R.A.S. Notices* for March 1876, and in the letter accompanying it Mr. Newall writes: 'The next view we got of the comet was on Tuesday, July 14, when a wonderful change had taken place. This is extremely well represented by fig. 2, which was made by Mrs. Newall, and was so exact that I did not touch it. Here the nucleus is still very distinct. The two streams of the tail have separated and become shorter, leaving a wider dark space between them, while from the two corners of the fan proceed two antennæ, which appear to be projections of the inner sides of the tails, and preceding these is a luminous cloud.'

M. Guillemin's description of the drawing is very exact, viz. the two parabolic arcs start from the nucleus, and being symmetrical with regard to the axis of the tail, resemble very closely the antennæ of a moth. In the drawing by Mr. Plummer of the comet on July 14,* as seen in a refractor of 10 inches aperture, the two arcs are also placed symmetrically with regard to the axis; and a sketch made by M. Dreyer at Copenhagen,† on July 13, with an 11-inch refractor, also shows the same arrangement.

The drawing of Mr. With ‡ represents a fan-shaped structure, with the apex at the nucleus and slightly inclined to the axis; and there is but a very slight want of symmetry shown in the drawings of Mr. Wilson § or of Mr. Huggins.‖ The drawing of Dr. Vogel for July 14¶ does not show the interior structure very clearly.

It is well known how greatly the drawings of astronomical appearances, as seen in different telescopes, may differ from

* *R.A.S. Notices*, Dec. 1874.

† *Id.*, May 1876.

‡ *Id.*, March 1876.

§ *R.A.S. Notices*, Dec. 1874.

‖ *Proc. Royal Soc.*, vol. xxiii., p. 159.

¶ *Ast. Nach.*, No. 2,018.

one another, (as, for example, is the case with the careful delineations of the great nebula in Orion, by Lord Rosse, Mr. Lassell, and Father Secchi), and, making some allowance for this, I have been struck with the generally close agreement in the representations of Coggia's comet, the most exceptional of which seems to me to be that of M. Rayet, shown in fig. 60. The duplicate structure, resembling two parabolas superposed the one over the other, is mentioned by several observers, and the chief difference in the drawings is that while M. Rayet inclines the two parabolas to one another at an angle, most of the other observers agree with Mr. Newall in making them cross one another symmetrically. The fact that Mr. Newall alone saw the plumes reaching right up to the nucleus is, doubtless, due to the great aperture of his telescope.

Coggia's comet was seen at the Observatory, Melbourne, by Mr. R. L. J. Ellery, on the morning of July 27. A series of drawings was obtained by means of the Great Reflector: the comet was very bright, and the nucleus very stellar. It had much diminished in brightness by August 10. On October 7 it was still visible, but was too faint for smaller telescopes than the Melbourne Reflector.

The comet was observed by Mr. Tebbutt at Windsor, New South Wales, from August 1 till October 7. It was a very conspicuous object during the first week of August, and was still faintly visible to the naked eye at the end of that month. It was also seen on July 27 by Mr. A. A. Anderson, in the eastern hemisphere (in lat. 23° 30′ S., long. 28° 54′ E.), when travelling to Barkly, Griqualand West, South Africa. It was very bright, and the tail was apparently short, but this was partially owing to the brightness of the moon. Mr. Anderson made observations with a sextant from July 27 to August 8.

It does not appear that any signs of the division or disruption of the comet were noticed in the southern hemisphere. Mr. Ranvard has remarked that when the comet became visible in the southern hemisphere 'the inner duplicate structure was still visible, but the outer arcs had been dissipated:' so that the comet does not seem to have undergone any marked changes in consequence of its passage near the sun.

CHAPTER XI.

THEORY OF COMETARY PHENOMENA.

SECTION I.

WHAT IS A COMET?

Complexity and extent of the question—The law of gravitation suffices to explain the movements of comets—Lacunæ in the theory; acceleration of the motion of the comets of Encke and Faye—Origin of comets; their systems—Questions relative to their physical and chemical constitution—Form of atmospheres; birth and development of tails.

LET us glance back for a moment at the contents of the preceding chapters.

We there find many facts accumulated, observations both interesting and instructive, phenomena whose variations suggest reflections without limit concerning the nature of the bodies to which they relate. Nevertheless, do these collected facts permit a clear and certain reply to the simple question: What is a comet?

I say a simple question, for so, as a rule, it is thought to be by non-scientific people; but in reality there is no question more complex. In order to attempt to reply to it, or at least to relate what is known for certain about comets, and to pass in review the most probable conjectures on doubtful points, we must proceed methodically, and thus as it were divide the difficulty.

A first natural division of the subject is at once apparent, it seems to us, from the exposition of cometary phenomena which has been made in the preceding chapters. This division includes the movements of comets, either apparent or real, all that relates to their orbits, and, in a word, the laws which govern them, not only as concerns what we may call the regular portion of their course, but in the vicissitudes and perturbations to which they are subjected by other celestial bodies. So far—in theory, at least—we find no difficulty in explaining the various facts, such as the periodicity of certain comets, the disappearance of some, the non-reappearance of others, the delay or too speedy arrival of those whose epoch of return has been calculated. Gravitation is the principle that renders an account of all these facts, of all these movements ; the theory of comets is, in this respect, the same as that of the planets; and, if there still remain difficulties and facts unexplained, neither the principle nor its application are for a moment doubted by any true astronomer.

There are difficulties, as we have already seen. For example, we ask ourselves, under the operation of what cause does Encke's comet continually shorten its period of revolution? Is this diminution due to the influence of a resisting medium or to the action of a repulsive force ? Opinions are divided; but this is no impeachment of the principle of gravitation, or the fact that the sun attracts a mass inversely as the square of its distance from his centre.

There are obscurities, as, for instance, the origin of comets. That all comets belong to the solar system cannot be supposed, as certain amongst them move in hyperbolic orbits. But have all these bodies come originally from beyond the limits of the solar system? Do they form, as M. Hoek believes, groups or systems ; and are we to consider the conversion of their original orbits into closed orbits as due to the disturbing action of the planetary masses? These questions are not yet decided ; but, whatever the reply that science may hold in store, it is certain that they affect in no respect either the cause of the cometary movements or their laws.

Lastly, there are unexplained facts, such as the non-reappearance of the comet of 300 years' period (that of 1264–1556), the non-return of some of the comets of short periods, and the division of Biela's comet into two distinct comets.

But this last and very curious phenomenon may perhaps have been due to the action of external forces, and in this case it would belong to the second or physical category of problems involved in the question before us.

The mass, the density, the physical constitution of the luminous nucleus, of the atmosphere surrounding it, and of the matter which streams out from it as the comet draws near the sun; the variations of form and volume of the nucleus and the nebulosity; the singular transformations which are revealed to us by the telescope, more especially those relating to the origin, development, and disappearance of the tail, are all facts that have been well and carefully observed, as the preceding chapters testify, but which are nevertheless difficult to co-ordinate into a logical whole and to reduce to a single principle, from which all the observed facts could be deduced as so many particular consequences. The phenomena presented to us by these bodies indicate that they have a special constitution of their own, as has been justly remarked by M. Roche, the author of some researches of the highest interest that we shall shortly proceed to analyse. In the meanwhile we will give M. Roche's views on the subject before us:—

'Comets are characterised less by the form and position of their orbits than by the changes they submit to during the times of their apparitions, and which sometimes succeed each other with wonderful rapidity. These changes denote a physical condition peculiar to comets, and mark important distinctions between them and other celestial bodies. Whilst the centre of gravity of the comet is describing its trajectory around the sun, under the influence of the solar gravitation and the disturbing action of the planets it may happen to approach, the comet itself experiences important changes, in which it is impossible not to recognise the action of the sun; for it is chiefly in the neighbourhood of the perihelion that these modifications are developed upon the grandest scale.'

M. Roche divides the phenomena in question into two kinds—those relating to the tail, its appearance, its varied form, its brightness and extent; and those which have reference to the variations of form or luminous intensity of the parts which constitute the head. The latter, as we have seen, are phenomena which have only been observed in comparatively recent times, whilst the formation of the tail has been long studied. To the explanation of these appendages, considered as the principal characteristic element of cometary bodies, astronomers have devoted many efforts. The hypotheses arising from these attempts are numerous; but they may be distinguished into four principal hypotheses, which we will now proceed to describe successively.

SECTION II.

CARDAN'S HYPOTHESIS.

Cometary tails considered as effects of optical refraction—Objections made by Newton and Gregory—New theory of Gergonne: ideas of Saigey on the subject of planetary tails—Difficulties and lacunæ in this theory.

PANÆTIUS, a philosopher of antiquity, held the belief that comets did not really exist, but were false appearances. 'They are,' he says, 'images formed by the reflexion, in the heavenly expanse, of the rays of the sun.' In the opinion of Cardan and some astronomers and physicists, Apian, Tycho Brahé, in the Renaissance, and Gergonne and Saigey, in our time, the tails of comets are simple optical appearances.

The following is the passage in Cardan's work (*De Subtilitate*) which relates to this question: 'It is, therefore, evident that a comet is a globe situated in the heavens and rendered visible by the illumination of the sun; the rays which pass through it form the appearance of a beard or tail.'* The Milanese doctor has entered into no particulars respecting the manner in which these appearances are formed, which, in his opinion, were doubtless analogous to the effects of refraction produced by the convergence of luminous rays passing

* 'Quo fit ut clarè pateat cometem globum esse in cælo constitutum, qui a sole illuminatus videtur, et dum radii transeunt; barbæ aut caudæ effigiem formant.' (*De Subtilitate*, lib. iv. 118; edition 1554.)

through a lenticular glass or globe filled with water; such as were formerly employed by artisans for concentrating the light upon their work.

'But, as Newton and Gregory have remarked,' justly objects M. Roche, 'the light is only visible in proportion as it reaches the eye; it would be necessary, therefore, that the solar rays, refracted by the head of the comet and collected behind in one convergent beam, should be sent towards the earth by material particles. Thus, sometimes when the sun is near the horizon and hidden by clouds, its rays, reflected by particles of air or vapour, are seen clearly defined against the sky like luminous jets.'

The fundamental idea upon which this explanation rests, and which still bears the name of Cardan, has since been several times taken up and modified. Following in the same order of ideas, we shall mention only a memoir by Gergonne, entitled *Essai analytique sur la nature des queues des comètes.* The origin of the phenomenon is there considered as purely optical, tails being only an appearance due to the most illuminated portion of the cometary atmosphere, or, more correctly speaking, to the caustic surface which is the envelope of the solar rays that are refracted whilst traversing the nucleus or its surrounding layers. These rays become visible when reflected upon the particles which compose the atmosphere of the comet. But, according to this theory, the atmosphere should have a radius at least equal to the length of the tail; and this constitutes an almost insurmountable objection, which the author himself does not conceal. We must not forget that certain comets have had tails many millions of miles in length. The atmospheres of comets being certainly much more limited, the reflexion, we must suppose, takes place upon the particles of an interplanetary medium, independent of the comet itself, and extending to distances far beyond the limits of the zodiacal light. Saigey, in his *Physique du Globe*, admits this explanation of cometary tails, and, according to him, planets have virtual tails, which would become real 'if the interplanetary spaces were filled with a matter similar to that which accompanies these last-mentioned bodies.' Saigey, it is evident by this last line, believes in the indefinite extension of cometary atmospheres, and the objection made above exists in full force. As regards the form of tails, their curvature, their multiplicity, and their oscillations, he explains them as follows: the curvature, by an effect of aberration due to the finite velocity of the propagation of light, the multiplicity by irregularities in the form of the nucleus, the oscillations by a movement of rotation, which causes these irregularities to be periodic.*

According to this system, now almost entirely abandoned, it is difficult to explain the phenomena of which we have already given an account, and which are seen by the telescope to vary from hour to hour; we mean the development of luminous jets in front of the nucleus, together with the envelopes and the lateral reflux of the luminous matter to form the edges of the tail. Nor is it less difficult to explain the formation of tails, which have sometimes been projected towards the sun, unless, not content with comparing comets to transparent and refracting globes, we are willing to make them perform at the same time the part of concave mirrors.

We have seen that the observations of occultations of stars by cometary atmospheres do not furnish proof of any refracting power whatever in the nebulosity of the head. It must,

* Speaking of the tail of the earth, he observes: 'The axis of the luminous sheaf ought to have, mathematically, the form of a spiral of Archimedes, the generating circle of which is 64,000 times greater than the terrestrial orbit; so that the most brilliant portion of this sheaf is slightly curved in the rear of the earth's movement of translation.'

therefore, be the refraction due to the nucleus alone that gives rise to the formation of caustics, and thus produces, at a distance, the illusion of cometary tails. We have said enough, however, of an hypothesis which has little chance of rising again from the discredit into which it has fallen.

SECTION III.

THEORY OF THE IMPULSION OF THE SOLAR RAYS.

Ideas of Kepler concerning the formation of tails—Galileo, Hooke, and Euler—Hypothesis of Kepler formulated by Laplace—Where does the impulsion come from in the theory of undulations?

KEPLER, who for a moment suffered himself to be led away by the idea of Cardan,* soon abandoned it, and substituted in its place that of the action of the solar rays. According to this theory cometary tails have substance and are formed of materials borrowed from the comet, its nucleus, or at least its nebulosity. 'The sun,' says Kepler, 'strikes upon the spherical mass of the comet with direct rays, which penetrate its substance, and carrying with them a portion of this matter, issue thence to form that trace of light which we call the tail of the comet.

* It appears that Galileo was also a partisan of the same theory. 'We find,' says Arago, in a work entitled *Il Trutinatore*, 'that Galileo gave it his approbation.'

[It is worthy of remark that Kepler seems to have abandoned Cardan's theory mainly because it failed to explain the curvature of tails. He remarks that the laws of optics teach us that the paths of light-rays are rectilinear, so that, if produced as supposed by Cardan, the tail could not be curved. But if we take into account the fact that the velocity of light is finite—a fact not known in Kepler's time—it is easily seen that the tail will appear curved except when the earth should happen to be in the plane of the comet's orbit. M. W. de Fonvielle has recently called attention to this point in the history of Cardan's theory. See Monthly Notices of the Royal Astronomical Society, vol. xxxv. p. 408. 1875.—ED.]

This action of the solar rays rarefies the particles which compose the body of the comet; it drives them away and dissipates them.'

Hooke, a contemporary of Newton, in order to explain the ascent of the light and tenuous matters which, emanating from the nucleus and flowing back in a direction opposite the sun, contribute to form the tail, assumes that these volatile matters are imponderable: to *gravitation* he opposes their *levitation*; according to him they have a tendency to fly from the sun. This amounts to assuming a repulsive force without explaining where this force resides.

The opinion of Kepler has been completed, extended, and modified. Admitted by Euler, and then by Laplace, it may be considered as the starting-point of the theory maintained by several contemporary astronomers, and notably by M. Faye—the theory according to which the solar rays exercise a repulsive action at a distance. We shall devote to it in its present form a separate section. In the meantime let us see how this theory has been formulated by Laplace in his *Exposition du Système du Monde*:—

'The tails of comets appear to be composed of the most volatile molecules which the heat of the sun raises from their surface and by the impulsion of his rays banishes to an indefinite distance. This results from the direction of these trains of vapour, which, always situated, as regards the sun, on the further side of the head of the comet, increase in proportion as the comet draws near to the sun, and only attain their maxima after the perihelion passage. The extreme tenuity of the molecules increasing the ratio of the surface to the mass, the impulsion of the solar rays becomes sensible, and causes nearly every molecule to describe a hyperbolic orbit, the sun being the focus of the corresponding conjugate hyperbola. The series of molecules moving in these curves form, beginning from the head of the comet, a luminous train in a direction opposite the sun, and slightly curved towards that region which the comet has just quitted, whilst advancing in its orbit; and this is what observation shows to us to be the case. The rapidity with which cometary tails increase enables us to judge of the extraordinary velocity with which these molecules ascend. The different volatility, size, and density of the molecules must needs produce considerable differences in the curves which they describe: hence arise the great varieties of form, length, and breadth observed in the tails of comets. If we suppose these effects combined with others which may result from a movement of rotation in the comet itself, and the apparent changes arising from the illusions of the annual parallax, we may partly conceive the reason of the singular phenomena presented by the nebulosities and tails of comets.'

This hypothesis, it is evident, supposes two modes of action of the emanations from the sun. The first, which Kepler vaguely indicated, is an effect of dilatation due to the calorific activity of the solar rays, an effect doubtless itself preceded by an evaporation throughout the liquid parts of the surface of the nucleus. The nebulosity thus becomes more voluminous and the layers which form it more and more attenuated. Up to this point there is no difficulty—the known physical effects of heat justify this portion of the theory. The difficulty begins when it is necessary to assume that the same rays which have hitherto acted as calorific rays are endowed with another property, hitherto unknown, that of giving an onward motion or propulsion to all molecules reduced to a state of suitable tenuity. Does such a force exist?

To Laplace, who was justified in adopting the theory of emission at the epoch when he wrote, this repulsive force was quite natural. The luminous molecules emitted by the sun, moving with enormous speed, communicated a portion of their momentum to the molecules emitted by the comet, to those which by the action of heat had been previously reduced to a sufficiently small tenuity, and hence the formation of the tail. But it is less easy to conceive of this repulsive force in the wave-theory of light, now universally adopted. Undulations are propagated with enormous velocity in the ether, but the matter is not transported forward.* It is difficult to see how the force that gives rise to the successive waves can produce the rectilinear movement of the molecules of the cometary atmosphere. Moreover, before assuming the existence of an actual repulsive force, it should, if possible, be demonstrated by experiment. Arago, when citing the experiments of Homberg, opposes to them the negative verifications of Bennet, and concludes thus : 'The fundamental idea of an impulsion due to the solar rays is, therefore, only an hypothesis, without real value.' The question, however, is not yet settled, as we shall see.

* M. Roche cites the following comparison, due to Euler, and by which that great mathematician, a partisan of the theory of luminous waves, justifies his adhesion to the hypothesis of the impulsion of the solar rays :—

'As a violent sound excites not only a vibratory movement in the particles of air, but also causes a real and perceptible movement in the light dust floating in the atmosphere, we cannot doubt that in the same way the vibratory motion caused by light produces a similar effect.'

This very vague comparison is not conclusive. Sonorous waves have an amplitude sufficient to produce visible agitation ; the fact that we have to prove is the existence of a progressive rectilinear movement, not the existence of oscillation.

SECTION IV.

HYPOTHESIS OF AN APPARENT REPULSION.

Views of Newton on the formation of the tails of comets—Action of heat and rarefaction of the cometary matter—The ethereal medium, losing its specific weight, rises opposite the sun, and carries with it the matter of the tail—Objections which have been made to the hypothesis of a resisting and ponderable medium.

NEWTON, in order to explain the formation of the tails of comets, had recourse to no other causes than the ordinary action of the calorific rays on the one hand, and that of gravitation on the other. But, although he does not introduce any new force, he is obliged to suppose that the comet during the whole time that its tail is developing is traversing a medium subject to the force of gravitation and tending towards the sun. Newton thus explains the theory:—

The tail is composed of vapours, that is to say, of the lightest parts of the atmosphere of a comet. These vapours are rarefied by the action of the solar heat, and in their turn heat the surrounding ether. Thus, the medium which surrounds the comet becomes rarefied; it consequently loses its specific weight, and instead of tending with the same energy towards the sun, it continues to rise in the same manner as layers of air heated at the surface of the soil rise in virtue of the principle of Archimedes. In rising it carries with it particles of cometary matter, which by their ascension produce the tail, rendered visible by the reflexion of light proceeding from the sun. In this manner smoke ascends in a chimney by the impulsion of the air in which it is suspended; this air is rarefied by the heat; it ascends because its gravity or specific weight has become less, and it draws along in its ascent columns of smoke. The ascension of cometary vapours further arises from the fact that they revolve about the sun, and for this reason have a tendency to fly from it. The atmosphere of the sun or matter of the heavens is at rest or turns slowly, having received its movement of rotation from the sun. Such are the causes which determine the ascent of cometary tails in the vicinity of the sun where the orbits are much curved, and where the comet, plunged in a dense and consequently heavier atmosphere, emits a longer tail.

This theory, which had been in the first place vaguely formulated by Riccioli, and then by Hooke (the latter, we may remember, inclines rather to the doctrine of a repulsive force), was adopted by different astronomers of the eighteenth century, Boscovich, Gregory, Pingré, Delambre, Lalande. Gregory, however, was not contented with the cause assigned by Newton for the ascent of cometary tails; he believed also in an active impulsion in addition to an apparent repulsive force. His system is a combination of the two systems we have just described.

Various objections of a serious kind have been urged against the theory of Newton. The existence of a resisting medium, gravitating towards the sun, of a solar atmosphere, in fact, would necessarily be limited to within a certain distance of the sun himself. Laplace has proved that for such an atmosphere to subsist it must be animated by a movement of rotation about the sun's axis, and that it could not extend beyond the distance at which the centrifugal force arising from that movement would become equal to the force of gravitation. In the plane of the solar equator the limit is seventeen hundredths of the mean distance of the earth; it corresponds to the radius of the orbit of a planet whose revolution would be equal in duration to the solar rotation, which is effected in twenty-five days and a half. Now, comets, before attaining such proximity to the sun, are provided with tails; and considerable tails have been exhibited by comets whose perihelion distance has even exceeded the radius of the terrestrial orbit, which is nearly six times as great as the extreme possible limit of the solar atmosphere.

Besides, this ponderable medium would be a resisting medium as well. In addition to the disturbing action that this resistance would exercise upon the head of the comet and likewise upon its orbit, it would act with much more intensity upon the tail of the comet, on account of its extreme rarity. Before the perihelion passage, in the first part of the comet's movement, the curvature and the drifting back of the tail would be easily explained by this resistance; but, after the perihelion passage, the tail continues to keep the same position relatively to the radius vector joining the nucleus to the sun, so that the comet appears to move its tail to a position in advance of itself, a phenomenon incompatible with the hypothesis of a resisting medium. The medium of which we speak has likewise been assimilated to the zodiacal light; and Mairan, who has thus explained the terrestrial Aurora Borealis, finds in this light the cause and origin of cometary tails. But the preceding objections and others, which would take too long to repeat here, have been justly opposed to this new theory.

SECTION V.

THEORY OF OLBERS AND BESSEL.

Hypothesis of an electric or magnetic action in the formation of tails—Repulsive action of the sun upon the cometary matter, and of the nucleus upon the nebulosity —Views of Sir John Herschel and M. Liais—Theory of Bessel—Oscillations of luminous sectors—Magnetic polar force.

WHETHER the cause which determines the production of cometary tails and their development, at once so immense and so rapid, be a force *sui generis*, or only an apparent force, it is none the less true that it has all the features of a repulsive action or force. Heat, the impulsion of the solar rays, gravitation, have all been variously combined in order to furnish the desired explanation; it evidently remained to try the intervention of the electric and magnetic forces.

From this point of view Olbers, Herschel, and Bessel have in turn applied themselves to the problem. We will give a brief analysis of the opinions held by these illustrious astronomers.

The comet of 1811 first drew the attention of Olbers to the subject. 'This astronomer,' says M. Roche, 'attributes to the proximity of the comet and the sun a development of electricity in both these bodies; hence arises a repulsive action of the sun and another repulsive action of the comet upon the nebulosity which surrounds it.' By the first of these forces Olbers has explained the formation and development of tails; by the second he has accounted for the formation of the luminous sectors or plumes of the comet, and also the successive envelopes similar to those which were observed in Donati's comet. Biot has given his adhesion to this theory.

Sir John Herschel's view is nearly the same. It is not improbable, he observes, that the sun is constantly charged with positive electricity; that as the comet draws near the sun and its substance becomes vaporised the separation of the two electricities takes place, the nucleus becoming negative and the tail positive. The electricity of the sun would direct the movement of the tail, just as an electrified body acts upon a non-conducting body electrified by influence.*

* In his *Outlines of Astronomy* Sir J. Herschel is less explicit in regard to the physical nature of the force which produces the tails, and he does not refer to electricity. But, speaking of the curious phenomena which were observed to take place in the head of Halley's comet during its apparition of 1835, he proceeds:—

'Reflecting on these phenomena, and carefully considering the evidence afforded by the numerous and elaborately executed drawings which have been placed on record by observers, it seems impossible to avoid the following conclusions:—

'1st. That the matter of the nucleus of a comet is powerfully excited and dilated into a vaporous state by the action of the sun's rays, escaping in streams and jets at those points of its surface which oppose the least resistance, and in all probability throwing that surface or the nucleus itself into irregular motions by its reaction in the act of so escaping, and thus altering its direction.

'2ndly. That this process chiefly takes place in that portion of the nucleus which is turned towards the sun, the vapour escaping freely in that direction.

'3rdly. That when so emitted it is prevented from proceeding in the direction originally impressed upon it by some force directed from the sun, drifting it back and carrying it out to vast distances behind the nucleus, forming the tail, or so much of the tail as can be considered as consisting of material substance.

'4thly. That this force, whatever its nature, acts unequally on the materials of the comet, the greater portion remaining unvaporised; and a considerable part of the vapour actually produced remaining in its neighbourhood, forming the head and coma.

'5thly. That the force thus acting upon the materials of the tail cannot possibly be identical with the ordinary gravitation of matter, being centrifugal or

M. Liais, in his work entitled *L'Espace Céleste*, has pronounced in favour of a repulsive force of an electric nature. According to this astronomer the calorific action of the solar rays causes a physical and chemical modification of the molecular condition of the nucleus, and thus gives rise to the two electricities. Whilst the nucleus is charged with the one electricity, the opposite electricity becomes developed to the more attenuated and lighter portions, and is carried by them to the limits of the cometary atmosphere. But, on the other hand, the sun himself is constantly in a state of strong electric tension. And of the two electricities that which he possesses in excess will attract, for example, the nucleus, if it should be charged

repulsive as respects the sun, and of an energy very far exceeding the gravitating force towards that luminary. This will be evident if we consider the enormous velocity with which the matter of the tail is carried backwards, in opposition both to the motion which it had as part of the nucleus and to that which it acquired in the act of its emission, both which motions have to be destroyed in the first instance before any movement in the contrary direction can be impressed.

'6thly. That unless the matter of the tail thus repelled from the sun be retained by a peculiar and highly energetic attraction to the nucleus, differing from and exceptional to the ordinary power of gravitation, it must leave the nucleus altogether; being in effect carried far beyond the coercive power of so feeble a gravitating force as would correspond to the minute mass of the nucleus; and it is, therefore, very conceivable that a comet may lose, at every approach to the sun, a portion of that peculiar matter, whatever it may be, on which the production of its tail depends, the remainder being of course less excitable by the solar action, and more impassive to his rays, and therefore, *pro tanto*, more nearly approximating to the nature of the planetary bodies.

'7thly. That, considering the immense distances to which at least some portion of the matter of the tail is carried from the comet, and the way in which it is dispersed through the system, it is quite inconceivable that the whole of that matter should be reabsorbed; that, therefore, it must lose during its perihelion passage some portion of its matter; and if, as would seem far from improbable, that matter should be of a nature to be repelled from, not attracted by, the sun, the remainder will, by consequence, be, *pro quantitate inertiæ*, more energetically attracted to the sun than the mean of both. If, then, the orbit be elliptic, it will perform each successive revolution in a shorter time than the preceding, until, at length, the whole of the repulsive matter is got rid of.'

with the opposite electricity, and will, in the same way, repel the molecules of the electrised atmosphere.

This expulsion, exercised over a portion of the atmosphere which already has a tendency to fly from the sun, originates a tail nearly opposite to the radius vector; but, acting upon the anterior molecules, it drives them back and causes the formation of a second tail. 'As the electric action of the sun,' he observes, 'exerted as an attractive force upon the nucleus, and as a repulsive force upon the nebulosity, somewhat disturbs the figure which the comet would assume under the sole influence of gravitation, and causes a much greater extension of the nebulosity in the rear of the nucleus than on the side near to the sun, the gravitation of the molecules towards the nucleus is less on the side opposite to the sun than on the other, added to which the electricity of a similar kind to that of the comet is there more abundant, and there consequently the repulsive force is manifested with the greatest energy. Besides, the currents of matter are directly in the direction of this force, and have not to curve themselves back like those that arise in front. The posterior tail is, therefore, more energetically repelled than the tail produced in the anterior region. For this reason it makes with the prolongation of the radius vector of the comet a less angle, which accords with what has been found by observation. For the same reason also the tail which is opposite to the sun is generally the first to appear and the last to disappear. M. Liais assumes that this double tail, which was very observable in the comet of 1861, exists in nearly all comets, and that the two parts of which the tail is composed appear distinct only when they are very long, unless, from the position of the earth in their common plane—the plane of the orbit—they should be projected one upon the other.'

The preceding will suffice to give a general idea of the

above theory, which differs in no essential particular from that of Olbers and Sir John Herschel. M. Liais has entered into details too long to be here reproduced, by means of which he proceeds to explain all observed anomalies, and to show, with perhaps a little too much confidence, that these apparent anomalies are legitimate consequences of the theory of cometary electricity. Tails multiple in number, straight or curved, plume-shaped or fan-shaped, luminous sectors and jets, tails directed towards the sun, the double curve of the tail of the comet of 1769, the accelerated movement of certain comets, the rapid and transient coruscations of tails, all correspond exactly, according to M. Liais, with this hypothesis.

In a memoir on the physical constitution of Halley's comet the illustrious astronomer Bessel has formulated a theory not very different to that of electricity. The end which he chiefly had in view was the explanation of the luminous *aigrettes* observed in 1835, a phenomenon which has since been repeated and observed in the head of the great comet of 1862. Bessel compares the axis of the comet to a magnet, one of whose extremities or poles is attracted to the sun, whilst the other extremity is repelled. To an equilibrium by turns broken and established under the action of internal forces, and the polar force proceeding from the sun, the observed oscillation is due. This polar force, in the part of its action which is repulsive, tends to form and develop the tail. 'As regards the physical origin of the force,' says M. Roche, in his analysis of Bessel's memoir, 'he attributes it to a particular action of the sun which accompanies the volatilisation of the cometary fluid, and tends at the same time to repel each molecule of matter and to direct the axis of the comet towards the sun. This force would not be proportional to the mass, but specific; that is to say, it would act with different intensity upon different matters. This specific character would explain the production of multiple tails. The action of the polar forces as conceived by Bessel is very complicated; their influence upon the nucleus of the comet is obscure, and renders impossible any kind of equilibrium in the atmosphere by which it is surrounded.'*

* The following is a *résumé* of Bessel's memoir, taken from the notice of it inserted in 1840 in the *Connaissance des Temps* :—

'The illustrious astronomer, in the first place, describes the appearances presented by the head of the comet between the 2nd and the 25th of October, 1835, dwelling more especially upon the movements of the *aigrette*, of "that effusion of luminous matter which issued from the nucleus and was directed towards the sun. The most curious phenomenon presented by the comet," he observes, "was unquestionably the movement of rotation or oscillation of the luminous cone." He then endeavours to determine in what manner and in what direction this movement was effected, and from observations and measures of the position of the *aigrette* he was led to consider probable the hypothesis of a pendulous motion of the luminous cone in the plane of the orbit of the comet and about an axis perpendicular to this plane. The duration of the period was found to be comprised between four and five days, and the amplitude of the oscillations had a mean value of 60 degrees.

'Now, how are we to explain this movement? Could not the attraction of the sun, acting unequally upon the particles of the comet, more or less distant from the sun, and combining with the movement of the comet in its orbit, produce in the nucleus a libration analogous to the libration of the moon?' To the question thus raised Bessel has given a reply in the negative, because in that case the oscillations arising from the attraction of the sun would have been of very long duration, whilst the period of the oscillations was found, on the contrary, to have been very short. This leads us to consider the possibility of a special physical force. He thus explains his ideas on the subject:—

'We must assume a polar force tending to direct one of the rays of the comet towards the sun, whilst at the same time it directs the opposite ray in a contrary direction: there is no *à priori* reason for rejecting the existence of such a force. The magnetism of the earth supplies us with an example of an analogous force, although it is not yet proved that it has any relation to the sun; if this were the case we should see the effect manifested in the precession of the equinoxes. This force once admitted, the oscillatory movement of the *aigrette* is easily explained; the duration of the oscillations depends upon the magnitude of this force, and the amplitude upon the initial motion of the molecules. I remark, moreover, that if the sun exerts upon one portion of the comet a force other than attraction, this force must be a polar force; that is to say, a force producing a contrary action upon another portion of the mass. For if this were not the case, the sum of all the actions which the sun exerts upon the mass of the comet would not be proportional to the mass, and consequently the movements of the comet according to the laws of Kepler would not correspond to the mass of the sun as determined by the movements of the planet. Now, observation has shown no deviation that could be attributed to this cause; if, therefore, we can prove that the sun does not act equally upon the whole mass of the comet, it will be a new argument in favour of a polar force.'

Bessel next investigates analytically the path followed by a particle which, emitted by the nucleus and escaping from the luminous *aigrette* in the direction of the sun, proceeds under the influence of the solar repulsion to turn back upon its path and to separate from the head of the comet in a direction opposite the sun; then, comparing the results of analysis with those of observation, he accounts for the appearances presented by the tails of different comets, their curvature, their more or less size, &c.

But all comets are far from presenting the same phenomena: for example, whilst the *aigrettes* of the comets of 1835 and 1744 issued from a particular region of the surface of the nucleus, in that of 1811 the luminous matter was emitted in all directions; in that of 1769 there were two distinct *aigrettes*, and the comet of 1807 had two tails. We have seen other examples of these differences. Bessel explains the first by a simple difference in the value of a constant; the others by assuming that the repulsive force of the sun is exerted with variable intensity, being dependent upon the nature of the different portions of the luminous matter.

We will terminate this analysis by quoting the passage in which Bessel explains the movement of oscillation which, by principally fixing his attention, gave rise to his theory:—

'I regard,' he says, 'the oscillatory movement of the luminous *aigrette* of Halley's comet as an effect of the same force which projects in opposite directions the particles emitted from the nucleus parallel to the radius vector. This is how I conceive that the force in question acts.'

'The total action of one body upon another may be divided into two parts, the one part acting equally upon all the particles of the other; and the second part of the action composed of different actions exerted over different parts. When two bodies are far separated from each other, and the action is small, it is only the first part which becomes sensible, in proportion as the distance diminishes; the second part can only have an appreciable value later. Thus, when a comet approaches the sun after having been very far removed, we perceive the general action which the sun exercises over all its parts. I suppose that this action consists of a volatilisation of particles, which, moreover, are so polarised as to be repelled by the sun. The second part of the sun's action may have for its effect the polarisation of the comet itself and a particular emission of luminous matter in the direction of the sun. The part of the surface from which issues the luminous *aigrette* has a polarisation such that it has a tendency to be attracted towards the sun; and consequently the particles which compose it having the same polarisation, tend also to approach the sun. But these particles are moving in a space filled with matter having an opposite polarisation, and which is constantly being replaced. Thus the two polarisations neutralise each other, and the particles which compose the *aigrette* will acquire the opposite property to that which they had previously, in proportion as they recede from the comet.'

SECTION VI.

THEORY OF COMETARY PHENOMENA.

Researches of M. E. Roche upon the form and equilibrium of the atmospheres of celestial bodies under the combined influence of gravitation, solar heat, and a repulsive force—Figure of equilibrium of a solid mass submitted to gravitation and the heat of the sun—Comets should have two opposite tails—Completion of the theory of cometary tides by the admission of a repulsive force, real or apparent—Accordance of the theory so completed with observation.

M. Edouard Roche has devoted a series of highly interesting memoirs to the discussion of the figure assumed by the atmosphere of celestial bodies under the action of the forces of the solar system. He has more particularly given his attention to the study of cometary atmospheres, and to all the phenomena which take place in and around cometary masses.

M. Roche begins by reducing the question to its simplest form. He assimilates a comet to 'an entirely fluid mass, sensibly homogeneous, and having no movement of rotation.' The forces which act upon it are the mutual attraction of its own particles and gravitation towards the sun. For such a mass to be in equilibrium under the action of these forces it must have the figure of a prolate spheroid with its centre at the centre of gravity, and its axis of revolution coincident with the radius vector from the sun.

Introducing then the motion of the comet towards the sun, M. Roche examines the modifications that would be occasioned in the figure of the atmosphere by a diminution of the distance between itself and the sun, only taking into account the mutual attraction of the two bodies. 'At first spherical when the comet is far off, its figure becomes ellipsoidal, and gradually lengthens as it draws near the sun.' But there is a limit to the amount of lengthening, a limit which depends upon the density of the fluid of which the cometary atmosphere is formed.

Although, according to M. Roche, there may not exist amongst the numerous comets of the solar system any one whose physical constitution accords exactly with the above hypothesis, nevertheless it is more natural to consider the general question of a central nucleus surrounded by an atmosphere of much greater rarity. In all the atmospheres of the celestial bodies the gaseous envelope is retained by gravitation towards the nucleus. Let us see what theory gives on this new hypothesis:—

'On following with care,' says M. Roche, 'the phenomena developed by a comet during its approach to the sun, we clearly see that they result, at least in part, from the increasing action of the solar gravity. The difference of the attractions exerted by the sun upon the nearest and the furthest portions of the cometary atmosphere must needs have the effect of lengthening the comet in the direction of the sun, and more and more in proportion as the distance of the comet from the latter continues to decrease. In a word, the cause that produces the terrestrial tides will here manifest itself in a similar manner, but on a much grander scale, in the neighbourhood of the perihelion. This has been confirmed by my mathematical investigations.' In fact, the surfaces of equilibrium,* originally spherical, lengthen more and more towards

* [A surface of equilibrium, or, as it is often called, a level surface, is a surface such that if a particle were placed at any point on it, the resultant of the action

the sun, as by the diminution of the distance from the sun its action increases; the comet also lengthens not only towards the sun, but also to an equal extent in the opposite direction. But the atmosphere cannot extend further than to the points where the attractions of the sun and the nucleus exactly balance one another. Any particle that escapes beyond this limiting surface of equilibrium, which is called the free surface, is then subject to the preponderating influence of the sun, and as it were abandons the comet. The exterior surfaces of equilibrium have no longer a spheroidal form: they open themselves as it were at the two poles, *A* and *A′*, and consist of sheets that extend to infinity. (Fig. 63.)

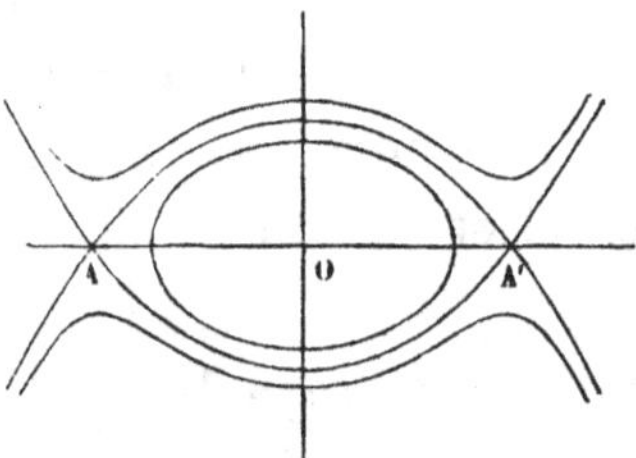

Fig. 63.—M. Roche's theory of cometary phenomena. Limiting atmospheric surface of equilibrium.

'If, from any cause, the cometary fluid should pass beyond the free surface, it will spread itself in all directions over the surfaces of equilibrium that are immediately exterior; and, as they are infinite in extent, the fluid in excess will stream off through the two conical summits or poles as through two openings and lose itself in space.' (Fig. 64.) So far gravitation is the only force whose action we have

of the external forces upon the particle would act in a direction perpendicular to the surface at the point. Thus, at any point within the free, or bounding, surface of the atmosphere the resultant of the external forces acting upon the particle of the atmosphere that is situated there is perpendicular to the surface of equilibrium passing through it, and this resultant is balanced by the pressure of the atmospheric layers underneath.—Ed.]

taken into account, and we have deduced from it the figure of equilibrium of a comet, regarded as a homogeneous fluid mass, or as a nucleus, either liquid or solid, surrounded by a ponderable atmosphere. We may go further still without the intervention of another force; for M. Roche finds that, as regards the free surface, its dimensions vary with the distance of the sun: 'this surface, as it were, contracts as the comet approaches perihelion, and the fluid layer that is thus left outside it flows away at the two poles, thus forming two opposite jets along the radius vector of the sun. If the fluid is elastic and behaves like a gas, the outflow will continue so long as molecules continue to come from the interior to re-

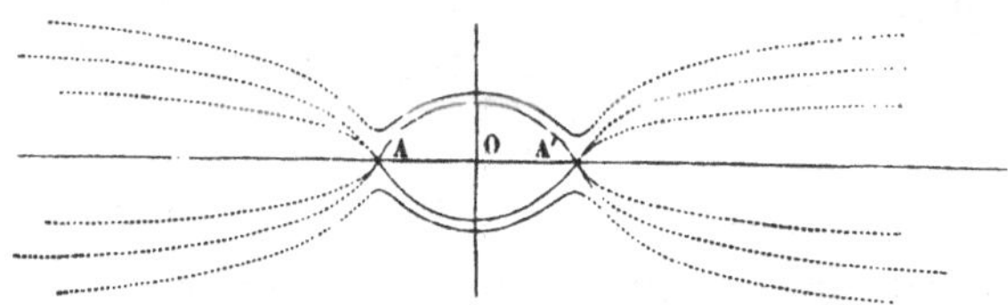

Fig. 64.—Flow of cometary matter beyond the free surface of the atmosphere. No repulsive force.

place those which escape. This is what must happen whilst the comet is drawing near the sun.' When the perhelion has passed this state of things no longer continues.

But another cause will have been contributing, and doubtless in a much more powerful degree, to produce this outflow of cometary fluids from the two extremities of the axis of the comets, as represented in fig. 64. This cause is the calorific action of the solar rays upon the nucleus, an action which continues rapidly to increase in the neighbourhood of the perihelion. The accumulated heat then gradually dilates and volatilises the cometary substance which rises by the diminution of its specific weight, attains the free surface, and, passing beyond it, flows off into space, as we have just said.

The whole of this matter which thus abandons, not only the nucleus, but the cometary atmosphere itself, becomes resolved into particles independent of the comet, 'into a cloud of dust composed of an infinite number of disconnected molecules;* if they fail to disperse and continue still agglomerated, it is because their motion is nearly the same, owing to their common initial velocity.'

Such, in its principal features, is the theory which M. Roche has named the *theory of cometary tides.* It involves only the action of gravitation and heat. We must next enquire if the phenomena actually observed correspond with the results of the analysis. As regards the appendages of the head and the growth and development of tails, does it suffice to furnish an explanation of these phenomena? To this question M. Roche replies in the negative. 'If this theory,' he remarks, 'were sufficient to explain all; that is to say, if the attraction of the sun and that of the nucleus were the sole influences concerned in the production of these phenomena

* This is the expression employed by M. Roche in his memoirs, and it is no doubt true, if we consider this dust, molecule by molecule; but is it right to say that these particles have no longer any connexion with each other? They are always under the influence of gravitation, both towards the nucleus and the sun; the latter being preponderant, the system of expelled molecules continues to exist, and for two reasons: firstly, because there is no reason to believe that the molecules have lost all mutual action; secondly, because they all continue to gravitate towards the sun, as is shown besides by the apparent solidarity of comets and their tails. Now, are we to suppose that the whole of this abandoned matter returns no more to the comet? It is both possible and probable, if it continues to be endowed with the property of gravitation (and how could this property have been lost?), that it may be attracted by the sun or some other planet encountered on its way, which benefits by the losses of the comet; but, on the other hand, as the velocity acquired causes it to follow to a great extent the movement of the comet in its orbit, it will happen that at a certain distance from the perihelion the action of the sun ceases to preponderate, and gives place to that of the nucleus; the disengaged matter will then regain the limits of the cometary atmosphere, since the free surface dilates in the second half of the orbit in proportion as the distance from the sun augments.

(there is heat as well), a constant agreement would exist between the phenomena observed and the theoretical results I have pointed out. Every comet would then have not one tail but two tails, directly opposite to one another. This circumstance, which was observable in the comet of 1824 and some others, is nevertheless exceptional; as a rule, only the tail opposite the sun is observed. The tail in front is more frequently replaced by a brilliant *aigrette*, which may be considered as a rudimentary tail. In the aspect of comets there is nevertheless so great a want of symmetry, that we are obliged to acknowledge the insufficiency of our theory of cometary tides; it only explains a part of the phenomena.'

He then asks himself if this want of symmetry may not be explained in part by the unequal action of the solar heat upon the anterior and posterior portions of the comet. To a certain extent, yes. But it still remains to explain the extraordinary length of the tail opposite the sun, the cessation in the development of the second tail, the figure of the *aigrettes*, which appear to be bent and driven back to contribute to the formation of the tail opposite. A necessity arises, therefore, for the intervention of a new force, either apparent or real, which shall exercise a repulsive action upon the matter of the comet.

When M. Roche published the memoirs above referred to, Donati's comet had recently attracted the attention of the astronomical world by the singular phenomena revealed to observation in the structure of its different parts, its nucleus, atmosphere, and tail. M. Faye, taking up Kepler's theory of an actual repulsive force inherent in the solar rays, discussed the whole of its consequences and compared them with the facts observed. At his suggestion M. Roche included this action also in his analysis. How M. Faye has defined the force in question we shall see further on; we will restrict ourselves to stating that, under the influence of this force, the figure of the surfaces of equilibrium ceases to be symmetrical with respect to the central nucleus. The free surface, convex and slightly flattened towards the sun, is terminated at the opposite extremity by a conical apex. The inner surfaces of equilibrium envelop the nucleus on all sides; but the

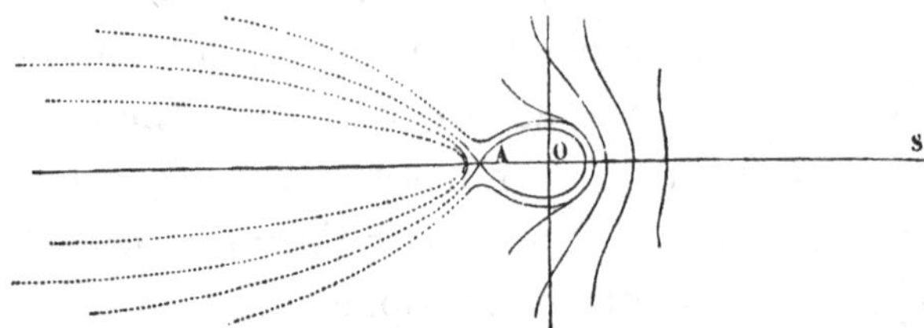

Fig. 65.—Development of cometary tails, on the hypothesis of an intense repulsive force. M. Roche's theory.

outer surfaces closed towards the sun are open on the opposite side, in the vicinity of the conical point A (fig. 65). By this single opening behind the nucleus the excess of cometary matter makes its escape. Hence we find regularly superimposed layers towards the sun, in the form of envelopes; on the other side these layers are traversed and as it were broken

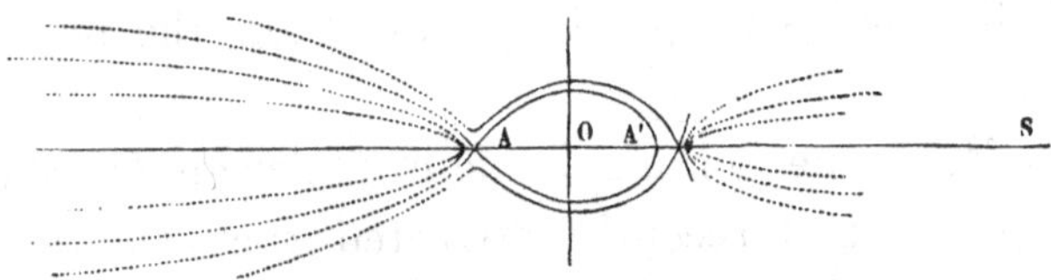

Fig. 66.—Development of cometary tails on the hypothesis of a feeble repulsive force. M. Roche's theory.

by emissions from the summit A, which is the origin of a tail opposite the sun; such is the aspect a comet would present in the new conditions to which we have supposed it subjected. Also the figure varies, if the intensity of the repulsive force varies with the nature of the particles upon which it acts. Hence the different forms which may co-exist in the same comet. The three theoretical figures that we have given correspond, the first (fig. 64) to the absence of a repulsive force, the second (fig. 65) to a repulsive force acting upon the whole of the fluid with great intensity. The third (fig. 66) supposes the force in question to be extremely feeble. The tail directed towards the sun exists then but as a very slightly elongated *aigrette*, which is speedily drifted back to unite with the opposite tail.

M. Roche then examines in detail several of the observed facts that we have described, and shows that in his theory they find a logical interpretation: the successive envelopes formed by the vapours raised from the nucleus by the influence of the solar heat, the formation of *aigrettes* towards the sun, and the drifting back of the emitted matter, the general form of the tail, more brilliant as a rule on its outer edge, where the repelled matter is accumulated, the dark sector behind the nucleus, &c.

There are, doubtless, in any particular comet complications of aspect which require special study. These may arise from the peculiar constitution of the comet itself, or quite as frequently from changes in the relative position of the observed comet and the earth, in which case they are only effects due to perspective.

The theory of M. Roche has received the adhesion of several French and foreign astronomers. And in our opinion it has one great merit: it is more than a simple hypothesis, because it rests upon two undisputed principles in astronomy, as it only depends upon the action of gravitation and solar heat; and when the new force—the repulsive force—is introduced no hypothesis is made with regard to its origin and nature. The sole hypotheses assumed by M. Roche consist (1) in the repulsive force, which he regards as directly proportional to the density of the matter submitted to its influence and inversely to the square of the radius vector, and (2) in a fact concerning which all contemporary astronomers appear to be agreed, viz. the extreme tenuity of the cometary matter in those portions of the nebulosity which are detached from the nucleus. 'This tenuity,' he observes, 'is so great that it exceeds anything that we can imagine, and renders the nebulosity of a comet comparable to the medium which occupies the vacuum of an air-pump. The repulsion which the comet appears to experience on the part of the sun would, therefore, seem to be a consequence of that singular condition of matter concerning which physical science does not as yet possess any certain data.'*

* 'From another point of view,' adds M. Roche, 'the study of a substance reduced to this state of extreme dilatation would be of not less importance.' In fact, as M. Radau has judiciously remarked, 'the matter of the tails of comets is so disseminated and rarefied that we are compelled to renounce the idea of expressing their density in figures. This extreme division of the matter is not so extraordinary as might be thought. It is comparable to the density of cosmical matter, the concentration of which, according to Laplace, has given birth to the planets and their satellites. Let us suppose the earth's radius increased till it becomes equal to the distance of the moon; it would then be so rarefied that it would become fifty times less dense than ordinary air; and if the sun were expanded till its radius was equal to that of the terrestrial orbit, its density would not be more than sixteen-millionths of that of atmospheric air. It is then a medium similar in its extreme rarity to cometary tails, that has constituted the atmosphere of the sun during the period of its condensation. The as yet unknown properties of this medium will no doubt ultimately throw light upon the primitive condition of the solar system.'

It is not only to the terrestrial orbit that we should suppose the radius of the sun or of the planets extended in order to reconstitute the primitive solar nebula; it should be extended at least to the orbit of Neptune. By supposing a homogeneous sphere of this radius filled with the matter of the sun, its density as compared with that of water would be 0·000 000 000 005 29, which is a little more than five billionths of the actual density of water. This is a density two hundred and fifty millions of times less than that of atmospheric air.

SECTION VII.

THE REPULSIVE FORCE A REAL PHYSICAL FORCE.

Theory of M. Faye—Rigorous definition of the repulsion inherent in the solar rays—Its intensity varies with the surfaces of the two bodies; it decreases inversely as the square of the distance—It is not propagated instantaneously—Discussion and accordance of the facts—Experiments in support of a repulsive force.

It was at the suggestion of M. Faye, as we have seen, that M. Roche introduced into his analytical researches upon cometary phenomena the hypothesis of a repulsive force which has, in fact, led to results more in conformity with what is observed. It should be remarked, however, that M. Roche has considered the matter rather from the point of view of a mathematician than of a physical astronomer; whilst, on the contrary, the physical bearing of the problem has more especially occupied the attention of M. Faye. This astronomer, after passing in review the different theories we have mentioned, and rigorously comparing their conclusions with the facts of recorded observations, in short, after the most exhaustive discussion, has finally decided in favour of an actual repulsive force inherent in the solar rays. This is the base of the theory known as Kepler's theory, and which has been distinguished by the adhesion of Euler and Laplace.

At the time when M. Faye made known his views, two great comets—that of Donati (1858) and that of 1861—had recently appeared. Both comets had been subjected to careful telescopic scrutiny, and it was necessary to explain the physical phenomena which had been daily followed in their details by observers in Europe and America, and also to account for a phenomenon of another kind, but equally important, viz. the accelerated movement of the comets of Encke and Faye.

Encke, as we shall see, was in favour of the hypothesis of a resisting medium, and regarded it as the cause of the known acceleration of the above-mentioned comets. Newton likewise, as we have seen, attributed the formation of cometary tails to the existence of an interplanetary medium. Here, then, is a connexion between two very different classes of phenomena. M. Faye, on the hypothesis of a repulsive force, proceeds to examine the cause of the formation and development of *aigrettes* and tails, and the accelerated movement of the comets above referred to.

Let us see, in the first place, how M. Faye defines the repulsive force. This is an important point, concerning which the partisans of this theory had hitherto neglected to be explicit; there was supposed to be an impulsion of the solar rays, and that was all. M. Faye's words are:—

'A repulsive force having its origin in heat. By means of it heat produces mechanical effects. It depends upon the surface and not upon the mass of the incandescent body. The action upon a body is proportional to the surface of the body, and not to its mass. It is not propagated instantaneously, like the attractive force of Newton. Nor does it act through intervening matter, like attraction. It is provisionally assumed that its intensity decreases inversely as the square of the distance, and that its velocity of propagation is the same as that of rays of light or heat.'

Now, the existence of such a force being admitted, how does M. Faye deduce from it the theory of cometary phenomena? How does he explain by it the formation of tails, simple or multiple, their curvature, their direction, the development of luminous or dark sectors, the disengagement of envelopes more or less parabolic? Upon all these points M. Faye gives the following explanations, which he finds confirmed point by point by the observations of the most brilliant comets which have recently appeared:—

'The action of the repulsive force upon a body in motion about the sun does not coincide with the radius vector, but is always exerted in the plane of the orbit, so that the figure which it tends to impress upon a body originally spherical, such as that of a comet very remote from the sun, will be symmetrical with respect to this plane; nor can this result be changed either by the sun's attraction or by that of the nucleus, or by the progress of the deformation itself. In the second place, the action of this force being in proportion to the surface, the effects produced depend upon the density of the matter of which the comet is composed; it follows, then, that, except in the plainly exceptional event of these materials being completely homogeneous, it must give rise to the formation of several tails, resulting from a sort of purely mechanical selection on the part of the repulsive force. But the axes of these multiple tails, which are longer in proportion as their curvatures are less, will always be situated in the plane of the orbit, as in the case of a single tail.

'According to the mechanical generation of these appendages, the matter of which is in a state of division, tenuity, and molecular independence whereof it is difficult to form an idea, each tail, in its regular portion, exhibits a simple curvature behind the line of motion of the nucleus. At its origin each of these tails is tangential to the radius vector, or rather is inclined to it at a small angle.

'With respect to the special form of any particular tail, it must be regarded as the envelope of matters of the same density which successively abandon the head of the comet, under the triple influence of the repulsive force, the power of solar attraction, and the general velocity, to which we must add, as Bessel has done, the small velocity of the nucleal emission. If we consider at a given moment the whole of the molecules thus driven from the narrow sphere of attraction of the comet, they will be found principally distributed over the circumference of a nearly circular section of the nebulosity; and if we follow the same series of particles for the next few moments, we shall see that, as a consequence of their motions in their independent trajectories, the nature of which can be assigned, they will occupy constantly increasing areas, the section continually lengthening in the plane of the orbit, while the transverse diameter increases in much less proportion. The tail of the comet, therefore, will be principally displayed in the plane of the orbit, more especially tails which are very much curved. But should they be viewed edgewise, they will appear straight, under the form of a narrow band, equally defined on the two edges, and more brilliant at the edges than in the middle. The two edges will be nearly parallel, or at all events but slightly divergent, unless the observer should be situated very near to a portion of the tail... Should there be several tails, they will appear projected one upon another, while the earth is crossing the plane of the comet's orbit, and as they are very far from being opaque, the narrowest of the tails will be seen defined in the midst of the largest, or that which is nearest to the observer. It is evident, therefore, that before they can be distinguished one from another the earth must have passed by a very considerable distance the plane of the comet's orbit.'

The above is M. Faye's explanation of the origin and development of tails, as well as of the varied appearances observable in cometary appendages. As a whole, this theory is certainly satisfactory, but we cannot affirm, in presence of the numerous and complex facts which we have described, that it is quite complete or free from objection. For example, we do not see very clearly how M. Faye would explain the appearance of the multiple fan-shaped tail presented by the great comet of 1861, on June 30, the day when the earth was situated exactly in the plane of the orbit. In this situation the tails of the comet should have been seen projected upon each other, as described. But these are minor difficulties, arising, doubtless, from the real complexity of the phenomena, further enhanced by the effects of perspective.

As regards nucleal emissions, the sectors and luminous envelopes, &c., their formation is considered to be wholly attributable to the forces of attraction and the increasing influence of the solar calorific radiations. It is here that M. Faye refers to M. Roche, and considers the theoretical diagrams given by the latter as the most faithful representations possible of the real phenomena.

'Thus,' observes M. Faye, in conclusion, 'the figure of a comet, and the more extended portion of the tail, are the result of a purely mechanical action of two forces: the Newtonian attraction and the repulsion due to heat. The attraction is exercised by the respective masses of the sun and the comet, the repulsion by the incandescent surface of the sun; but it is further necessary to take into consideration the repulsive force which the heat belonging to the comet, or rather that which it receives in approaching the sun, develops amongst its molecules. From this cause arises an expansion more or less analogous to that of terrestrial bodies when brought to a gaseous state, an expansion which occurs in the phenomenon of the double nucleal emission. It is, thus, this expansion which enables the solar repulsion to take effect, and which, dilating the matter of the nucleus more and more, renders it of extreme tenuity, as in the already mentioned case of the envelopes of Donati's comet. The question is, therefore, very simple in principle, notwithstanding the enormous complexity of the phenomena involved; and as in the universe there is a relation between all things, it will be perceived more and more that in the sidereal universe as well as in the terrestrial domain there exist many other manifestations of this repulsive force, the effect of which comets present to us upon so gigantic a scale.'

However well these deductions of an able theory may accord with each other and with the fundamental principle, they none the less rest upon an hypothesis : that of a repulsive force inherent in the solar radiations. The theory of M. Faye, therefore, stood in need of the decisive and indispensable sanction that experiment alone can give. This was fully appreciated from the beginning by M. Faye, who, in 1861, in order to obtain the confirmation which the former experiments of Bennet had seemed to promise, made, in conjunction with M. Ruhmkorff, a series of experiments upon the action which metallic plates, when heated to a state of incandescence, exercise in vacuo upon the stratification of the induction spark. 'In all these experiments,' says he, 'the repulsive action of the incandescent surface was very decided, but I was more especially struck with it in the case of arsenic and sulphur.' The influence of the heat is manifest, if it be true that the repulsion increased with the temperatures of the plates heated to incandescence. The influence of the density is not less, if the indications of the repulsion were weak, in proportion to the amount of air admitted within the globes that were used in the experiments.

The direct verification of the solar repulsion, according to M. Faye, is impossible on the surface of the earth, 'as in all probability it exhausts itself upon the upper strata of our atmosphere.' If this be the case, there is no reason, it is clear, why the earth and all the planets that have atmospheres should not be provided with tails after the manner of comets. It remains to be determined whether they are sufficiently extended to be visible, but that they must exist is certain, unless the repulsive force should be here exhausted without producing its effect.

[* During the last two years the subject of attraction and repulsion as resulting from radiation has been the subject of much discussion and investigation, in consequence of the experiments of Mr. Crookes and the invention by him of his radiometer. As the question of whether the solar emanations are accompanied by a repulsive action is one of the highest importance in regard to the motions of comets and the explanation of their tails, and as this is a matter which has been recently the object of the most searching and thorough examination by means of instruments of extraordinary delicacy, I think it desirable to give a brief account of what has been effected.

I commence by giving in Mr. Crookes's own words a short summary of a historical summary of the investigations prior to 1873:—

'The Rev. A. Bennet recorded the fact that a light substance delicately suspended in air was attracted by warm bodies: this he ascribed to air-currents. When light was focused, by means of a lens, on one end of a delicately suspended arm, either in air or in an exhausted receiver, no motion could be perceived distinguishable from the effects of heat.

'Laplace spoke of the repulsive force of heat. Libri attributed the movement of a drop of liquid along a wire heated at one end, to the repulsive force of heat, but Baden Powell did not succeed in obtaining evidence of repulsion by heat from this experiment.

'Fresnel described an experiment by which concentrated solar light and heat caused repulsion between one delicately suspended and one fixed disk. The experiment was tried in air of different densities; but contradictory results were obtained under apparently similar circumstances at different times, and the experiments were not proceeded with.

'Saigey described experiments which appeared to prove that a marked attraction existed between bodies of different temperatures.

'Forbes, in a discussion and repetition of Trevelyan's experiment, came to the conclusion that there was a repulsive action exercised in the transmission of heat from one body into another which had a less power of conducting it.

'Baden Powell, repeating Fresnel's experiment, explained the results otherwise than as due to repulsion by heat. By observing the *descent* of the tints of Newton's rings between glass plates when heat was applied, Baden Powell showed that the interval between the plates increased, and attributed this to a repulsive action of heat.

'Faye introduced the hypothesis of a repulsive force of heat to account for certain astronomical phenomena. He described an experiment to show that heat produced repulsion in the luminous arc given by an induction-coil in rarefied air.'

Mr. Crookes's own experiments showed that a heavy metallic mass when brought near a delicately suspended light ball attracts or repels it under the following circumstances :—

I. When the ball is in air of ordinary density—
  (*a*) If the mass is colder than the ball it repels the ball ;
  (*b*) If the mass is hotter than the ball it attracts the ball.

II. When the ball is in a vacuum—
  (*a*) If the mass is colder than the ball it attracts the ball;
  (*b*) If the mass is hotter than the ball it repels the ball.

And in an experiment in which the rays of the sun, and then the different portions of the solar spectrum, were projected on to a delicately suspended pith-ball balance, he found that *in vacuo* the repulsion was so strong as to cause danger to the apparatus, and resembled that which would be produced by the physical impact of a material body.

The application of these facts to the question of a solar repulsive action is obvious; and Mr. Crookes himself, after discussing the explanations which may be given of the phenomena and showing that they cannot be due to air-currents, thus referred to the evidences of a repulsive action of heat and attractive action of cold in nature. 'In that portion of the sun's radiation which is called heat we have the radial repulsive force possessing successive propagation, required to explain the phenomena of comets and the shape and changes of the nebulæ. To compare small things with great—to argue from pieces of straw up to heavenly bodies—it is not improbable that the attraction now shown to exist between a cold and a warm body will equally prevail, when, for a temperature of melting ice is substituted the cold of space, for a pith-ball a celestial sphere, and for an artificial vacuum a stellar void.'

All this is taken from Mr. Crookes's abstract of his paper (*Proc. Roy. Soc.*, vol. xxii. pp. 37–41). The paper itself was printed in the *Philosophical Transactions*, vol. clxiv. pp. 501–527.

By these experiments Mr. Crookes was led to examine more fully the action of radiation upon black and white surfaces. He found that at the highest exhaustion heat appeared to act almost equally on white and on lampblacked pith, repelling them in about the same degree, but that the action of luminous rays was different. These were found to repel the black surface more energetically than the white surface. Taking advantage of this fact, Mr. Crookes was led to invent the instrument now so well-known as the radiometer or 'light-mill.' It consists of four arms of very fine glass, supported in the centre by a needle-point, so that it is capable of revolving horizontally. To the extremity of each arm is fastened a thin disk of pith, lampblacked on one side, the black surfaces all facing the same way. The arms and disks are delicately balanced, so as to revolve with the slightest impetus. The whole is enclosed in a glass globe, which is then exhausted to the highest attainable point and hermetically sealed. This instrument revolves under the influence of radiation, the rapidity of revolution being in proportion to the intensity of the incident rays.

The speed with which a sensitive radiometer will revolve in full sunshine is almost incredible, the number of revolutions per second being several hundreds. One candle will make the arms spin round forty times a second. Mr. Crookes found that the action of dark heat (as, *e.g.* from boiling water) was to repel each surface equally, and the movement of the radiometer is therefore arrested if a flask of boiling water is brought near it. The same effect is produced by ice.

In a brief notice of his subsequent experiments read before the Royal Society on June 15 of the present year, Mr. Crookes attributes the repulsion caused by radiation as shown by the radiometer to the action of the residual gas, *i.e.* to the very small amount of gas (the gas employed was generally dry atmospheric air, but the effect only differed in degree with other gases) still remaining in the almost perfect vacuum. 'In the early days,' says Mr. Crookes, 'of this research, when it was found that no movement took place until the vacuum was so good as to be almost beyond the powers of an ordinary air-pump to produce, and that as the vacuum got more and more nearly absolute, so the force increased in power, it was justifiable to assume that the action would still take place when the minute trace of residual gas which theoretical reasoning proved to be present was removed.

'The first and most obvious explanation, therefore, was that the repulsive force was directly due to radiation. Further consideration, however, showed that the very best vacuum which I had succeeded in producing might contain enough matter to offer considerable resistance to motion. I have already pointed out that in some experiments where the rarefaction was pushed to a very high point the torsion-beam appeared to be swinging in a viscous fluid; and this at once led me to think that the repulsion caused by radiation was indirectly due to a difference of thermometric heat between the black and white surfaces of the moving body, and that it might be due to a secondary action on the residual gas.'

Mr. Crookes having contrived an apparatus by means of which the viscosity of the residual internal gas, as well as the force of the radiation, could be measured, found that up to an exhaustion at which the gauge and the barometer were sensibly level there was not much variation in the viscosity of the internal gas, and that on continuing to exhaust, the force of radiation commenced to be apparent, the viscosity remaining about the same. The viscosity next commenced to diminish, the force of radiation increasing. After long-continued exhaustion the force of radiation approached a maximum, but the viscosity began to fall off: at a still higher exhaustion the force of repulsion diminished. In a radiometer exhausted to a very high degree of sensitiveness the viscosity of the residual gas is almost as great as if it were at the atmospheric pressure. Mr. Crookes concludes as follows: 'The evidence afforded by the experiments of which this is a brief abstract is to my mind so strong as almost to amount to conviction, that the repulsion resulting from radiation is due to an action of thermometric heat between the surface of the moving body and the case of the instrument, through the intervention of the residual gas. This explanation of its action is in accordance with recent speculations as to the ultimate constitution of matter and the dynamical theory of gases.' It will thus be seen that although the radiometer at one time seemed to afford experimental evidence of the direct repulsive action of the light-rays, Mr. Crookes after a remarkable series of experiments extending over four years has come to the conclusion that at all events the repulsion is only an indirect effect of the action of light; so that the evidence in favour of a real solar repulsive force, after being submitted to a very severe test, has been found wanting.

It is to be noticed that although Mr. Crookes seems to have in his first paper inclined to the belief that his experiments tended to establish a direct action of radiation, he has throughout refrained from adopting this as a theory, and the final conclusion quoted above is I believe the only explanation he has at any time offered of the action of the radiometer.

The existence of a solar repulsive action has not, I think, ever found much favour with mathematical physicists, and recent investigations do not seem to have afforded experimental evidence in favour of it; but at present, while the details of the chief experiments of Mr. Crookes remain still unpublished, it would be premature to attempt to decide under what circumstances repulsion may result directly or indirectly from radiation. In any case it now may be considered as certain that the matter is one that can be satisfactorily determined by experiments in the laboratory, and in consequence of the interest that has been excited there is little doubt that before very long much more will be known upon the subject than at present.

Abstracts of Mr. Crookes's papers have been published in the *Proceedings of the Royal Society*, vol. xxii. p. 37, vol. xxiii. p. 373, vol. xxiv. pp. 276 and 279, and vol. xxv. p. 136. Two papers only have as yet been printed *in extenso* in the *Philosophical Transactions*, (vol. clxiv. p. 501 and vol. clxv. p. 519.)

Mr. Bennet's paper was published in the *Philosophical Transactions* for 1792.

I may add in conclusion, that in my own opinion the solar repulsive force seems to me still to be merely a hypothesis, and I cannot feel that any explanation of cometary phenomena that is dependent upon it is satisfactory.—ED.]

SECTION VIII.

THEORY OF THE ACTINIC ACTION OF THE SOLAR RAYS.

Experiments and hypotheses of Tyndall—Originality of his theory; objections and omissions—Is this theory incompatible with that of a repulsive force?

A NEW theory of cometary phenomena which has been proposed by Professor Tyndall, one of the most distinguished of contemporary physicists, in our opinion merits special attention. In the first place, because we believe it to be altogether new and original; and, in the second place, because it is derived, not from *à priori* conceptions, like so many other theories in astronomy and physics, but from accurate experiments and their interpretation.

The study of the action of radiations upon very rarefied media of gaseous matter first led Professor Tyndall to consider the mode of production of the phenomena presented by the heads and tails of comets. Of the undulations proceeding from any luminous source, such as the sun, some have a purely calorific action; these are those which have the greatest amplitude or are least refrangible; the undulations which constitute or produce light come next in the order of length of wave or refrangibility; the shortest waves are those which manifest themselves exclusively by chemical action. We now proceed to explain Professor Tyndall's views on the subject of these modifications, and his manner of accounting for the fact that the rays of shortest wave-length are endowed with the property of acting upon chemical substances, decomposing them and separating the atoms of which their molecules are composed, whilst the larger and mechanically more powerful waves are, on the contrary, ineffectual to perform any such decomposition.

'Whence, then, the power of these smaller waves to unlock the bond of chemical union? If it be not a result of their strength, it must be, as in the case of vision, a result of their periods of recurrence. But how are we to figure this action? I should say thus: the shock of a single wave produces no more than an infinitesimal effect upon an atom or a molecule. To produce a larger effect the motion must *accumulate*; and for wave-impulses to accumulate they must arrive in periods identical with the periods of vibration of the atoms on which they impinge. In this case each successive wave finds the atom in a position which enables that wave to add its shock to the sum of the shocks of its predecessors. The effect is mechanically the same as that due to the timed impulses of a boy upon a swing. The single tick of a clock has no appreciable effect upon the unvibrating and equally long pendulum of a distant clock; but a succession of ticks, each of which adds, at the proper moment, its infinitesimal push to the sum of the pushes preceding it, will, as a matter of fact, set the second clock going.'

After having thus explained the chemical action of light, Professor Tyndall proceeds to study its action upon the vapours of different volatile substances, sometimes employing a beam of electric light, and at other times the solar light. He fills a tube of certain length with a mixture of air and the vapour of nitrite of amyl, of nitrate of butyle, or of iodide of allyl, after having taken the requisite precautions for the exclusion of all foreign matters, and more especially of particles floating in the air—dust, organic germs, mineral matters, &c. When thus filled the tube remains dark, and the mixture it contains is absolutely invisible. But should a luminous beam of light, such as that given by the flame of a lamp, be rendered convergent by a lens and allowed to fall upon the interior of the tube, the following will be observed: the space for an instant after the introduction of the beam will remain dark; but this brief moment passed, a white luminous cloud will be seen to invade that portion of the tube occupied by the beam of light. How has this change been effected? The action of the waves has decomposed the nitrite of amyl and precipitated a rain of particles which from that moment are capable of reflecting and diffusing in all directions the light of the beam. 'This experiment,' says Tyndall, 'illustrates the fact, that however intense a beam of light may be, it remains invisible until it has something to shine upon. Space, although traversed by the rays from all suns and all stars, is itself unseen. Not even the ether which fills space, and whose motions are the light of the universe, is itself visible.'

We may see by this last remark the capital objection which forces astronomers to reject the theory of Cardan, according to which the tails of comets are simply the effect of refraction. This theory we have already mentioned.

It is to be remarked that the end of the experimental tube most distant from the lamp is free from cloud. Now, the nitrite of amyl vapour is there also, but it is unaffected by the powerful beam passing through it. Why? Because the very small portion of the beam competent to decompose the vapour is quite exhausted by its work in the frontal portions of the tube; it is the longer waves that continue their course; but these waves are powerless to produce a chemical decomposition. Thus can the able physicist find in the detail of facts the confirmation of his ingenious hypotheses. But let us now proceed to the theory of cometary phenomena.

The substance of this theory has been embodied by Professor Tyndall in the seven following propositions, which we will reproduce in the author's own words:—

'1. The theory is, that a comet is composed of vapour decomposable by the solar light, the visible head and tail being an actinic cloud resulting from such decomposition; the texture of actinic clouds is demonstrably that of a comet.

'2. The tail, according to this theory, is not projected matter, but matter precipitated on the solar beams traversing the cometary atmosphere. It can be proved by experiment that this precipitation may occur either with comparative slowness along the beam, or that it may be practically momentary throughout the entire length of the beam. The amazing rapidity of the development of the tail would be thus accounted for without invoking the incredible motion of translation hitherto assumed.

'3. As the comet wheels round its perihelion, the tail is not composed throughout of the same matter, but of new matter precipitated on the solar beams, which cross the cometary atmosphere in new directions. The enormous whirling of the tail is thus accounted for without invoking a motion of translation.

'4. The tail is always turned from the sun, for this reason: two antagonistic powers are brought to bear upon the cometary vapour—the one an *actinic* power, tending to effect precipitation; the other a *calorific* power, tending to effect vaporisation. Where the former prevails, we have the cometary cloud; where the latter prevails, we have the transparent cometary vapour. As a matter of fact, the sun emits the two agents here invoked. There is nothing whatever hypothetical in the assumption of their existence. That precipitation should occur behind the head of the comet, or in the space occupied by the head's shadow, it is only necessary to assume that the sun's calorific rays are absorbed more copiously by the head and nucleus than the actinic rays. This augments the relative superiority of the actinic rays behind the head and nucleus, and enables them to bring down the cloud which constitutes the comet's tail.

'5. The old tail, as it ceases to be screened by the nucleus, is dissipated by the solar heat; but its dissipation is not instantaneous. The tail leans towards that portion of space last quitted by the comet—a general fact of observation being thus accounted for.

'6. In the struggle for mastery of the two classes of rays a temporary advantage, owing to variations of density or some other cause, may be gained by the actinic rays, even in parts of the cometary atmosphere which are unscreened by the nucleus. Occasional lateral streamers, and the apparent emission of feeble tails towards the sun, would be thus accounted for.

'7. The shrinking of the head in the vicinity of the sun is caused by the breaking against it of the calorific waves, which dissipate its attenuated fringe and cause its apparent contraction.'

This very brief exposition of an hypothesis which might be termed the *physico-chemical theory* is taken from the new edition of Tyndall's work upon Heat. It is unaccompanied by explanation or commentary of any kind, and to us at least seems to be wanting in completeness, and to contain some obscurities which we shall briefly notice, in the form of questions and objections, and on the subject of which we should be glad to receive additional elucidation from the author.

Professor Tyndall defines comets without making mention of the nucleus. Comets, for him, would appear to be simple masses of vapour rendered visible by the actinic action of the solar rays. Further on, nevertheless, he considers the nucleus as endowed with the property of absorbing the calorific waves, whilst the efficacy of the chemical waves is in no respect impaired. In those comets where observation has proved that a nucleus exists, is that nucleus solid or liquid, or a simple gaseous mass of greater density than the other portions of the comet?

He considers the chemical and calorific rays as antagonistic powers. Nevertheless they are regarded by all physicists as undulatory movements differing in no essential respect from each other, or in other words differing only in amplitude and length of period. In what respect, then, are the calorific and chemical waves opposed ?

What is this dissipation by the solar heat of the tail no longer screened by the nucleus ? Is it the effect of a repulsive force inherent in the calorific rays which are no longer absorbed by the nucleus? Are the particles precipitated by the decomposing action of the actinic rays re-composed so as to resume their original state of transparency?

Professor Tyndall makes no mention of the curvature of the tail, of that disposition which causes it to be displayed in the plane of the orbit, or of the production of multiple tails. Do the accidental lateral currents of which he speaks afford a sufficient reply to this last question?

Finally, according to this theory, it follows that comets are agglomerations of matter of extreme tenuity, certain parts only of which are rendered visible by the solar action. The precipitation takes place in the interior of the mass, in a determinate but continually varying direction, so that the mass would have to be considered as having in every direction a diameter equal to the enormous length of the comet's tail. These spheres of vapour, therefore, millions of miles in diameter, thus travel in known orbits in the midst of the interplanetary spaces. Is it gravitation towards the nucleus which thus keeps together the constituent molecules of these attenuated masses; or if not, if the vapours thus formed are incessantly abandoned in space, how are they incessantly replaced? Are they an emission proper to the nucleus, or the effect of a repulsive action of the solar heat? If Professor Tyndall should admit this last hypothesis, of what use would be the actinic action of the solar rays in explaining the development of the tail? His theory would in that case be grafted upon the theory of M. Faye, and would have no other *raison d'être* than to explain the visibility of a matter so attenuated as that of which the tails of comets are composed.

In our opinion all these questions require elucidation; but they are rather questions than objections. We must not forget that the forces called into play in phenomena of this kind—that is to say, the recognised forces—are gravitation and the ethereal radiations. It has been supposed by some that gravitation itself is due to the waves of ether; these last are revealed to us by their triple manifestations—calorific, luminous, and chemical. Professor Tyndall invokes chemical action for the purpose of explaining cometary phenomena; M. Roche and M. Faye appeal to gravitation and heat. It remains to be seen whether the repulsive force may not be explained as a component of the solar radiations. If so, by far the greatest obstacle to the reconciliation of these different theories would be removed, and the cause of all those curious and diverse phenomena, the movements of comets, their perturbations, as well as their physical transformations, whether apparent or real, would be reduced to the one principle which Lamé believed to be the universal connecting link between all the phenomena, viz. the undulatory movement of the ether.

SECTION IX.

COMETS AND THE RESISTANCE OF THE ETHER.

Accelerated motion of Encke's comet; its periods continually diminish—It describes a spiral, and will ultimately fall into the sun—Hypothesis of a resisting medium; how does the resistance of a medium increase the rapidity of motion?—The nature of this supposed medium, according to Arago, Encke, and Plana—Objections of M. Faye; the acceleration of motion explained by the tangential component of the repulsive force.

In our account of the periodic comet of Encke we gave, together with the dates of its successive apparitions, the durations of the revolutions comprised between these dates. If the reader will turn back to the table on p. III he will readily perceive that these durations are unequal, and that the period is continually decreasing, and has suffered a diminution of a little more than two days, or exactly of 2·06 days. As the table includes twenty-two revolutions of the comet it is at most a diminution in each revolution of two hours twenty-two minutes, a quantity small in itself, but which, incessantly accumulating, is capable of producing changes of very great importance in the course of time.

The discovery of this acceleration is due to the astronomer whose name the comet bears. Since 1824 Encke had remarked the diminution of the period, and he was unable to account for the result observed, even by admitting very considerable errors in the masses of the planets, whose disturbing influence upon the comet he had himself calculated with the utmost care; but, by assuming the existence of a resisting medium, Encke found that the major axis of the orbit would decrease as well as the eccentricity, and that the mean motion would increase, while the inclination and the longitude of the node would remain unchanged. As this agreed with observation he was led to attribute the acceleration of the motion of the comet to the resistance of a medium now generally spoken of as the resistance of the ether.

Encke continued his researches on the subject on each return of the comet; he calculated with care, taking into account the disturbing influence of the planets, the epoch of the perihelion passage of the comet; lastly, he published, in 1858, the memoir from which we have extracted the table, which exhibits in a striking manner the acceleration of its movement.

In order to explain this diminution of period Encke, as we have already seen, regards the interplanetary spaces, not as a vacuum, as assumed by Newton and astronomers of his school,* but as filled with a medium of sufficient density to oppose to the movements of bodies circulating therein a resistance capable of producing in the course of time modifications in their orbits. The mean augmentation of the motion of the comet, which has been found by observation, arises, according to him, from a tangential force acting in a direction opposite to that of the comet's motion, 'which accords entirely,' he observes, 'and in the most simple manner, with the hypothesis of a resisting medium in the universe. Proofs of the

* Newton, though regarding the interplanetary space as a vacuum, explains nevertheless, as we have seen, the formation of tails by an ascensional movement of the cometary particles in the midst of a ponderable medium which surrounds the sun to a certain distance, and whose density increases in proportion as this distance decreases. It is really, therefore the hypothesis of Newton which in this case has been adopted by Encke.

existence of such a medium appear so evident that there can no longer be any doubt about the matter.' We shall see, however, that the evidence is not considered satisfactory by all astronomers; and, in particular, M. Faye has raised objections which it is difficult to ignore. But, from whatever cause may arise the acceleration of Encke's comet, and whatever may be the physical nature of the force in question, it has always been regarded by mathematicians and astronomers as a resistance applied to the comet, and exerted in a direction opposite to that of its motion.

On this point we must enter into some details, in order to explain what always seems strange to those persons who have but a slight knowledge of the principles of mechanics, and of celestial mechanics in particular. Such persons find it difficult to understand that a resistance experienced by a body in motion, describing a certain curve around a centre, should produce an acceleration; it appears to them that the reasoning is false, and that a retardation must be the necessary result of such a resistance. They would be right if the body in question were moving in a determinate and invariable line from which it could never swerve. A railway-train, for example, moving against a strong wind experiences a resistance which reduces its speed. This is not the case with a comet or any other body which, animated by a determinate velocity, and free to take any direction whatever in space, necessarily follows a course dependent, on the one hand, upon the velocity it has at any instant, and, on the other, upon the force with which the sun attracts it.

Now, if the velocity of the comet undergo a diminution, such as would be caused by a resisting medium, then, as the force to the sun remains the same, the comet would, as it were, be pulled in towards the sun, so that instead of describing CC', it would in the same time describe CC''. But, by Kepler's third law, which Newton proved by means of the principle of gravitation, there is a fixed relation between the major axis of an orbit and the periodic time;* so that if the mean distance from the sun be diminished the time of revolution must also be diminished.

The effect, therefore, of a resisting medium upon a comet would be to diminish the size of its orbit, and consequently to shorten its period of revolution. As, moreover, the same cause would be unceasingly in operation, and even—if we suppose that the medium gradually increases in density

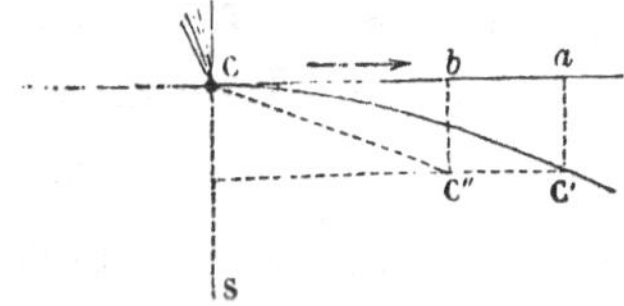

Fig. 67.—Influence of a resisting medium upon the orbit of a comet.

towards the sun—would act with increasing intensity, the acceleration itself would gradually become more considerable. The comet, therefore, would describe a curve continually and steadily approaching the sun; in other words, it would describe a spiral, and at the end of a certain time would be precipitated upon the sun itself.

What we have observed of Encke's comet applies equally to all other comets, and to the various celestial bodies, planets or satellites, which compose the solar system; only as regards the planets, whose masses, or rather densities, are very great in comparison with those of comets, the effect of this resistance has been up to the present time imperceptible. It is but a question of time, however; and, as Arago observes, 'mathematically speaking, if no cause should be discovered which will compensate for the resistance experienced, it will be certain

* The squares of the periodic times are as the cubes of the major axes. See note, p. 71.

that, after a sufficient lapse of time—consisting, perhaps, of several thousands of millions of years—the earth itself will be united to the sun.'

But let us leave the planets, which are only remotely concerned in the question, and return to the resisting medium. What is this medium? According to Arago it is ether; 'that is to say, the ethereal matter which fills the universe, and whose vibrations constitute light.' Such is not the opinion of Encke, who considers the resisting medium as a kind of atmosphere enveloping the sun on all sides to a certain distance, and whose density increases inversely as the square of the distance. The origin of this medium would be either a primitive atmosphere of the sun or the *débris* of atmospheres left in space by the planetary and cometary masses. Further, Plana, in a memoir upon the subject treated of by Encke, thus expresses himself: 'The resisting medium to which the formulæ have been applied is not the imponderable and universal ether which propagates light, but a kind of atmosphere surrounding the sun.'

We have already had occasion to speak of objections that have been made to this hypothetical medium. M. Faye, after remarking that 'the analysis of those mathematicians who assume its existence proves that they regard this kind of ponderable atmosphere as immovable,' proceeds: 'Now, this immobility is impossible; no ponderable particle can exist in the solar system without precipitating itself towards the sun or circulating about him; there can be no other alternative.' M. Faye shows that the second supposition is alone possible; but that if there were a resisting medium circulating about the sun the effect would be somewhat different. Instead,' he proceeds, 'of forcing the comet to describe a spiral, so as to approach the sun, and finally precipitate its mass upon him, the action of such a medium would chiefly affect the eccentricity. If this element be sufficiently diminished, the orbit would become more and more circular, but the major axis would cease to diminish, and the comet would not be precipitated upon the sun. In the case of a direct comet, such as that of Encke, the action of a medium circulating in the same direction would depend only upon the relative velocity of the comet and the layers it encounters. There would be alternate periods of acceleration and retardation. The first would predominate until the orbit became circular; the influence of the medium would then cease. A retrograde comet—Halley's comet is the sole instance amongst periodic comets—would, on the contrary, experience a much greater resistance, and the acceleration would be very considerable.'

This objection relates to an event requiring for its determination a longer series of observations than we yet possess. It is very possible that the result may be as indicated by M. Faye, and that the acceleration of Encke's comet will have a limit. The hypothesis of a resisting medium appears, then, in no respect invalidated.

But M. Faye, as we should expect, has sought in his own theory for an explanation of the observed acceleration. The repulsive force, by the aid of which he accounts for the phenomena of tails, furnishes him quite naturally with the means. In fact, the action of this force, as we have seen, is not instantaneous; it is propagated with the velocity of rays of light or heat. A kind of aberration ensues—a deviation in the direction along which the repulsive force acts. We may, therefore, conceive of this force as a compound of two parts, viz. the radial component, *CF'* (Fig. 68), in the direction of the radius vector, which causes the formation of tails; the tangential component, *Cb*, opposite to the direction of motion of the comet. It is the latter that plays the part of a resisting medium, and causes, whilst allowing the attraction of the sun to preponderate, the diminution of distance from the sun and the accelerated motion of the comet.

Further, every repulsive force, whatever be its nature, so long as its velocity of propagation is not infinite, will produce the same effect and explain in the same manner the acceleration of motion. Bessel, in the memoir that we have referred to, thus expresses himself on this subject: 'The luminous *aigrette* of Halley's comet gave it nearly the aspect of a rocket. Consequently it must have exercised an effect similar to that which is observed in the movement of rockets. It is not the centre of gravity of the comet only, but the centre of gravity of the comet and the *aigrette*, which describes

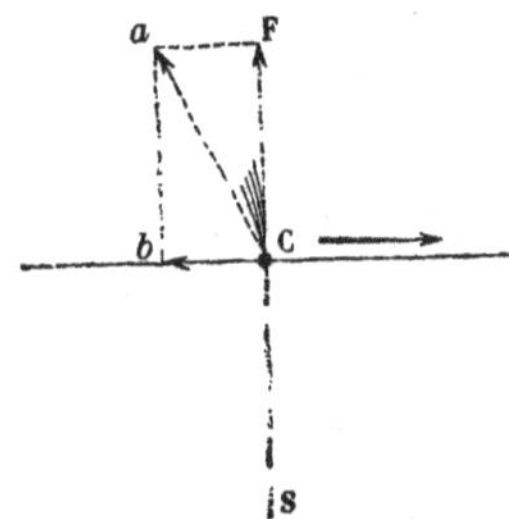

Fig. 68.—Radial and tangential components of the repulsive force, according to M. Faye.

a conic section according to the laws of Kepler; the luminous matter, therefore, which issues from the comet to form the *aigrette* must exercise upon the centre of gravity a repulsive action, which, being continuous, produces an accelerating force. From the brilliancy of the *aigrette*, which gives the apparent proportion of its mass to that of the nucleus, we may imagine that this disturbing force might very appreciably change the elliptic movement. In the comet of 1811 the delicate researches of M. Argelander appear to indicate a deviation arising from an analogous cause; the more exact observations of Halley's comet will allow a closer investigation of the subject.'

Biot came to a similar conclusion, taking into account the loss of substance the comet would experience on each of its revolutions.

Such are the hypotheses that have been proposed, in order to explain the accelerated motion of Encke's comet. But is this the only case of acceleration? No. It appears from the researches of M. Axel Möller, a Swedish astronomer, that Faye's comet presents a similar phenomenon; this follows from an examination of the three successive revolutions it has accomplished since the date of its discovery in 1861. 'The motion of the comet,' he remarks, 'cannot be accounted for by attraction only.'

Why, then, is it that these two comets alone have manifested the operation of a cause whose action, whether produced by a resisting medium or a repulsive force, must be general? On Encke's hypothesis the short-period comet, being that which, of all known periodic comets, or rather of comets that have returned, approaches the nearest to the sun, would necessarily experience most strongly the influence of a medium which is denser in the regions nearer the sun. On the hypothesis of a repulsive force the explanation is the same, since the intensity of that force being greater at less distances from the sun, its tangential component is likewise greater. But this reasoning no longer holds good when we turn from the consideration of Encke's comet to that of Faye, whose perihelion distance, and even mean distance, are, on the contrary, among the greatest. Perhaps these apparent divergencies are due to nothing more than a want of accuracy in the calculated movements of these bodies and their perturbations. In any case much remains to be done before we shall be enabled to decide between the theories that have been proposed.

CHAPTER XII.

COMETS AND SHOOTING STARS.

SECTION I.

WHAT IS A COMET ?

The ancients were unacquainted with the physical nature of comets—False ideas entertained by astronomers of the eighteenth century respecting the physical constitution of comets; comets regarded by them as globes, nearly similar to the planetary spheroids—Views of Laplace upon comets, compared by him to nebulæ—Contemporary astronomers have confirmed these views and rectified the errors of the ancient hypotheses—Desideratum of science; the rencontre of the earth with a comet or the fragment of a comet.

THE question, *What is a comet?* examined in the preceding chapter, and which we reproduce as the heading of this Section, has been the subject of numerous hypotheses. It cannot, however, yet be considered as answered. But it has lately been attempted in an entirely new manner, and by a method least of all to be expected—that of direct investigation. The exposition of this method, and the considerations which have led to it, will be the object of this new chapter.

Let us commence by recapitulating the substance of what our previous enquiries and researches have already taught us.

The ancients, as we have seen at the commencement and in the course of this work, held notions concerning the nature of comets that were entirely hypothetical, and moreover contradictory. On passing their conjectures in review it is surprising, no doubt, to meet with ideas, to some extent, in conformity with the accepted facts of modern science. But the astronomers of the Middle Ages and of the Renaissance, up to the time of Newton, and even later, were not more advanced than the ancients: the coincidences of which we speak, therefore, are purely accidental. If we fabricate hypotheses entirely conjectural, we may occasionally chance upon propositions so much in unison with the truth that they might be for a moment regarded as the result of a marvellous divination; but it is not so. Xenophanes, for example, has called comets *wandering clouds*, which is doubtless true; but what a difference between the meaning attached by the philosopher to such an expression and that given to it by science at the present day!

We have already said that the progress of cometary astronomy from the time of Newton, a progress essentially mathematical or mechanical, caused the philosophers of the eighteenth century to entertain very erroneous views concerning the physical constitution of comets. Seeing only in these bodies planets making longer voyages than others, and having orbits more inclined to the ecliptic, they regarded them almost as planets of the solar system. For these philosophers they were globes surrounded with denser atmospheres, which were subjected to extreme vicissitudes of heat and light, and consequently differed greatly from those of the planets; in fact, their climates were different, but that was all.

Laplace, guided by his own more profound views of the origin of the solar and planetary world—views nevertheless propounded with reserve—was the first to see clearly that there must be a difference of origin between comets and the planets, as both the smallness of the masses and the optical appearance of the former denote an essential difference of structure. Not only does he look upon comets as nebulæ, but as wandering nebulæ—visitors for a time to our planetary world, wandering from star to star or from system to system. Some few only, conquered for a time by the effect of the planetary perturbations, become temporary satellites of the sun.

But did Laplace himself, whose views seem to gain credit in proportion as science advances, attach to the word nebula a well-defined physical signification? Evidently he neither could nor would assimilate a comet to a resolvable nebula; that is to say, to a mass composed of a multitude of little stars. In his opinion it was a confused mass of elements analogous to the proper nebula of Herschel, in which that celebrated astronomer saw matter being condensed to form centres of light, or suns; or rather it was, as it were, a portion of the primitive solar nebula or of some other similar agglomeration.

What have science and observation added to these naturally somewhat vague notions? This we have seen in the chapters which treat in detail of the physical and chemical constitution of comets, and from which it follows that a comet is altogether differently constituted to the globes more or less similar to the earth which form the planets. It is a mass in a state of unstable equilibrium, whose form is modified with extreme rapidity, according as it receives the solar radiations at a greater or less distance, and which is subjected to the attraction of the sun and of the planets. The transformations, physical, calorific, luminous, and perhaps chemical, of the nebulous portion are so extraordinary and rapid, that nothing in other bodies can furnish an adequate idea of them. Everything leads us to believe that cometary nuclei, even when they may be regarded as solid masses, or composed of multiple solid masses, more or less aggregated, are themselves the seat of transformations quite as singular.

Spectral analysis, in spite of the few results it has as yet furnished concerning the light of comets, gives us, nevertheless, reason to suspect that the matter of cometary nebulosities is chemically composed of but two or three elements at most:

while the matter of the nucleus, whether it be an incandescent liquid or solid, or only solid matter reflecting the light of the sun, quite eludes our research, nor do we know anything of the chemical elements of which it is composed.

To improve our knowledge on these doubtful points of cometary astronomy it would be necessary that one of those events so much dreaded by the timid and superstitious should come to pass. It would be necessary that our globe should encounter a comet on its path, or let us say, as a more inoffensive hypothesis, the fragment of a comet. Would not the penetration of the cometary matter into the atmosphere, its fall upon the earth, by permitting men of science to contemplate *de visu* and touch with their hands the substance of a comet, put an end to all uncertainty?

Up to the present time there has been nothing in the history of mankind, nothing in the past history of the earth, geologically considered, to indicate that such an event has ever taken place. But even were it otherwise, and had such an event actually occurred, supposing it to have been attended with no disastrous consequences, science, which then did not exist, was in no condition to profit by the occurrence, and we should be no better informed than we are at present respecting the true constitution of a comet.

Is it true that in 1861, on June 30, the earth passed through the tail of the great comet which appeared at that time? We shall see in a subsequent Section that it is possible that such an event took place. Several astronomers who have carefully studied the question affirm that it did. And what was the result? At the utmost a phosphorescent glimmer; certainly nothing was either seen or felt. From what we know of cometary tails, their mass and excessively small density, nothing could have been seen.

We should have to penetrate to the midst of the cometary nucleus, or at least to the matter of the vaporous head, the luminous *aigrettes* and envelopes, in order to witness the singular phenomena whose development the telescope has revealed to us already. But the probabilities of such a rencontre are so slight that there is little chance that we shall ever add to our knowledge by this means. The ether is so vast an ocean, its abysses are so profound, so prodigious in extent compared with the insignificant volumes of the stars, of our pigmy earth and even of the larger comets, that the chance of a collision between our earth and any of these bodies, whose paths are so well regulated, is very small.

But have we not other reasons for hoping for a rencontre, if not between the earth and a comet, at any rate between the earth and fragments of cometary matter? This would supply the desideratum of science. We have seen comets divide into two; we have seen their nuclei project, under the influence of unknown forces, waves and jets of matter to enormous distances in space. Does the matter thus distributed finally regain the centre from which it was projected; or does it, on the contrary, abandon it for ever, so that comets, in each of their revolutions, lose a portion of their substance? This is the question which we shall now examine.

SECTION II.

IS THE MATTER OF COMETS DISPERSED IN THE INTER-PLANETARY SPACES?

At a distance from the sun the nebulous agglomerations which constitute a comet preserve a spherical or globular form, a certain indication that their molecules obey the preponderating action of the nucleus. This form would be preserved if no foreign influence interfered to derange their mutual positions or to disturb the general equilibrium.

But the comet, when approaching its perihelion, is subjected more and more to the attractive power of the sun, whose enormous mass suffices to change the spherical form of the cometary nebula, to render it more and more ellipsoidal, and finally to carry away beyond the sphere of the attraction of the nucleus whole strata of the nebulosity. This is proved beyond a doubt, as we have seen, by the analysis of M. Roche. In addition to the action of the solar mass there is likewise the action of radiated heat from the sun, which determines changes of great importance: the emission of vaporous matter from the nucleus, luminous jets, *aigrettes*, and successive concentric envelopes. If the tails of comets, as everything leads us to believe, are material realities, and not simple visual effects; if they are molecules detached from the nebulosity and projected far beyond it by a repulsive force, we may say that, having passed beyond the preponderating action of the nucleus, they have for the moment become foreign to the comet itself, which has thus suffered a portion, however small, of its matter or its mass to escape.

We have said *for the moment.* In fact, the molecules, although no longer retained by the cometary nucleus, but abandoned and repelled, have not for that reason lost the orbital movement, or movement of translation, which they possessed in common with the others. They continue, therefore, to describe around the sun orbits which carry the whole of them together through the same regions of space as the comet itself. No sooner have they passed the points which form their respective perihelia than, like the comet, they recede more and more from the sun, the focus of their movements. Now, in proportion as the distance from the sun increases, the causes which have led to their dispersion diminish in intensity, and the nucleal attraction gradually resumes the influence it originally possessed. We do not see, then, why the whole of the dispersed molecules should not ultimately return, if not to resume their original places in the system, at least to reconstitute the nebulous agglomeration, not in the same form, no doubt, but so as to reproduce the primitive mass.

In point of fact, however, for matters to happen exactly in this way we should have to suppose that no cause of perturbation could arise to derange the system or disturb the re-constitution. Now, we know that comets traverse the planetary system, and project to enormous distances across it the matter of their tails. The nuclei themselves are disturbed in their orbits by the masses of Jupiter, Mars, and the Earth. This same disturbing action must exert its influence upon the detached nebulosities of comets and carry away for ever from the latter the outlying portions of their atmospheres and tails. Such an intervention doubtless divided in two the comet of Biela, and has since dispersed one or other of the fragments. What, in fact, has become of these *débris* which failed to reappear at the time when their regular period should have brought them within sight of our globe?

Thus everything inclines us to believe that scattered here and there upon the planetary shores, or floating upon the waves of the ethereal ocean, shattered and broken-up comets exist; remains of shipwrecks suffered by millions of comets; waifs of aërial barks unable to accomplish the voyage without paying tribute to the element in which they float. Nevertheless such fragments, more or less disintegrated, do not wander at will in space; they move in orbits whose forms are dependent upon the modifications which the disturbing influence has brought to bear upon the original motions. We may suppose that these orbits continue to be closed curves of which the sun is the focus, or that they have become changed to hyperbolas. We may even suppose the disturbing influence to have been such that these disintegrated fragments have become satellites of the disturbing planet. Perhaps all these results and many more have been brought about in the course of time.

The number of comets which penetrate into our world is, in all probability, so immensely great, that during the hundreds of millions of years which we may assume to have elapsed since the beginning of the world the interplanetary spaces must have been furrowed by prodigious multitudes of currents of matter from disintegrated comets, fragments of comets, which the planets, in their regular course about the sun, cannot fail frequently to encounter. It can hardly be otherwise. And indeed at the present day there is nearly certain evidence that this actually is the case; and, although originally the proof of the existence of these material currents may have been arrived at in another manner, the fact of their existence is certain. Their origin alone can be matter of doubt.

SECTION III.

COMETS AND SWARMS OF SHOOTING STARS.

Periodicity of the meteor-swarms; radiant points; number of swarms recognised at the present day—Periodical maxima and minima in certain meteoric currents; thirty years period of the November swarm—Parabolic velocity of shooting stars; the swarms of shooting stars come from the sidereal depths.

THESE considerations bring us to the theory recently elaborated by the learned Italian astronomer M. G. V. Schiaparelli, Director of the Observatory of Brera, at Milan.

According to this theory there exists between comets and shooting stars a connexion and community of origin, which henceforth we may regard as certain, as it is supported both by logical deduction and observation. We shall now explain by what train of ideas this assimilation between phenomena which at first sight appear so foreign to each other has passed from the phase of simple hypothesis into that of a theory which observations of great value permit us at the present time to consider demonstrated.

Let us first of all pass in review the facts upon which the theory is based.

The shooting stars which may be observed on any clear night throughout the year are notably more numerous at certain times, the dates of which are nearly fixed, as, for example, August 10, November 13 or 14, and April 20. They then appear in sufficient numbers to be considered, not as isolated meteors, but as groups or swarms of meteors. This connexion soon became still more manifest when it was found that the stars of each swarm were moving in trajectories, not distributed at random over the celestial vault, but so disposed that if prolonged backwards all or nearly all passed, if not through the same mathematical point, at least through one very circum-

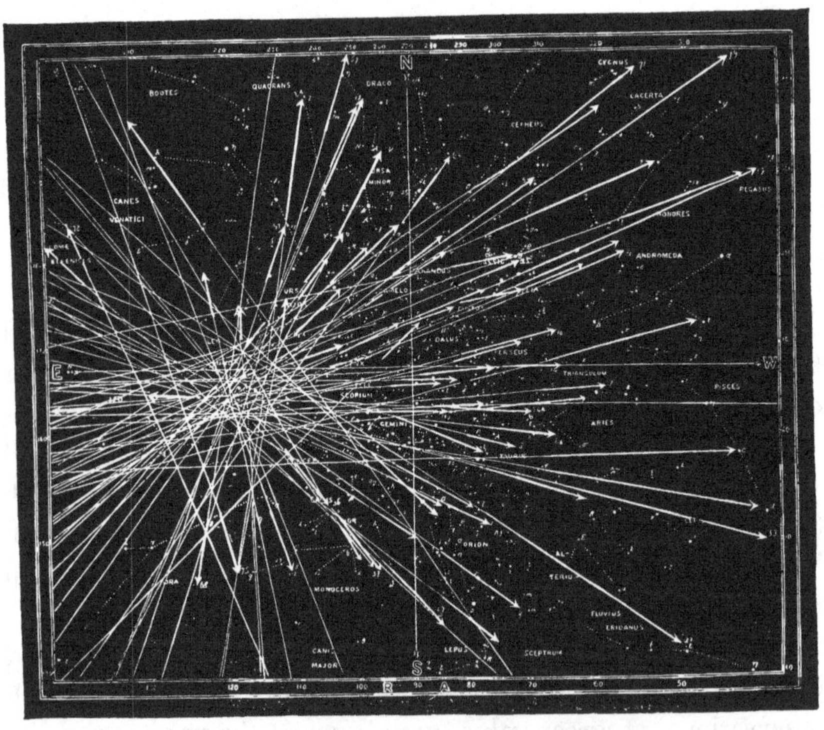

Fig. 69.—Shooting Stars of November 13–14, 1866. Convergence of the tracks, according to A. S. Herschel and A. MacGregor.

scribed region of the heavens. To obtain an idea of this remarkable conjunction of the apparent trajectories of a stream of shooting stars the reader has only to glance at fig. 69, representing the tracks among the stars of a certain number of these meteors which appeared on the night of November 13–14, 1866.

The point of emanation or convergence of the trajectories of a swarm is called the *radiant point.* Now, not only have the meteors which appear at the fixed dates we have mentioned, both on the same night and on several successive nights, the same radiant point, but this radiant point does not vary in position, or varies very little, in the course of years for successive apparitions of the same meteor-current.

At the time when researches were first commenced respecting these singular phenomena the known streams of periodical return were few in number; those of August 10 and November 14 were at first alone recognised. It was thought that the shooting stars of ordinary nights were scattered without any apparent connexion. But more careful observations showed that in reality these sporadic shooting stars obeyed laws similar to those of the other swarms; a great number of streams were thus distinguished, and the positions of their radiant points determined. At the present time 102 are known.*

But another fact of great importance was established, partly by historical researches in regard to the previous apparitions of similar phenomena, and partly by the continuous observations of contemporary astronomers. The fact in question is this:—The periodicity of meteoric swarms is not only annual, so that the same nights of each succeeding year are remarkable for displays of meteors sufficiently abundant to distinguish these nights from those immediately adjacent, but it also happens that, as regards certain streams, the display varies in a manner which clearly enables us to distinguish the recurrence of periodical maxima and minima. We will cite a remarkable instance of this periodicity, viz. that of the swarm of the middle of November. In 1799 the stream of shooting stars was of

* [See addition to this chapter.—Ed.]

wonderful intensity. The same phenomenon was repeated thirty-four years later; that is to say, in the year 1833. On tracing back the records of similar phenomena the display of 1766 was found to have been not less abundant at the same date. Similar instances were likewise found at more remote epochs, and it appeared that the intervals between these extraordinary apparitions were always either thirty-three or thirty-four years, or a multiple of these numbers. It seems certain, then, that for this particular swarm there is a periodical maximum which recurs about every third of a century. Thus Olbers did not hesitate to predict, long beforehand, the recurrence of a maximum for the year 1867. Professor Newton, of America, furnished more precise data, and defined the period with greater accuracy, assigning to it a duration of thirty-three years and a quarter, and fixing the date of its next return for the night of November 13–14, 1866.

The meteor-swarm was true to the prediction.

It soon became clear that these singular phenomena could only be explained by considering the different swarms of meteors as of extra-terrestrial or cosmical origin. Other circumstances, which we shall enlarge upon in a separate work, contributed, in conjunction with those which we have just related, to prove that shooting stars form currents of celestial particles which circulate independently in space and describe regular orbits like comets. These numerous currents traverse the interplanetary spaces in all directions, and by their occasional rencontres with the earth give rise to the production of shooting stars. Our globe, or simply its atmosphere, penetrating more or less deeply to the centre of one of these groups, a meeting takes place, the velocity being sometimes equal to the sum and sometimes to the difference, or to any intermediate amount, of the respective velocities of the two bodies. In any case there is a loss of *vis viva*, or rather a transference of *vis viva* into heat, which generally produces incandescence.

But, if these swarms are currents of meteoric matter, what law governs their circulation in space? What are the elements of their orbits? And last, and most interesting of all, what is their origin? These questions called forth various hypotheses in reply. The swarms, it was supposed, were closed rings more or less elliptic and more or less eccentric in regard to the sun, the focus of their movements. In this way the facts of observation were accounted for.

In this state of the question, M. Schiaparelli, by some bold speculations, succeeded in throwing a new light upon an obscure point in the theory. He believed it could be shown from different observations that the velocity of meteors at the moment of their entrance into the terrestrial atmosphere was at least equal to what is termed cometary velocity, and consequently nearly half as much again as the velocity of the earth's translation. He showed that this hypothesis accounted for a fact* which at first appeared to be inconsistent with the cosmical origin of shooting stars, but which was, on the contrary, a striking confirmation of it. Provided with this important element, the Italian astronomer was enabled to calculate the elements of the orbits of certain swarms, and he found that they describe in space excessively elongated curves, parabolas or hyperbolas.

* That the number of shooting stars observed on any night varies with the time of observation. The maximum horary number takes place in the hours 2 or 3 A.M. [for the November meteors.]

SECTION IV.

COMMON ORIGIN OF SHOOTING STARS AND COMETS.

Transformation of a nebula which has entered into the sphere of the sun's attraction; continuous parabolic rings of nebulous matter—Similarity between the elements of the orbits of meteor streams and cometary orbits—The August stream; identity of the Leonides and the comet of 1862—Identity of the Perseids and the comet of 1866 (Tempel)—The shooting stars of April 20 and the comet of 1861—Biela's comet and the December stream—Did the earth encounter Biela's comet on November 27, 1872?

It still remains to explain the origin of meteor swarms or streams, and the reason of their annual periodicity and the maxima which appear at dates separated by intervals of several years. For this purpose it will be necessary for a moment to quit the domain of fact and consider some theoretical speculations.

The swarms of shooting stars appear to be constituted, as it were, of aggregations of particles separated from one another by some distance. But if, instead of seeing them on their arrival in the proximity of the earth, in contact with its atmosphere, it were possible to contemplate them from a distance in the heavens, the whole of these myriads of particles, whether illuminated by the sun's rays or shining by their own light, would appear to the observer like a cloud or nebulosity. And as the supposed velocity of meteoric swarms in their orbits is that of cometary velocity, it follows that the nebulosities of which we speak come from the depths of space, from regions far beyond the sun and the planets. Never theless, it is certain that these nebulæ, which have, perhaps, quitted some other sidereal system, only make their entrance into ours under the influence of a momentary preponderance of the attraction of our sun.

It was, doubtless, from considerations of this kind that Schiaparelli was led to propound the following problem: Given a nebulosity situated at a very great distance, but nevertheless such that the attraction of the sun determines its movement towards our system, what will be the form of this aggregation (supposed to be spherical at starting) when it arrives at its perihelion? Resolving this problem by analysis, and according to the principle of gravitation, M. Schiaparelli proves that the nebulous mass, though globular at first, will be gradually transformed, so as to be, at the time of its passage in the vicinity of the sun, converted into a continuous stream or current of parabolic form, very much more dense than it was originally, and taking hundreds and even thousands of years to effect completely its perihelion passage.

Hence it will be understood that the earth encountering this stream at one point of its orbit, and passing, after each of its revolutions, through this same point of interplanetary space, a periodic apparition of meteors will take place; particles of the stream traversing the higher regions of the atmosphere will shine each for a moment as a shooting star, some to be totally destroyed by their combustion, others to pursue their course, after having thus manifested for a moment the presence of the nebula of which they form a part. These long parabolic trains explain the yearly-recurring streams of meteors: according to the greater or less thickness of the portion of the stream traversed by the earth will the number of shooting stars seen be more or less considerable.

As regards the longer periods which give the maxima at

regular intervals of several years, M. Schiaparelli accounts for them as follows:—'In the same manner as the long parabolic currents are comparable, as far as their movements are concerned, to comets of infinite orbits, intermittent periodic streams are analogous to periodic comets of regular return. Incidental circumstances—the planetary perturbations, for example—may transform an endless stream into an elliptic closed ring.' The meteors of November 13 and 14 are, very probably, according to Schiaparelli, a case in point.

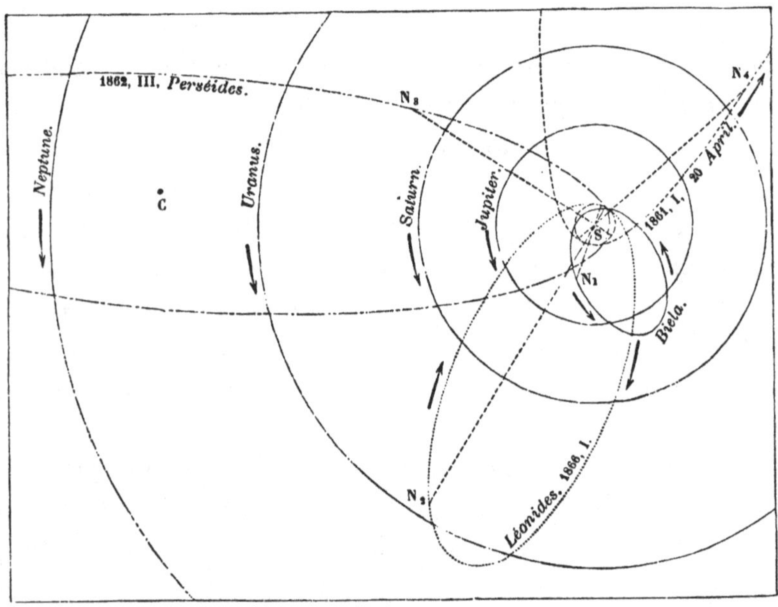

Fig. 70.—Orbits of the Meteoric Streams of November, August, and April, and of the Comets of 1866 and 1861.

We shall not enter more particularly into the details of this remarkable theory, but shall confine ourselves to pointing out the analogy existing between the nebulous currents which give birth to the meteoric swarms and the nebulosities of comets. The velocity of translation, the inclinations of the planes of the orbits at all angles, the movements in all directions, are elements common alike to the comets and to the meteor swarms.

One most essential sanction was still wanting to this theory, that given by observation, alone capable of actually demonstrating the connexion between the two kinds of nebulæ. Now at the present time this sanction exists; it soon followed the theory, and appears evident enough to defy all contradiction. M. Schiaparelli, having collected the observed elements of the meteoric swarm of August 10, was enabled to calculate its orbit as if it had been that of a celestial body or a comet. On examining the parabolic elements of comets which had been catalogued he recognised the almost identity of the elements of one of these comets with those of the meteor swarm. The following table of elements will exhibit this comparison:—

	Elements of the orbit of the meteoric stream of August 10, calculated by Schiaparelli	Elements of the orbit of the comet of 1862, calculated by Oppolzer
Perihelion passage . .	August 10·15.	August 22·9, 1862.
Longitude of perihelion .	343° 28′	344° 41′
Longitude of node . .	138° 16′	137° 27′
Inclination	64° 3′	66° 25′
Perihelion distance . .	0·9643	0·9626
Direction of motion . .	Retrograde.	Retrograde.

A similarity so complete could hardly be an accidental coincidence. But M. Schiaparelli did not confine himself to this one comparison. He calculated in the same manner the elements of the orbit of the November stream, and found that they were almost identical with the elliptic elements of Tempel's comet (1866 I.), calculated by Oppolzer. The following table affords the means of comparing these elements:—

	Elliptic elements of the orbit of the meteoric stream of November 13, calculated by Schiaparelli	Elliptic elements of the orbit of Tempel's comet (1866, I.), calculated by Oppolzer
Perihelion passage . .	November 10·092	January 11·160, 1866
Longitude of perihelion .	56° 25′ 9″	60° 28′ 0″
Longitude of node . .	231° 28′ 2″	231° 26′ 1″
Inclination	17° 44′ 5″	17° 18′ 1″
Perihelion distance . .	0·9873	0·9765
Eccentricity . . .	0·9046	0 9054
Semi-axis major . .	10·340	10·324
Period	33·250 years.	33·176 years.
Direction of motion . .	Retrograde.	Retrograde.

After having pointed out this new and important coincidence, M. Schiaparelli observes: 'It is very worthy of remark that the two well-known meteor streams, those of August and November, have each their comet. Are we to suppose that it is the same with all the others? If so, we should be forced to regard these cosmical streams as the result of the dissolution of cometary bodies. But it would be at least premature to extend this conclusion to all shooting stars. It is possible, as I have shown, that the whole of these bodies, great and small, may form systems in space bound together solely by their own attraction, and afterwards destroyed by the action of the sun. Perhaps, also, that which we term a comet is not a single body, but a collection of very numerous and minute bodies, attached to a principal nucleus.'

We cannot fail to notice the connexion existing between these views and those which led M. Hoek to his study of the theory of cometary systems; we must also perceive how completely these new views on the subject of cometary physics accord with the facts of observation that have been mentioned in a preceding section respecting the duplication of Biela's comet and the division and shattering of ancient comets, phenomena which have been handed down to us by tradition, but which had hitherto been generally denied and regarded as fables by astronomers.

In conclusion, and before leaving so vast a domain, open alike to new researches and conjectures, we must not forget to mention two more cases of identity between meteoric swarms and comets. The first relates to the meteors of April 20. According to MM. Galle and Weiss the orbit of this swarm has the same elements as the orbit of the comet of 1861. D'Arrest and Weiss have likewise found an accordance between Biela's comet and the shooting stars of the end of November and the first days of December. We have already said that the remarkable shower of shooting stars which distinguished the night of November 27, 1872, appears certainly to have been due to the rencontre of the earth either with one of the two comets, fragments of that of Biela, or with a stream of matter which originally belonged to that comet, and which followed in space nearly the same course.*

If these views—which would have appeared so strange half a century ago—should be confirmed, we have a new and quite unexpected means of putting ourselves in direct communication with comets, since the earth every year—every night in the year, indeed—comes in contact with nebulosities which have been comets. A new light would be thrown upon the physical constitution of these bodies, and we might then consider as highly probable that granulated structure of cometary nuclei, formed of isolated particles, which Babinet was led to suspect upon very different grounds.

* [See note, p. 265.—Ed.]

ON THE CONNEXION BETWEEN COMETS AND METEORS.

BY THE EDITOR.

The intimate connexion now known to exist between comets and meteors is perhaps the most striking and novel discovery of a purely astronomical kind that has been made in our time. To those who are aware how few years have elapsed since the apparitions and tracks of meteors seemed to be so arbitrary and capricious that in the opinion of many it was scarcely worth while to record them, it cannot but be matter of wonder to consider how great has been the advance in our knowledge, and how rapid has been the progress of ideas on this subject. On account of the importance of the results found upon the nature of comets, I, therefore, add here several details, chiefly historical, which will serve to show more fully how remarkable is the connexion that has been established.

It is less than ten years since the orbit of the first stream of shooting stars—that of the middle of November—was calculated, the circumstances being as follows :—

Professor H. A. Newton, of the United States, collected and discussed thirteen historic showers of the November meteors between the years 902 and 1833.

The following table exhibits these displays, and the earth's longitude at each date, together with the same particulars for the shower of 1866 :—

A.D.	Day and Hour.	Earth's longitude.
902	October 12·17	24° 17′
931	" 14·10	25° 57′
934	" 13·17	25° 32′
1002	" 14·10	26° 45′
1101	" 16·17	30° 2′
1202	" 18·14	32° 25′
1366	" 22·17	37° 48′
1533	" 24·14	41° 12′
1602	" 27·10 (Old style)	44° 19′
1698	November 8·17 (New style)	47° 21′
1799	" 11·21	50° 2′
1832	" 12·16	50° 49′
1833	" 12·22	50° 49′
1866	" 13·13	51° 28′

From these data Professor Newton inferred that these displays recur in cycles of 33¼ years, and that during a period of two or three years at the end of each cycle a meteoric shower may be expected. He also concluded that the November meteors belong to a system of small bodies describing an elliptic orbit about the sun, and extending in the form of a stream along an arc of that orbit which is of such a length that the whole stream occupies about one-tenth or one-fifteenth of the periodic time in passing any particular point. He further showed that the periodic time must be 180 days, 185 days, 355 days, 377 days, or 33¼ years, and suggested that by calculating the secular motion of the node for each one of the five possible orbits, and by comparing the values with the observed motion (about 52″ annually or 29′ in 33¼ years) it would be possible to decide which of these five orbits was the correct one.

Soon after the remarkable display of the November meteors in 1866, Professor Adams, of Cambridge, undertook the examination of this question. Beginning with the orbit of 355 days, which Professor Newton considered to be the most probable one, Professor Adams found the motion of the node would only amount to 12′ in 33¼ years; that for the orbit of

377 days the value would be nearly the same, while if the periodic time were a little greater or a little less than half a year, the motion of the node would be still smaller. It therefore only remained to examine the orbit of 33¼ years, and Professor Adams found that, for this orbit, the longitude of the node would be increased 20′ by the action of Jupiter, nearly 7′ by the action of Saturn, and about 1′ by the action of Uranus. The other planets produce scarcely any sensible effects, so that the entire calculated increase of the longitude of the node in the above-mentioned period is about 29′. As already stated, the observed increase of longitude in the same time is 29′, and this remarkable accordance between the results of theory and observation left no doubt as to the correctness of the period of 33¼ years.*

Subsequently, however, to the commencement, but before the publication of Professor Adams's results, M. Schiaparelli had been led, on totally different grounds, to conclusions which first suggested a probable connexion between meteors and comets. These related to the August meteors, or Perseids as they are called from the constellation which usually contains their radiant point. A comparison of the average hourly increase in the frequency of meteors, throughout the year, from evening until daybreak, with a mathematical formula for the same variation in terms of their velocity, led M. Schiaparelli to conclude that the real average velocity of shooting-stars in their orbits round the sun did not differ much from that of comets moving in parabolic orbits, which is greater than the earth's mean orbital velocity at the same distance from the sun in the proportion of 1·414 to 1. Discussing, then, the origin of meteoric currents M. Schiaparelli remarked that in all respects shower-meteors resembled comets rather than planetary

* *Monthly Notices of the Roy. Ast. Soc.*, vol. xxvii., p. 250 (April 1867).

bodies indigenous to the solar system; and shortly afterwards supposing groups of meteors to have entered our system originally as cosmical clouds, he formulated his theory in a series of nine propositions, of which I extract two. 'IV. Whatever may be the shape and size of a cosmical cloud, it can rarely enter the central parts of the solar system without being transformed into a parabolic current, which may occupy years, centuries, or thousands of years in completing its perihelion passage, in the form of a stream extremely narrow in comparison with its length. Of such streams those which the earth encounters in its annual revolution present themselves as a shower of shooting-stars diverging from a common centre of radiation.' 'VIII. Shooting-stars and other like celestial bodies, which in the last century were regarded as atmospheric, which Olbers and Laplace first maintained might be projected from the moon, and which afterwards came to be regarded as planetary bodies, are in reality bodies of the same class as the fixed stars; and the name of *falling stars*, applied to them, simply expresses the real truth. They bear the same relation to comets which the planetoids between Mars and Jupiter bear to the larger planets, the smallness of the masses, in both cases, being compensated for by the greatness of their number.' *

Subsequently M. Schiaparelli showed that if the Perseids be supposed to describe a parabola, or a very elongated ellipse, the elements of their orbit calculated from the observed positions of their radiant point agree closely with those of the orbit of Comet II., 1862, as calculated by Dr. Oppolzer; so that it seemed probable that the great Comet of 1862 was part of the same current of matter as that to which the August meteors belong: he also gave approximate elements of the orbit of the November meteors (or Leonids), calculated upon the supposition

* *Bullettino Meteorologico dell' Osservatorio del Collegio Romano*, vol. v.

that the period of the revolution was 33¼ years: but as the calculations were founded on an imperfect determination of the radiant point, he failed to find any cometary orbit that could be identified with that of the meteors. A few weeks later M. Le Verrier gave more accurate values of the elements of the November meteors, his calculations being based upon a better determination of the radiant point, and M. Peters, of Altona, pointed out that these elements closely agreed with those of Tempel's comet (I., 1866); an agreement which M. Schiaparelli, who had recalculated the elements of the orbit of the November meteors, also remarked independently.

Thus we see at very nearly the same time M. Schiaparelli was enabled to identify the orbit of the August and November meteors with those of two known comets, and Professor Adams placed almost beyond doubt the fact that the November meteors did actually describe an orbit round the sun in 33¼ years; so that almost simultaneously it was shown that the November meteors did describe their orbit in 33¼ years; and that assuming this, the orbit resembled very closely that of Tempel's comet of 1866.

The reader will readily see from fig. 69 (p. 426), that the tracks of the meteors, as laid down upon a celestial chart, intersect in a very restricted region of the sky, but that this region is very far from being an exact point. The same will be seen in fig. *A*, which shows the tracks of certain of the meteors observed at Greenwich, on the night of the great star-shower of November 1866. This want of precision is partly due to the extreme difficulty of noting down quite accurately the track of a meteor from a rapidly made eye-observation, and partly to the fact that the meteors do not all proceed from the same mathematical point. Thus the determination of the radiant point, and therefore of the orbit, of the meteors is always a matter subject to some slight amount of uncertainty. It will be remarked that in fig. *A*, the tracks are not prolonged backwards as in fig. 69, so that the apparent lengths of the meteor-tracks are shown, and that these are shorter in proportion as they are nearer to the radiant point. This is merely an effect of perspective, for a meteor seen at the radiant point would be directly approaching us, and therefore would merely appear like a

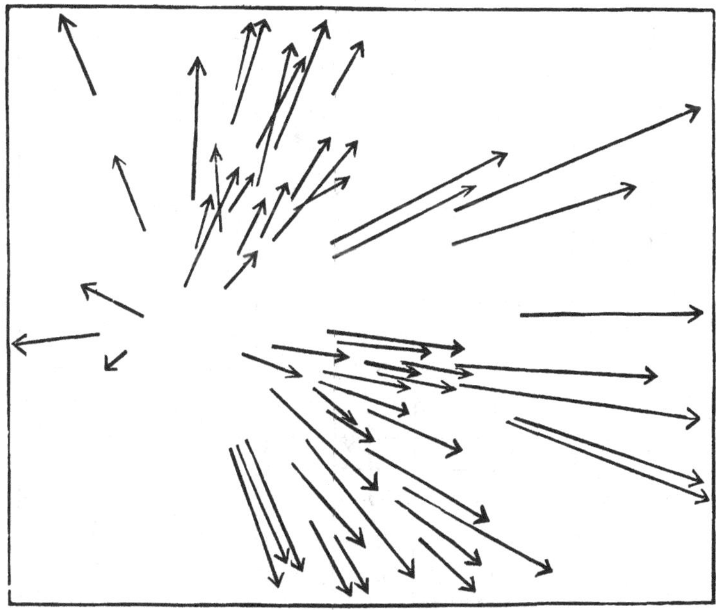

Fig. *A*.—Tracks of Meteors observed at the Royal Observatory, Greenwich, on the night of November 13—14, 1866.

motionless star, destitute of train, appearing and then disappearing; and the tracks would be longer the further removed they were from the radiant point.

In fig. *B*, is shown the rate of frequency of the meteors in the same remarkable star-shower. Eight of my observers, at the Royal Observatory, Greenwich, were engaged in counting the number of shooting-stars per minute, each taking a different portion of the heavens. The maximum number was therefore attained between 1 and 2 A.M. on the morning of the 14th, and the display then rapidly subsided. The curve in fig *B* is interesting as it may be regarded as representing

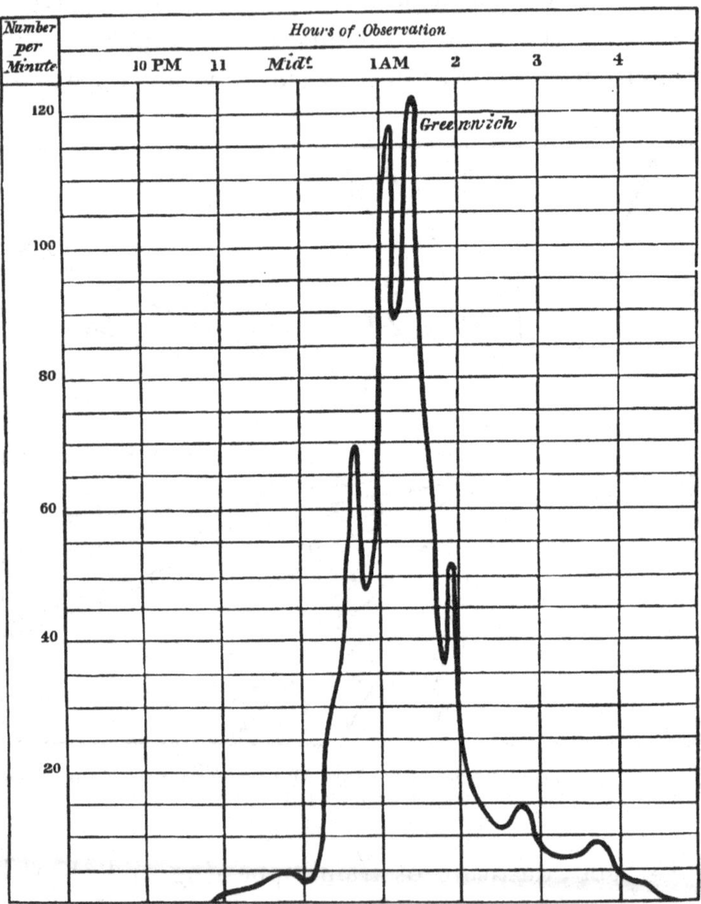

FIG. *B*.—Showing the rate of frequency of Meteors seen per minute at the Royal Observatory, Greenwich, on the night of November 13—14, 1866.

the density of the meteors in the stream that we were crossing.

The August meteors (the Perseids) are much less numerous than those of November, but they appear every year, and large meteors are frequent among them. Fig. *C* gives a number of tracks of these meteors observed by Professor Tachini in 1868.

It was this meteor-swarm that M. Schiaparelli first identified with a comet (Comet II. 1862).

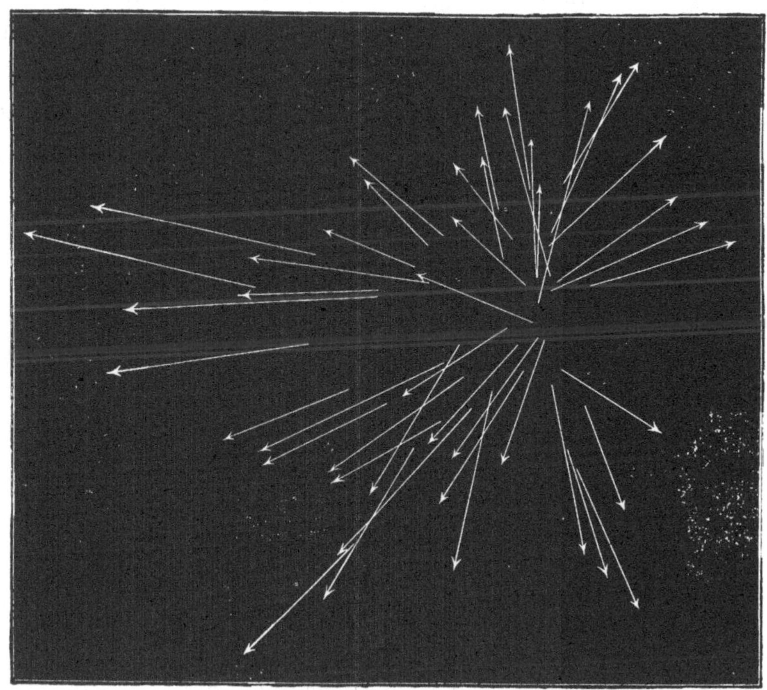

FIG. *C*.—Tracks of Meteors recorded by Professor Tachini at Palermo on August 8 to August 12, 1868.

The number of known radiant points has in the last few years—now that the subject of meteoric astronomy has begun to excite attention—been greatly increased, and now exceeds seven hundred.*

* References to 695 radiant points will be found in *Monthly Notices of the Royal Ast. Soc.*, vol. xxxv., p. 249 (1875).

The fact that there is a close connexion between comets and meteors is all that can be considered established with certainty; but M. Schiaparelli has developed his views on their relation to one another at some length. According to him, comets and meteoric groups are portions of the nebulous matter in space in two different states of condensation, which may arise either together or apart, according to the tenuity of the matter which produced them. Such differences are observable in the nebulæ of which some are resolvable and others not resolvable in the telescope. Comets with two or

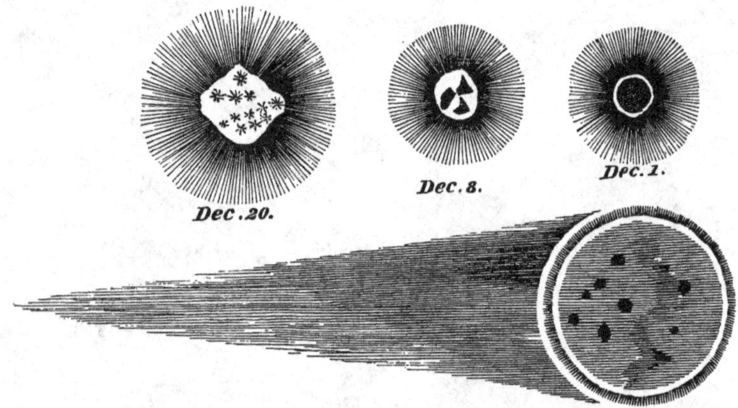

Fig. *D.*—(1) Nucleus of the Comet of 1618, observed telescopically by Cysatus. (2) Comet of 1652, as seen December 27, according to Hevelius.

more nuclei have several times been observed. Figs. 61 and 62 (p. 341) exhibit the multiple nuclei of the comets of 1618 and 1661, according to Hevelius, and fig. *D* shows the nucleus of the comet of 1618 according to Cysatus, and the comet of 1652 which, according to Hevelius, consisted of a disk of pale light of the apparent diameter of the full moon, and in the telescope appeared to be filled with points of light. M. Schiaparelli considers that the various characters of all these nebulous bodies can only be explained upon the hypothesis that they have been condensed from a state of highly heated vapour, gradually undergoing a process of cooling and condensing of its parts. 'Not only the various features of star-showers and comets, but even the mineralogical structure of aërolites appear to be explained on this supposition. The theory of Faye that they are developed from the nuclei, and of Secchi that they are the remnants of the tails, and of Erman that they are particles detached from comets by a resisting medium, are not so immediately referable to the known laws of gravitation as the hypothesis that all classes of luminous meteors, like comets themselves, are drawn from the sun by its attraction from the regions of intra-stellar space, which the telescope declares to be empty, but which, in all probability, are strewed with cosmical clouds, containing in one order of phenomena both meteoroids and clouds.' *

It would require too much space to enter further into M. Schiaparelli's theory and explain how he accounts for accidental or sporadic meteors, not apparently belonging to any stream, &c. The reader will, however, see that in a great many different respects the close relation of comets to meteor-swarms seems to be established.

In a note at the end of Section iii., Chapter VII. (p. 265), I have given an account of the remarkable circumstances connected with the meteor-shower of November 27, 1872, and Biela's comet; and in relation to that note the following particulars, due to Professor H. A. Newton, will not be found uninteresting. The line of nodes, or the place of the earth's nearest approach to the comet's track, being at N (fig. *E*), it appears that in the year 1798, when the earth encountered at that point the great meteor-shower of December 6 of that

* *British Association Report*, 1868, p. 414.

year, Biela's comet was in the position marked C, somewhat nearer to the earth than on the next occasion, when a similar occurrence was observed in 1838. The comet was in 1838 at the point A, about 300 millions of miles distant, measured, along its orbit, from the earth. At the apparition of the star-shower on November 27, 1872, the comet should have occupied the position B, at about 200 millions of miles along the comet's path from the place of the earth's intersection with the meteor-stream at N. It would thus appear that a

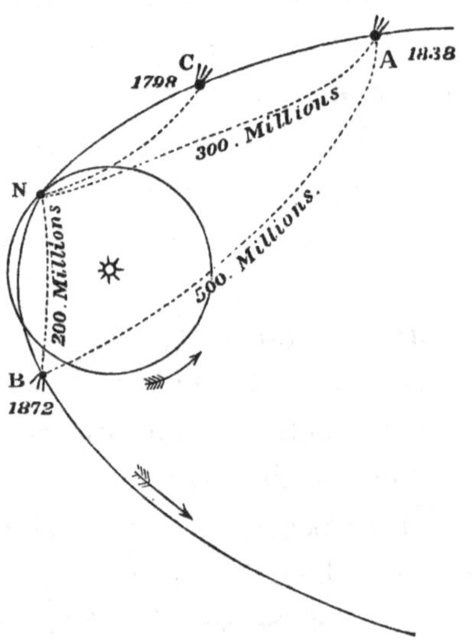

FIG. *E.*—Positions of Biela's Comet relatively to the Earth in 1798, 1838, and 1872.

long-extended group of meteor-particles must accompany the comet in its periodical revolution, preceding it to a distance of 300 millions of miles in front, and following it to a length of 200 millions of miles in the rear; so that, as there is no reason to suppose this elongated meteor-current discontinuous, it occupies fully 500 millions of miles in its observed length along the comet's path.*

* *British Association Report*, 1875, p. 224.

But, as has been shown in the note already referred to, there is considerable doubt with regard to the actual motion of the comet, so that these conclusions must be regarded as somewhat uncertain.

Professor Tait's Theory of the Constitution of Comets.

Professor P. G. Tait has recently published his theory of the constitution of comets, which he assimilates to swarms of stones or meteors, which are partly illuminated by the sun, and also give out a light of their own through the numerous and violent collisions which are always taking place among them, especially near the nucleus, where the swarm is densest. After stating that the researches of Mr. Huggins have shown the presence of some hydrocarbon in the comet, Professor Tait proceeds:—

'Now this is a most remarkable phenomenon. It at once suggests the question—How does the hydro-carbon get into this incandescent state in the head of a comet? A word or two on that subject may be of considerable interest, but we must lead up to it gradually. A great astronomical discovery of modern times is, that meteorites, the so-called falling stars, especially those of August and November, as they are called, follow a perfectly definite track in space, and that this track is, in each case, the path of a known comet; so that—whether, as Schiaparelli and others imagine, the meteorites are only a sort of attendants on the comet; or whether, as there is, I think, more reason to believe, the mass of meteorites forms the comet itself — there is no doubt whatever that there is at least an intimate connexion between the two. The path of the meteorites is the path of the comet. Well, let us consider a swarm of such meteorites (regarded each as a fragment of stone), like a shower in fact, of Macadamised stones, or bricks, or even boulders. What would be the appearances presented by such a cloud? It must in all cases be of enormous dimensions, because the earth takes two or three days and nights to pass through even the breadth of the stratum of the November meteors. Consider the rate at which the earth moves in its orbit, and you can see through what an enormous extent of space these masses are scattered. Now if you think for a moment what would be the aspect of such a shower of stones when illuminated by sunlight, you will see at once that, seen from a distance, it would be like a cloud of ordinary dust; and an easy mathematical investigation shows that it should give when sufficiently thick, except in extreme cases, a brightness equal to about half that of a solid slab of the same material similarly illuminated. The spectrum of its reflected or scattered light should be the spectrum of sunlight, only a great deal weaker. It is easy without calculation, but by simply looking at a cloud of dust on a chalky road in sunshine, to assure oneself of the property just mentioned of such a cloud of dust or small particles. Remember that in cosmical questions we can speak of masses like bricks or even paving-stones, as being mere dust of the solar system, and we may suppose them as far separated from one another, in proportion to their size, as the particles of ordinary dust are. Whether, then, it be common terrestrial dust, or cosmical dust, with particles of the size of brickbats or boulders, does not matter to the result of this calculation. Spread them about in a swarm or cloud as sparsely as you please, and only make that cloud deep enough, and illuminate it by the sun, then it can send back one half as much light as if it had been one continuous slab of the material.

'Now, look at the moon. You see there a continuous slab of material, and you know what a great amount of brightness that gives. And a shower of stones in space at the same distance from the sun as the moon, and of the same material as the moon, could, if it were only deep enough, however scattered its materials, shine with half the moon's brightness. Now no comet's tail has ever been seen with brightness at all comparable to that of the moon; and therefore it is perfectly possible and, so far as our present means enable us to judge, it is extremely probable, that the tail of the comet is merely a shower of such stones. But now we come to the question, How does the light from the head of the comet happen to contain portions obviously due to glowing gas, in addition to other portions giving apparently a faint continuous spectrum of sunlight and perhaps also light from an incandescent solid? The answer is to be found—at least so it appears to me—in the impacts of those various masses upon one another. Consider what would be the effect if a couple of masses of stone or of lumps of native iron such as occasionally fall on the earth's surface from cosmical space, impinged upon each other even with ordinary terrestrial, not with planetary velocities. In comparison with these latter, of course, the velocity of the shot of any of the big guns at Shoeburyness would be a mere trifle; yet we know that when a shot from one of them impinges upon an iron plate there is an enormous flash of light and heat, and splinters fly off in all directions. Now mere *differences* among the cosmical velocities of the particles of a comet, due to different paths round the sun, or to mutual gravitation among the constituents of a cloud, may easily amount to 1,400 feet per second, which is about the rate of a cannon-ball. Masses so impinging upon one another will produce several effects; incandescence, melting, development of glowing gas, the crushing of both bodies, and smashing them up into fragments or dust, with a great variety of velocities of the several parts. Some parts of them may be set on moving very much faster than before; others may be thrown out of the race altogether by having their motions suddenly checked, or may even be driven backwards; so that this mode of looking at the subject will enable us to account for the jets of light which suddenly rush out from the head of a comet (on the whole forwards) and appear gradually to be blown backwards, whereas in fact they are checked partly by impacts upon other particles, partly by the comet's attraction. Therefore so far as can be said until we get a good comet to which to apply the spectroscope, this excessively simple hypothesis appears easily able to account for many even of the most perplexing of the observed phenomena. I must warn you, however, that this is not the hypothesis generally received by astronomers.' (*Lectures on some recent Advances in Physical Science*, pp. 254–257, 1876.)

Sir William Thomson in his Address before the British Association in 1871, referring to Prof. Tait's theory—not then published—explained that according to it, the comet, 'a group of meteoric stones, is self-luminous in its nucleus, on account of the collisions among its constituents, while its tail is merely a portion of the less dense part of the train illuminated by sunlight, and visible or invisible to us according to circumstances, not only of density, degree of illumination, and nearness, but also of tactic arrangement, as of a flock of birds or the edge of a cloud of tobacco-smoke.'

It seems not at all improbable that this may be the real explanation of the constitution of comets; but it is clear that only a complete mathematical investigation of the motion and appearance of such an assemblage of particles, so illuminated, can decide whether this theory will account for the observed phenomena. It is clear that the mechanical difficulties which the motion of a comet's tail presents on the hypothesis of its consisting of matter moving with the comet have to be met; and this can only be effected by a thorough discussion of the complicated dynamical problem involved.

CHAPTER XIII.

COMETS AND THE EARTH.

SECTION I.

COMETS WHICH HAVE APPROACHED NEAREST TO THE EARTH.

The memoir of Lalande and the panic of the year 1773--Letter of Voltaire upon the comet—Announcement in the *Gazette de France* and the Memoirs of Bachaumont—Catalogue given by Lalande of comets which up to that time had approached nearest to our globe.

In the spring of the year 1773 a singular rumour, soon followed by a strange panic, obtained in Paris and rapidly spread throughout France. A comet was shortly to appear upon the earth's track, to come into collision with our planet, and thus infallibly bring about the end of the world. The origin of this rumour was a memoir which Lalande was to have read before the Academy of Sciences on April 21; it was, however, not read, but the title alone was sufficient to create a popular ferment. The work of the learned astronomer was entitled *Réflexions sur les Comètes qui peuvent approcher de la Terre.* It was speedily imagined, and without the smallest foundation—for nothing of the kind was to be found in the memoir—that a comet predicted by the author was about to dissolve the earth on May 20 or 21, 1773.

So great was the panic that Lalande, before publishing his work, caused the following announcement to be inserted in the *Gazette de France* of May 7: 'M. de Lalande had not time to read a memoir on the subject of the comets, which by their approach to the earth may occasion disturbance to it; but he observes that it is not possible to fix the date of these events. The next comet whose return is expected is that which is due in eighteen years, but it is not amongst the number of those which can harm the earth.' This notice, it appears, did not allay the public uneasiness, for, under the date of May 9, we read the following in the *Mémoires de Bachaumont:*—

'The cabinet of M. de Lalande is still besieged by the curious, anxious to interrogate him upon the memoir in question, and doubtless he will give to it a publicity which is now necessary, in order to reassure those whose heads have been turned by the fables to which it has given rise. So great has been the ferment that some *dévots*, as ignorant as they are foolish, solicited the archbishop to have a forty hours' prayer, in order to arrest the enormous deluge threatened; and that prelate was on the point of ordering the prayer, when some Academicians made him sensible of the absurdity of such a proceeding. The false announcement in the *Gazette de France* has created a bad effect, for it is believed that the memoir of the astronomer must have contained terrible truths, since they were thus evidently disguised.'

We see that a century ago *communiqués* were not more efficacious than at the present day, and were just as much believed. But the expressions which Bachaumont uses in regard to the *dévots* are not more misplaced than the abuse which he finds means further on to lavish upon Lalande. How much more to the point is the refined irony of Voltaire in his *Lettre sur la prétendue Comète!* Let the following short extract speak for itself:—

'Grenoble, May 17, 1773.

'Some Parisians, who are no philosophers—and, if they are to be believed, will not have time to become so—have informed us that the end of the world is approaching, and that it will infallibly take place on the 20th of this present month of May.

'On that day they expect a comet which is to overturn our little globe and reduce it to impalpable powder, according to a certain prediction of the Academy of Sciences which has not been made.

'Nothing is more probable than this event; for James Bernoulli, in his "Treatise upon the Comet," expressly predicted that the famous comet of 1680 would return with a terrible crash on the 17th of May, 1719. He assured us that its wig would signify nothing dangerous, but that its tail would be an infallible sign of the wrath of heaven. If James Bernoulli has made a mistake in the date, it is probably by no more than fifty-four years and three days.

'Now, an error so inconsiderable being looked upon by all mathematicians as of no account in the immensity of ages, it is clear that nothing is more reasonable than to expect the end of the world on the 20th of the present month of May, 1773, or in some other year. If the event should not happen, what is deferred is by no means lost.

'There is certainly no reason to laugh at M. Trissotin when he says to Madame Philaminte (*Femmes Savantes*, act iv., scene 3):—

> Nous l'avons en dormant, madame, échappé belle :
> Un monde près de nous a passé tout du long,
> Est chu tout au travers de notre tourbillon ;
> Et, s'il eût en chemin rencontré notre terre,
> Elle eût été brisée en morceaux comme verre.

'There is no reason whatever why a comet should not meet our globe in the parabola which it is describing; but what then would happen? Either the force of the comet would be equal to that of the earth or it would be greater or less. If equal, we should do it as much harm as it would do us, action and reaction being equal; if greater, it would take us along with it; if less, we should take it along with us.

'This great event can be managed in a thousand ways, and no one can affirm that the earth and the rest of the planets have not experienced more than one revolution from the embarrassment of a comet encountered in its way.'

Lalande's memoir was published in the course of the year 1773. It appeared moreover in the *Comptes rendus* of the Academy, and the prediction which had never been made was soon forgotten. 'The Parisians will not desert their city,' says Voltaire, in concluding his letter; 'they will sing their *chansons*, and the "Comet and the End of the World" will be played at the Opéra Comique.'

What, then, after all, was the purpose of Lalande's work? To find by calculation the distances of the nodes of sixty-one comets from the earth's orbit, as well as the distances of these comets from the ecliptic, when the comet's radius vector was equal to unity. By means of these elements it was possible to determine which among known comets could most nearly approach the earth, and, consequently, occasion or undergo the greatest perturbations. The table which he gave was perfected by a Swedish astronomer, Prosperin. The following extract contains the most curious of the results:—

Comets which have Approached Nearest to the Earth.

	Minimum distance between the comet and the earth's orbit		Date of arrival at the nearest points	
	In radii	In miles	Earth	Comet
Comet of 1472 . .	0·0434	3,980,000	January 19	January 22
" 1680 . .	0·0053	480,000	December 22	November 20
" 1684 . .	0·0092	850,000	June 18	June 29
" 1702 . .	0·0304	2,780,000	April 22	April 20
" 1718 . .	0·0449	4,110,000	January 27	January 10
" 1742 . .	0 0141	1,290,000	November 9	December 13
" 1760 . .	0·0536	4,910,000	January 16	December 31
" 1770 . .	0·0183	1,680,000	July 1	July 1

COMETS WHICH HAVE APPROACHED NEAREST TO EARTH.

Of these comets two have approached the orbit of the earth to a distance less than the hundredth part of the distance of the earth from the sun, viz. the comets of 1680 and 1684. The first passed within 500,000 miles, and the second within 850,000 miles of the earth's orbit; but both these comets were in reality much more distant from the earth, as it was not at the time of these passages at the nearest point of its orbit. This was not the case, however, as regards the comet of 1770, which passed on the same day as the earth through a point only 168,000 miles distant from the spot occupied by our globe five hours later.

SECTION II.

COMETS AND THE END OF THE WORLD.

Prediction of 1816; the end of the world announced for July 18—Article in the *Journal des Débats*—The comet of 1832; its rencontre with the orbit of the earth—Notice by Arago in the *Annuaire du Bureau des Longitudes*—Probability of a rencontre between a comet and the earth—The end of the world in 1857 and the comet of Charles V.

THE terrors of the year 1773 create a smile at the present day; but similar fears, it ought not to be forgotten, have been renewed from time to time in our own century; as, for example, in 1816, 1832, and 1857.

In 1816 a report was current of the approaching end of the world; July 18 was the date fixed for the fatal event. Some days after there appeared in the *Journal des Débats* a satirical article by Hoffmann, in which that critic ridiculed in the following manner the hypothesis * of the earth coming into collision with a comet:—

'A great mathematician (Laplace), to whom we owe the complete exposition of the system of the world, and whose work is law, has been kind enough to reassure us a little concerning the uncivil comets of Lalande; but that he is far from having banished all cause of alarm we may judge from the following passage, which I will literally transcribe: "The small

* The quotation is taken from a curious little work of M. Maurice Champion, *La fin du Monde et les comètes au point de vue historique et anecdotique.* Paris, 1859.

probability of such a rencontre may, by accumulating during a long series of years, become very great." Now, for many centuries no comet has come in contact with our globe. .... [Here follows an enumeration of the effects produced by the collision of a comet, which we shall reproduce further on from Laplace.] .... 'As it is a very long time,' continues Hoffmann, 'since this catastrophe has taken place, and as the possibility of the disaster increases with time, according to our great mathematician, it seems to me prudent to set about arranging our affairs; for, in three or four thousand years at the latest, we shall see a new representation of this great tragedy.'

In France wit, which in this case is the flower of good sense, never fails to assert its rights, as Bayle and Voltaire have already proved. Hoffmann in 1816 furnishes new testimony of the same truth; but not the less slow are superstitious ideas to relax their hold, and, assisted by the popular ignorance of astronomy, we see that the fear of comets, so vivid in the Middle Ages, continues to reappear from time to time amongst all classes of the people. There is, nevertheless, an essential difference between the superstitious beliefs of former times and the credulity of the present day; in former times every apparition of a comet passed for a kind of supernatural event, a warning from above, and the fatal consequences arising from the passage of the terrible visitor were so many decrees of Providence. In our day a comet is more especially feared from the widely-spread impression that a fortuitous meeting between a comet and the earth is a fact which may arise in the natural order of events. It may also happen, as in 1773, that a scientific announcement wrongly interpreted gives rise to chimerical fears, which find in ignorance and the remains of mystic beliefs elements favourable to their propagation. In proof of which we subjoin the two following examples.

The first of these is afforded by Biela's comet of six years and three-quarters period, owing to the astronomical prediction of its passage for the year 1832.

Olbers had just given the elements and the ephemeris of the comet discovered in 1826 by Biela, and calculated by Damoiseau to return in 1832. On October 29, before midnight, the new comet was expected to pass through its node —that isto say, to cut the plane of the ecliptic—at a point a little within the earth's orbit, the distance of the node

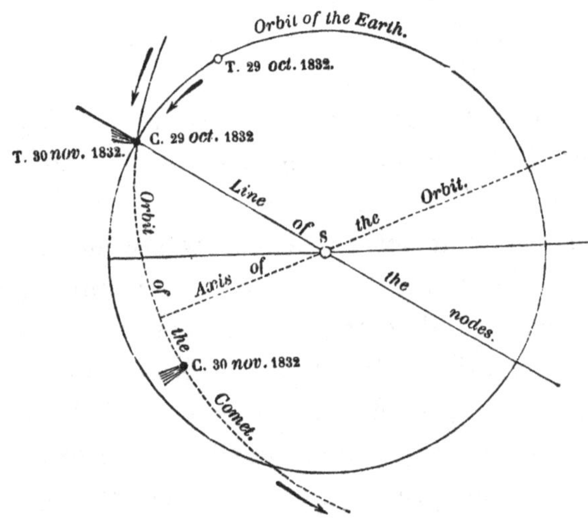

Fig. 71.—The Orbit of the Earth and that of Biela's Comet in 1832. Relative positions of the two bodies.

from the orbit, in fact, not exceeding 4·66 radii of our globe. Now, four radii and two-thirds are equivalent to rather less than 18,600 miles; so that however insignificant the dimensions of the nucleus and coma—according to the observations made by Olbers in 1805 the diameter of the atmosphere was equal to about 5 radii—it was evident that on October 29 the terrestrial orbit would intersect some portion of the atmosphere of the comet.

Nothing more was needed, these details having transpired,

to originate the report of an approaching rencontre between the comet and the earth. Our globe, violently struck, would be shattered to pieces; the end of the world was evidently at hand. One point alone had been forgotten by the alarmists, and Arago, who undertook to draw up for insertion in the *Annuaire du Bureau des Longitudes* a notice which should be calculated to allay the public uneasiness, alludes to it in the following terms. After concluding, as we have seen above, 'that on the 29th of October next a portion of the orbit of the

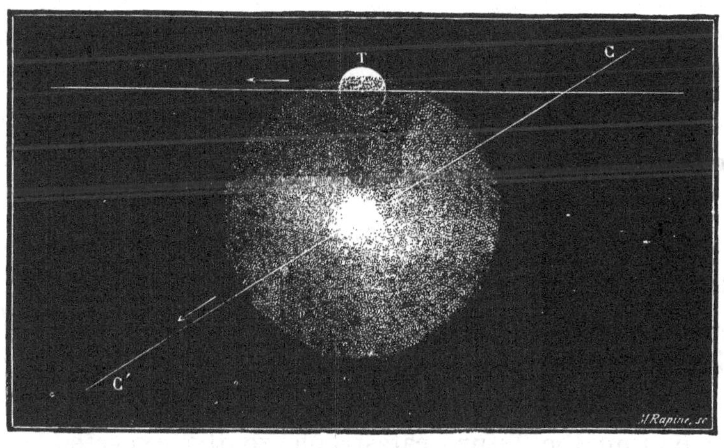

Fig. 72.—Biela's Comet at its node, October 29, 1832. Supposed position of the Comet at its least distance from the Earth.

earth will be comprised within the nebulosity of the comet,' the illustrious astronomer continues thus :—

'There remains but one more question, which is this: at the time when the comet is so near our orbit that its nebulosity will envelope some portion of it, *where will the earth itself be situated?*

'I have already said that the passage of the comet very near to *a certain point* of the terrestrial orbit will take place on the 29th of October, before midnight; the earth, however, only arrives at this point on the morning of the 30th of November, that is to say, *more than a month later.* We have, then, only to remember that the mean velocity of the earth in its orbit is 1,675,000 miles per day, and a very simple calculation will suffice to prove that the comet of six and three-quarter years period will be, at least during its apparition in 1832, *always more than* 49 *millions of miles from the earth!'*

Arago did well to reserve the question of future apparitions; for this same comet of 1832, forty years later, if it did not come into collision with the earth, must at least have grazed its surface.

Finally, before leaving these imaginary rencontres of comets and the end of the world, let us remark that the same chimerical fears were current in Europe in 1857, *à propos* of the predicted return of the comet of Charles V. This time the mystification came from Germany; according to a fantastic prediction the world was to be destroyed by fire, burnt by the terrible comet of June 13, in the year of grace 1857! The prediction at first related to the rencontre of an imaginary comet only; but afterwards the serious expectancy with which astronomers awaited the return of the comet of 1264 and 1556 suggested the idea of attributing the future catastrophe to the expected comet, although nothing in the elements of its orbit justified the probability of such a rencontre.*

In his notice upon the comet of 1832 Arago raises the following question: Can a comet come into collision with the earth or any other planet? By reasoning based upon the fact that the orbits of comets intersect the heavens in all directions, that they constantly traverse our solar system, and penetrate

* Perhaps it had been ascertained that the comets of 1264 and 1556, in Lalande's table, approached the earth's orbit to about 0·08 of the distance of the earth from the sun; and that the earth and the comet passed on the same day (March 12, 1556) through the points of their nearest mutual approach—at 7,200,000 miles, however, from each other.

to the interior of the orbits of the planets, even to the regions comprised between Mercury and the sun, he comes to the conclusion 'that it is not at all impossible for a comet to encounter the earth.' But, after having stated the possibility of the fact, he hastens to examine its probability. For this purpose he supposes a comet of which nothing more is known than that at its perihelion it will be nearer to the sun than we are ourselves, and that its diameter is equal to a fourth part of the earth. By the theory of probabilities Arago then finds that the odds are 281 millions to one against a rencontre with the earth. This probability, it is true, should be increased at least tenfold if, instead of our rencontre with the cometary nucleus, we were to substitute the entire nebulosity, the volume of which is comparatively much more considerable.

To illustrate the significance of the numerical results to which considerations of this kind lead, Arago continues: 'Let us for a moment suppose that if a comet were to strike the earth with its nucleus, it would annihilate the whole of the human race. The danger of death to which each individual, in that case, would be exposed from the apparition of an unknown comet would be exactly equal to the danger he would incur, supposing that in an urn containing a total number of 281 millions of balls there should be only one white ball, and that his condemnation to death were the inevitable consequence of this same white ball presenting itself at the first drawing. Anyone who will consent to use his reason, however attached to life he may be, will laugh at so trifling a danger; and yet the moment that a comet is announced, before it has been observed or its course determined, it is for each individual of our globe the white ball of the urn above-mentioned.'

These calculations are indisputably correct as regards their general application. But, when a particular comet is in question, whose elements are known, all considerations of probability are out of place. Arago with reason calls attention to this fact in reference to the comet of 1832, and its apparition in that year. With respect to ulterior apparitions it is different. There is always some uncertainty; the orbit of the comet and its elements may be modified by the planetary perturbations, and the return of the comet to its node may take place so that both the earth and the comet may arrive, if not at the same point together, at least sufficiently near one another to cause the contact of some of their parts. We have already said that this is what probably took place in 1872, as regards the comet of 1832, during the night of November 27 The meeting was absolutely inoffensive.

SECTION III.

MECHANICAL AND PHYSICAL EFFECTS OF A COLLISION WITH A COMET.

Opinions entertained by astronomers of the last century: Gregory, Maupertuis, Lambert—Calculations of Lalande; comets move too rapidly in the vicinity of the earth for the effects of their attraction to come into play—Opinion of Laplace—The collision of a comet with the earth; its effect according to the mechanical theory of heat.

It is interesting to note the opinions formed by savants a century ago respecting the probable effect of a collision between a comet and the earth. Further on we shall speak of the theological romance invented by Whiston for the scientific explanation of the Deluge. According to Whiston the famous comet of 1680, after having, 4000 years ago, produced the universal deluge, is destined to accomplish the destruction of the world, and our globe will be ultimately set on fire by the same comet which had previously inundated it.

Whiston wrote at the end of the seventeenth century. In the middle of the eighteenth century theological speculations engaged but very slightly the attention of astronomers; but that a very exaggerated idea continued to prevail respecting the amount of injury which the proximity of a comet or its collision with the earth would be capable of producing is undoubted.

In 1742 Maupertuis, in his *Lettre sur la Comète*, writes as follows: 'With their variety of movements it is clearly possible for a comet to encounter some planet or even our earth upon its way; and it cannot be doubted that terrible results would ensue. On the mere approach of these two bodies great changes would be effected in their movements, arising either from their mutual attraction or from the action of some fluid compressed between them. The least of these movements would suffice to change the position of the axis and the poles of the earth. That portion of the globe which had been previously near the equator would be situated, after such an event, near the poles, and that which had been near the poles would be situated near the equator.

'Some comets passing near the earth,' he observes elsewhere, 'might so alter its movement as to cause it to become a comet itself. Instead of pursuing its course, as it now does, in an uniform region, having a temperature suitable to man and the animals which inhabit it, the earth, exposed to the greatest vicissitudes, sometimes scorched at its perihelion, sometimes frozen by the cold of the remotest regions of the heavens, would be perpetually passing from one extreme to another, unless some comet should again come near enough to change its course and re-establish it in its original uniformity.'

Lastly, some large comet might, if Maupertuis is to be believed, divert the earth from its present orbit and subject it to its own attraction; in a word, make a satellite of our globe, which, henceforth compelled to follow the movements of the comet, would be exposed to the same vicissitudes as on the preceding hypothesis. 'A comet could in like manner deprive us of our moon, and if nothing more happened to us we should have no reason to complain. But the severest accident of all would be if a comet were to come in contact with the earth and break it into a thousand pieces. Both bodies would doubtless be destroyed; but from the fragments gravity would speedily form a new planet or several new planets.'

No one could be more accommodating. Maupertuis, it is plain, is of the same way of thinking as the mathematician so amusingly described in the *Lettres Persanes*, who regarded natural accidents of the most disastrous kind and the most formidable catastrophes simply as matters for calculation and opportunities for the furtherance of scientific knowledge.

So much for the mechanical effects due to the mass of a comet, which Maupertuis evidently regards as of an order of magnitude comparable to that of the earth. Now for the physical effects.

'The approach of a comet,' he observes, 'might be attended with results yet more fatal. I have not yet spoken to you of the tails of comets. These have been the subject of opinions not less curious than comets themselves. According to the most probable conjecture they are immense torrents of exhalations and vapours driven forth from the body of the comet by the heat of the sun. The strongest proof of this is that these appendages make their appearance only when the comet has approached to within a moderate distance of the sun; that they increase in proportion as the comet draws near to him, and diminish and become dispersed as it withdraws.

'A comet accompanied by a tail might pass so near the earth that we should find ourselves drowned in that torrent which it draws along with it, or in the atmosphere of the same nature which surrounds it. The comet of 1680, which approached near the sun, was subjected to a heat twenty-eight thousand times greater than that which the earth experiences in summer. Mr. Newton, from different experiments which he has made respecting the heat of bodies, having calculated the degree of heat which the comet would have acquired, finds that it must have exceeded two thousand times that of red-hot iron, and that a mass of red-hot iron as large as the earth would require 50,000 years to cool. What, then, could have been the heat which still remained in the comet when, coming from the sun, it crossed the orbit of the earth? Had it passed nearer, it would have reduced the earth to ashes or vitrified it; and if only its tail had reached us, the earth would have been inundated by a river of fire, and all its inhabitants killed—scalded to death like a colony of ants in boiling water.'

This, however, is the bad side of the rencontre. For, according to Maupertuis, who is afraid of having been too severe upon comets, they are also able to procure for us certain advantages; to raise, for instance, the axis of our globe, now too much inclined, and 'to maintain the seasons at a perpetual spring; to diminish the eccentricity of the orbit, and so cause a more equable distribution of light and heat. Lastly, instead of taking away our moon, it might itself be condemned to revolve about our earth, to illuminate our nights; to become to us, in short, a second moon. Who knows but that in former times we obtained in this manner possession of our own moon, originally, perhaps, some little comet, which, in consequence of having too nearly approached the earth, has been made captive by it?'

This opinion, which Pingré quotes, is, in his estimation, all the more probable, as it is based upon a tradition universally spread amongst the Arcadians. According to the testimony of Lucian and Ovid these people believed themselves to be more ancient than the moon. But the physical constitution of our satellite is altogether different from that of known comets, and Arago with reason calls attention to the fact that 'the almost total absence of atmosphere about the moon, far from being favourable, is rather adverse to the opinion which supposes the moon to be an ancient comet.'

Before leaving the speculations of Maupertuis let us quote the following passage from his letter, which shows the ideas then current respecting the physical and chemical constitution of comets, and the nature of the substances of which they are composed:—

'However dangerous might be the shock of a comet, the comet itself might, nevertheless, be so small as to be fatal only to that part of the earth with which it happened to come in contact, so that we might, perhaps, be compensated for the destruction of some kingdom by the enjoyment which the rest of the world would derive by the rarities which a body coming from so vast a distance could not fail to bring with it. The *débris* of those masses which we despise might prove, to the surprise of everyone, to be formed of gold and diamonds. But who would be the most astonished, ourselves or the inhabitants precipitated by the comet upon our earth? What should we think of each other?'

After all, the idea of Maupertuis is not so strange as might be thought. If no very small comet has yet fallen upon the earth, we frequently receive *débris* which have belonged to some celestial body; if neither gold nor diamonds fall upon the earth, it is certain that other minerals fall, iron and nickel, for example; but this is a matter that relates rather to shooting stars.

Gregory, a learned astronomer of the eighteenth century, observes, in his work *Astronomiæ physicæ et geometricæ elementa* (liv. v. prop. iv. coroll. 2): 'If the tail of a comet should extend as far as our atmosphere, or if a part of the sparse matter in the heavens of which that tail is composed should fall under the influence of our attraction, the exhalations of which it is formed, by mixing with the air that we respire, would occasion changes particularly sensible to animals and plants. Vapours, indeed, brought from distant and foreign regions, and excited by an intense heat, would be fatal, perhaps, to living beings upon the surface of the earth, and give rise to events similar to those which the testimony of all nations and universal consent concur in representing as a consequence of the apparition of comets, and which it does not become philosophers to assume too hastily to be ridiculous fables.'

Further on we shall return to examine the justice of these and similar views on the subject of cometary influences. Let us at the present moment confine ourselves to the opinions of astronomers concerning the probable effect of a mutual approach or rencontre.

Lambert, in his *Lettres Cosmologiques* (1765), thus expresses himself on the subject of these effects:—

'If comets,' he observes 'produce neither war, nor famine, nor mortality, nor the fall of empires, what are these evils in comparison with the catastrophes with which they menace the entire globe ? When we consider the movement of these bodies, and reflect upon the law of gravitation, it is not difficult to perceive that their near approach might be the occasion of the most dire catastrophes to our earth, cover it again with the waters of a deluge or cause it to be consumed by fire, crush it to atoms, or at least divert it from its orbit, carry away its moon, or worse still, carry away itself, and, bearing it off into the regions of Saturn, compel it to endure a winter of several centuries, which neither man nor animals would be capable of resisting.' This passage suffices to show that Lambert shares the ideas of Maupertuis respecting the probable influences of comets. But these events, though possible are not in his opinion probable, and he founds his opinion upon considerations based upon final causes, upon the order and harmony of the universe, the necessity for the preservation of those vast bodies, whose duration must needs be proportional to their masses.

He even considers 'that all these bodies have exactly the mass, the weight, the position, the direction and velocity requisite, to enable them to avoid dangerous collisions. A comet, for example, passing in the near vicinity of Jupiter, might be turned aside by that great planet, either to the right or left, with the express design of preventing one of these rencontres. This again is purely imaginary, and entirely opposed to the facts. The comet of Biela has been shattered; suns, like the stars of 1572, 1664, and 1866, have blazed forth suddenly, and have been extinguished almost as suddenly, under the eyes of observers. Revolutions and catastrophes happen in the physical world, which they seem to disturb for a moment as in the social world; so far from being departures from natural laws, science, which establishes their existence, has no other task than to show how they result from them.

The following calculation has been made by Lalande: 'If a comet * were five or six times nearer to us than the moon, that is to say, if it were passing at a distance of 13,000 leagues (40,000 miles) from the earth, it would suffice to raise the waters of the sea two thousand fathoms above the ordinary level, which would, in all probability, submerge the continents of the four quarters of the world.' To this calculation, Dionys du Séjour makes an objection, which destroys its value. In virtue of a principle, demonstrated by D'Alembert, he shows that 'if we suppose the globe entirely covered with water, to the depth of one league, the comet would occupy ten hours and fifty-two minutes in producing its effect, whatever that might be, upon the tides; this duration would depend neither upon the size, nor density, nor proximity of the comet, but solely upon the depth of the fluid. If the fluid were two leagues deep, the

[* The mass of the comet appears to be supposed equal to that of the earth. —Ed.]

elevation of the waters would continue for eight hours, twenty-five minutes, eleven seconds. Now, the action of the comet upon any one portion of the sea is far from being of so long a duration. Let the comet, at its perigee, be 13,000 leagues from the earth, one hour after it will be at least at a distance of 16,549 leagues, and will be vertically over a point of the earth, 23° 14′ distant from the place it occupied at the time of perigee. At the end of the second hour, the point to which it corresponds will have changed 27° 36′, and the distance of the comet from the earth will be 24,768 leagues. We have here chosen the most favourable case possible for the action of the comet. On another hypothesis, in one half hour only, the comet would be removed 32,569 leagues from the earth, and its corresponding point upon the earth would have changed 81° 27′ 30″.' All other possible hypotheses give a result intermediate to these two extremes. We may judge, from these considerations, whether comets have time to produce the great disturbances in the tides, which they would produce if they remained for a longer time vertically over the same point of the sea.

This objection has an important bearing upon the question now before us; for it does away entirely, so to speak, with the danger of a very near approach between a comet and the earth. This caused Laplace to say, thus confirming the conclusions of Dionys du Séjour: 'Comets pass with such rapidity in the vicinity of the earth, that the effects of their attraction need occasion no alarm. It is only by striking against the earth that they can be the cause of serious or fatal injury to our globe.'

The illustrious author of the *Mécanique Céleste* does not regard a collision as impossible, but he considers the probability of such an event as extremely slight. Although he points out that the cometary masses, as indicated by their inappreciable influence upon the planetary movements, are so small that in all probability only local disturbances would be produced; still he has depicted in the following terms the effects of a rencontre, on the supposition that the mass of the comet were comparable to that of the earth (this is the passage referred to in Hoffmann's article).

'The axis and the movement of rotation would be changed; the seas would abandon their ancient positions, in order to precipitate themselves towards the new equator; a great portion of the human race and the animals would be drowned in the universal deluge, or destroyed by the violent shock imparted to the terrestrial globe; entire species would be annihilated; all monuments of human industry overthrown; such are the disasters which the shock of a comet would produce, if its mass were comparable to that of the earth' (*Exposition du Système du Monde*). Laplace, moreover, does not seem far from believing that such a catastrophe has taken place, and the geological revolutions, the cataclysms which the ideas of Cuvier, then dominant, tended to refer back to a not very remote date, appear to him capable of being explained by such an event. 'We see then, in effect,' he continues, 'why the ocean has receded from the high mountains, upon which it has left incontestable marks of its sojourn. We see how the animals and plants of the south have been able to exist in the climate of the north, where their remains and imprints have been discovered; in short, it explains the newness of the moral world, certain monuments of which do not go further back than five thousand years. The human race reduced to a small number of individuals, and to the most deplorable state, solely occupied for a length of time with the care of its own preservation, must have lost entirely the remembrance of the sciences and the arts; and when the progress of civilisation made these wants felt anew, it was necessary to begin again, as if man had been newly placed upon the earth.'

Laplace, at the present day, would discard this explanation of the geological facts of past times, which modern science has since very differently interpreted. But it is interesting to know the opinion of the great mathematician with regard to the probable results of a collision between a comet and the earth ; an opinion differing very little from that of the savants of the eighteenth century, and consequently very far removed from that of certain contemporary astronomers, who, like Sir John Herschel and Babinet, regard comets as visible nonentities.

SECTION IV.

CONSEQUENCES OF A COLLISION BETWEEN A COMET AND THE EARTH ACCORDING TO THE MECHANICAL THEORY OF HEAT.

THE mathematicians and astronomers who have alluded to the effects of a collision between our earth and a comet, have more especially considered the event from a mechanical point of view; the two bodies were for them simply two projectiles which, both animated with enormous velocities, could not fail to encounter each other with a violence dependent upon their respective masses, velocities, and directions of motion. They foresaw only the dislocation or rupture of two gigantic masses, a catastrophe which would inevitably cause the destruction of the human race, and of all living beings upon the surface of the earth.

Some philosophers, believing a comet to be an incandescent mass, or at least to have become heated to an intense degree during its passage in the near vicinity of the sun, have conceived that it would inevitably set fire to our globe; in which case we should perish both by the shock and by fire.

But it was not then possible to view the phenomenon in its true light, since the great principle of the conversion of mechanical energy* into heat had not at that time been dis-

[*. The kinotic energy, or vis viva, of a body in motion is measured by the product of its mass and the square of its velocity. There are different forms of

covered. Let us therefore continue the same hypothesis of a comet of solid nucleus, of a mass comparable to that of our globe, and coming into collision with the earth from any direction whatever. The maximum effect, it is clear, would take place if the two bodies, travelling in opposite directions, were to encounter each other, so as to annihilate their respective motions, which could only happen in the event of the masses and velocity of the two bodies being equal. Under other conditions the result would vary as regards the amount of the effects produced, but not as regards their nature.

Now, the new principle, which has been established both by experiment and theory, is that even should all movement be apparently annihilated by the shock, it is in reality integrally preserved; only, it is transformed by molecular movement into heat.

The comet and the earth, by coming into collision with each other as we have just supposed, would be therefore stopped in their movements about the sun, and the sum of the energies of their motions, at the moment of collision, would be entirely converted into heat. Now, to show the enormous amount of heat that would be generated by the mere stoppage of the earth, we shall quote the following passage from Tyndall:—

'Knowing, as we do, the weight of the earth and the velocity with which it moves through space, a simple calculation enables us to determine the exact amount of heat that would be developed, supposing the earth to strike against a target strong enough to stop its motion. We could tell, for ex-

energy, of which heat is one; and the principle of the conservation of energy asserts that the total energy of a system of bodies cannot be increased or diminished by any mutual action between them, although it may be transformed into any other form of which energy is susceptible. Thus, when a moving body is brought to rest, the kinetic energy of its motion is transformed into some other form of energy, and is, in fact, expended in the generation of a definite quantity of heat.—Ed.]

ample, the number of degrees which this amount of heat would impart to a globe of water equal to the earth in size. Mayer and Helmholtz have made this calculation, and found that the quantity of heat which would be generated by this colossal shock would be quite sufficient, not only to fuse the entire earth, but to reduce it in great part to vapour.'

The catastrophe, as we thus see, would not be less enormous than was at first supposed. The two bodies, by their collision, would be converted into one mass, one part of whose elements would be in a state of fusion, whilst the rest would form an envelope of vapour. Nothing could more nearly resemble the idea we form at the present day respecting certain comets; but on our hypothesis all movement of translation would be annihilated. The new star, abandoned to the influence of the solar gravitation, would necessarily fall upon the surface of the sun, and the amount of heat generated by the blow, would be equal to that developed by the combustion of 5,600 globes of solid carbon, each having a volume double that of the earth.

Thus, even when considering the matter from the same point of view as the savants whose opinions have been quoted in the preceding section, we arrive at very different results. But it must be remembered that it is all but certain that comets have masses so very inferior to that of the earth, and that their physical constitution is so different, that a meeting would have little resemblance to a blow or shock such as would result from the collision of two solid globes.

SECTION V.

THE COMET OF 1680, THE DELUGE, AND THE END OF THE WORLD.

Ancient apparitions of the comet of 1680, on the hypothesis of a revolution of 575 years—Their coincidence with famous events—Whiston's theory of the earth: our globe is an ancient comet, whose movements and constitution have been modified by comets—The catastrophe of the Deluge caused by the eighth anterior apparition of the comet of 1680—Final catastrophe: burning of the earth—Future return of our globe to the condition of a comet.

THE comet of 1680 is one which has been made the subject of considerable discussion. If we adopt the calculations of Halley, confirmed in the first instance by Newton, it would be that which appeared in the years 531 and 1106 of our era, announced in the year B.C. 43, the death of Cæsar, presided at the taking of Troy, and eleven or twelve centuries earlier, was the direct cause of the deluge mentioned in the Mosaic records.

In 1106, the apparition of this celebrated comet did not, it is true, coincide with any great historical event; but, according to the chroniclers, it was of great brilliancy, 'resembling a flaming torch, covering with its rays a great portion of the heavens, and filling all minds with terror.' 575 years before, that is to say in 531, 'during twenty days was seen to the westward, a very large and fearful comet; it extended its rays, that is to say its tail, towards the most elevated portion of the heavens; on account of its resemblance to a burning lamp it received the name of Lampadias' (Theophanes, *Chron.*, quoted by Pingré). Other apparitions would have taken place in the year B.C. 619, that is, at the date of the destruction of Nineveh; and again in B.C. 1769, or, following Fréret, under the reign of Ogyges, who, according to the Greek legends, was contemporary with another deluge.

Whiston, an Englishman of the eighteenth century, a contemporary of Newton, and alike theologian and astronomer, published in 1696, 'A New Theory of the Earth,' in which he proposed to explain, by the action of a comet, the geological revolutions recorded in the Book of Genesis. His theory was at first entirely hypothetical, and not framed with reference to any comet in particular; but when Halley assigned to the famous comet of 1680 an elliptic orbit, with a period of 575 years, and Whiston identified the dates of two of its former apparitions with the years 2344 and 2919, that is to say, with two of the years fixed by chronologists as the date of the Mosaic deluge, he hesitated no longer; he developed his theory, and made the comet of 1680 appear, not only as the destroyer of the human race by flood, but also as the destroyer of the world by fire in future ages.*

Let us briefly give an idea of this curious theory, confining ourselves at present to that part of the hypothesis which concerns the deluge.

According to Whiston, the earth is an ancient comet whose perihelion was formerly very near the sun; the excessive temperature to which the comet was subjected at each of its perihelion passages explains the central, and continually main-

* The 28th of November, 2349, or 2348, according to the modern Hebrew text; December 2, 2926, according to the Samaritan text; the year 3308 according to others, would correspond to the exact chronological date of the Biblical deluge. Between 2349 and 2926 there are 577 years, only two years more than the period calculated by Halley for the comet of 1680.

tained, heat of the globe. When the earth was to be rendered habitable, one operation alone sufficed; the centrifugal force of the comet was diminished, and its orbit thus rendered less eccentric; this transformation effected, the eccentricity, nevertheless, was still sufficiently great to allow the hemisphere, destined to serve as the dwelling-place for man and animals, to enjoy the presence of the sun for a period of nine or ten months. Thanks to these changes, the thick atmosphere of the ancient comet became purified, and the air, the soil, and the water gradually found their equilibrium. The sun and the moon looked down upon the earth, and man and animals appeared.

When man had sinned, a small comet, passing very near the earth, cutting obliquely the plane of its orbit, impressed upon it a movement of rotation. No doubt we must attribute to this same comet the perfect circularity of the terrestrial orbit which, according to Whiston, was the case before the deluge. Pingré remarks, in citing Whiston, that: 'God had foreseen that man would sin, and that at length his crimes would demand a terrible punishment; consequently, he had prepared from the beginning of the creation a comet which he designed to make the instrument of his vengeance. This comet is that of 1680.' How was the catastrophe accomplished? Briefly as follows, according to Whiston.

On Friday, November 28, 2349, or on December 2, 2926, the comet was situated at its node, and cutting the plane of the earth's orbit at a point from which our globe was separated by a distance of only 3,614 leagues, of twenty-five to a degree. The conjunction took place at the hour of noon under the meridian of Pekin, where Noah, it appears, was dwelling before the flood. Now, what was the effect of this comet, to which Whiston assigns a mass equal to a quarter of the earth's mass? It caused a prodigious tide, not only in the waters of the sea, but also in those underneath the solid crust, disrupted by the movement of rotation and in the densest parts of the fluid portion of the terrestrial nucleus. The mountain chains of Armenia, the Gordian mountains, which were nearest to the comet at the moment of conjunction, were cleft and shaken to their foundations, and thus 'were all the fountains of the great deep broken up.' The disaster did not end here. The atmosphere and the tail of the comet coming in contact with the earth and its atmosphere, loaded the latter with aqueous and terreous particles, and 'thus were opened all the cataracts of heaven.' The depth of the waters of the deluge was, according to Whiston, six English miles, one mile of which was due to the eruption of the interior fluid, about five miles to the atmosphere or coma of the comet, and some little to its tail.

It was, then, as we see, a real inundation, an universal deluge which, according to this remarkable theory, was caused by the passage of the comet of 1680 to its node in the close vicinity of the position occupied by the earth, 4,223 years ago, according to some, 4,800 years according to others. The shock might perhaps have been sufficient to accomplish the work of destruction, but unquestionably a depth of six miles of water all over the earth was a more certain means of annihilation.

Now, how is this comet, which in the first instance drowned all the living beings upon the earth, to cause on its return the destruction of the earth's inhabitants by fire? Whiston is equal to the occasion. A second passage in the vicinity of the earth, but behind or to the west of it, will retard the movement of our globe, and change its nearly circular orbit into a very eccentric ellipse. The earth, at the time of its perihelion passage, will be situated in close proximity to the sun; it will experience an intense degree of heat, and enter into combustion.

But the comet may have a direct action as well; it may

meet the earth and strike against it. We know that the comet of 1680 approaches very near to the surface of the sun, so that, following Pingré's summary of Whiston's views, 'hardly can the mouth of a volcano vomiting forth lava liquified by the interior consuming heat give an idea of the fiery atmosphere of this comet. The air will then interpose no obstacle to the activity of the central fire; on the contrary, the inflamed particles with which our atmosphere will be charged, carried down by their own weight into the half-open bowels of the earth, will powerfully second the action of the central fire. This comet might even separate the moon from the earth, and affect the diurnal and annual motion of the earth by rendering both these movements equal, and by destroying besides the eccentricity of the terrestrial orbit, which would again become circular as before the deluge. Lastly, after the saints have reigned a thousand years upon the earth, itself regenerated by fire, and rendered habitable anew by the Divine will, a comet will again strike the earth, the terrestrial orbit will be excessively elongated, and the earth, once more a comet, will cease to be habitable.'

Such is the romance conceived by Whiston, a man of great erudition and science, but who shared the fault of his time in wishing to make his conceptions accord both with theology and astronomy. We are here only concerned with the scientific side of the question; and it is certain, and it was so a hundred years ago, that Whiston's theory is untenable. We will only notice two vital objections: in the first place, the enormous mass we are compelled to assign to the comet of 1680, and which no astronomer of our time would admit as probable; in the second place, even assuming such a mass, its action would be of so short duration, by reason, as we have seen, of the relative velocities of the comet and the earth, that the supposed effects would not have time to manifest themselves. But geologists, we believe, would have other objections to make to an hypothesis which we have only recorded because it is celebrated in science; and because the part it assigns to comets is truly curious.

A last and capital objection is this: The discussion of the elements of the comet of 1680 made by Encke with more accurate data than Halley possessed, has entirely overthrown the supposed chronological coincidences with its anterior apparitions. According to the new elements, the period of the comet is not 170 years (Euler), nor 575 years (Halley), nor 5,864 years (Pingré), but 8,814 years.

SECTION VI.

PASSAGE OF THE EARTH THROUGH THE TAIL OF A COMET IN 1861.

Possibility of our globe passing through the tail of a comet—Has such an event ever taken place?—The great comet of 1861—Relative positions of the earth and one of the two tails of that comet—Memoir of M. Liais and the observations of Mr. Hind.

THUS far, in treating of the possibility of a rencontre between a comet and the earth, we have more especially had in view the nucleus, or rather that portion of the comet's nebulosity which constitutes the coma. The effects of the rencontre have been studied on certain hypotheses respecting the mass and physical constitution of the comet whose nucleus we have supposed to be solid; this is far from certain, and, in any case, seems to be exceptional, as it is only in certain comets that the head is sufficiently condensed to exhibit a luminous nucleus.

A rencontre, of much greater probability, is that which would arise from the passage of the earth through the voluminous nebulosity of which the tail is formed. In all probability the masses of these appendages are all but inappreciable. Whatever opinion we may form of their nature, whether we regard them with Cardan and certain savants of our day as purely optical effects without material reality, or see in them the most tenuous portions of the atmosphere of the comet projected by a repulsive force, it appears certain that they consist of quantities of matter of extremely slight mass, and of even less density. It would be ridiculous to speak of a shock or any other mechanical effect; but it is not altogether evident that the matter of a comet might not produce some perceptible modification of the atmosphere of our globe.

Before considering what would happen in the event of the earth passing through the tail of a comet, we are naturally led to inquire if such an event has ever actually occurred. Now, according to several contemporary astronomers, the earth was, in fact, on June 30, 1861, plunged for some time in the nebulosity forming the large tail of the great comet of that year. M. Valz, in giving the elements of the comet, has observed: 'It follows that the comet having passed its node on June 28, at 9.50 p.m., at the distance of 0·132 from the orbit of the earth, the latter being less than 2° in advance of the node, must have been situated within the nebulosity of the tail, which was itself in the plane of the ecliptic. M. Loewy, in the *Bulletin de l'Observatoire* of July 12, likewise observes: 'It is probable that about June 28, the earth touched the tail of the comet.' M. Pape, of Berlin, was of a different opinion, his calculations leading him to conclude that an interval of more than two millions of miles separated the tail of the comet from the earth; but, according to M. Valz, this arises from the German astronomer having estimated the apparent breadth of the tail at 3°, whereas he himself estimated it at 6°, and Father Secchi at as much as 8°. M. Le Verrier, in giving the elements calculated by M. Loewy and Mr. Hind, adds the following remark: 'Did the earth pass through the tail of the comet? This question, apparently so simple, is, in reality, very complex. The calculations are complicated, and the data fail to determine this point with certainty.'

That the earth did pass through the tail of the comet was

the opinion of Mr. Hind from the very first. The following is an extract from the letter written on this subject by the English astronomer to the editor of the *Times*:—

'Allow me to draw attention to a circumstance relating to the present comet, which escaped my notice when I sent a communication on the 3rd instant, but which is now possessing some interest. It appears not only possible, but even probable, that in the course of Sunday last the earth passed through the tail, at a distance of, perhaps, two-thirds of its length from the nucleus.

'The head of the comet was in the ecliptic at 6 p.m. on June 28, distant from the earth's orbit 13,600,000 miles on the inside, its longitude, as seen from the sun, being 279° 1′. The earth, at this moment, was 2° 4′ behind that point, but would arrive there soon after 10 p.m. on Sunday last. The tail of a comet is seldom an exact prolongation of the radius vector, or line joining the nucleus with the sun ; towards the extremity it is almost invariably curved ; or, in other words, the matter composing it lags behind what would be its situation if it travelled with the same velocity as the nucleus. Judging from the amount of curvature on the 30th, and the direction of the comet's motion as indicated by my orbit already published, I think the earth would very probably encounter the tail in the early part of that day, or, at any rate, it was certainly in a region that had been swept over by the cometary matter shortly before.'

M. Liais, who observed the same comet in Brazil, speaks with still more certainty. He bases his assertions upon observations of his own made before and after the perihelion passage, upon the breadth and direction of the tail of the comet, as well as upon the elements of the orbit calculated by M. Seeling. We shall not enter into the details of the calculation and the discussion given by this savant in *l'Espace Céleste*, but content ourselves with stating the results. According to him, not only the earth, but the moon also, entered the tail of the comet on the morning of June 30, and at 6.12 p.m. on

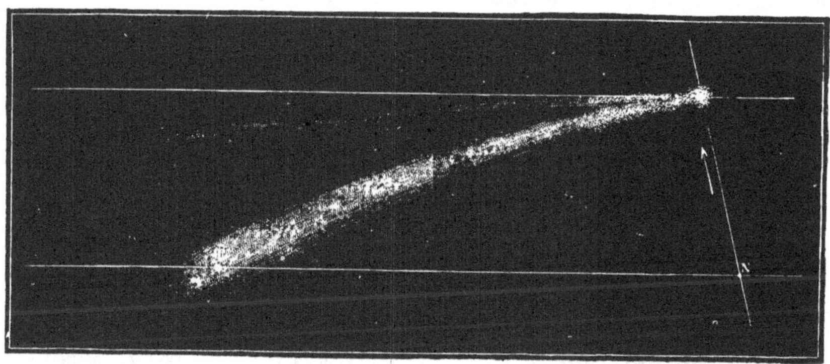

Fig. 73.—Passage of the Earth through the tail of the comet of 1861, on June 30.

that day our globe was plunged in it to a depth of 273,000 miles. Figures 73 and 74 give, the first, the position of the

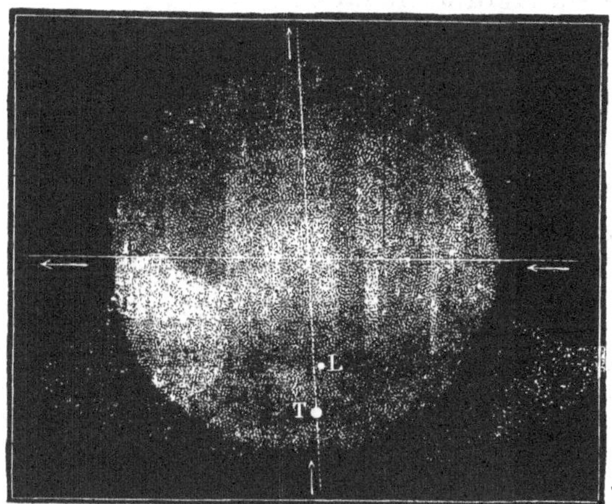

Fig. 74.—Positions occupied by the Earth and the Moon in the interior of the second tail of the comet of 1861.

comet in the plane of its orbit, at the moment of the passage of the axis of the second tail across the terrestrial orbit; the second, a section of the tail perpendicularly to its axis. In the latter are shown the positions occupied by the earth and its satellite in the midst of the nebulous appendage.

Now, assuming as a positive fact the passage of our planet through the tail of the comet of 1861, were any special phenomena observed which could be attributed to this singular rencontre? The reply to this question is probably to be found in the concluding observations of Mr. Hind's letter: 'I may add,' he observes, 'that on Sunday evening, while the comet was so conspicuous in the northern heavens, there was a peculiar phosphorescence or illumination of the sky, which I attributed at the time to an auroral glare; it was remarked by other persons as something unusual, and considering how near we must have been on that evening to the tail of the comet, it may, perhaps, be a point worthy of investigation, whether such effect can be attributed to our proximity thereto. If a similar illumination of the heavens has been remarked generally on the earth's surface, it will be a significant fact.'

The following note, to a similar effect, appears in the journal of another English savant, Mr. E. J. Lowe, of Highfield House, near Nottingham: 'June 30: A singular yellow phosphorescent glare, very like diffused aurora borealis, yet being daylight, such aurora would scarcely be noticeable.'* This is evidently the phenomenon described by Mr. Hind; but as both the observations were made in the same country, it may refer to a merely local appearance.

The fan-like figure presented by the tail of this comet during the night common to June 30 and July 1, that is to say, at the precise moment when the passage was taking place is connected, according to M. Liais, with this fact; but in our opinion the divergence of the rays of the tail might be ex-

[*Mr. Lowe's letter, containing this extract from his journal, was published in the *Times*, July 9, 1861. Mr. Hind's letter appeared in the *Times* of July 6. —Ed.]

plained as a simple effect of perspective. The comet projecting its tail towards the earth, it is evident that the form of the appendage, starting from the nucleus, would appear to enlarge considerably, even if its real form were cylindrical and not conical. On this subject M. Liais writes: 'Those divergent rays, which lasted so short a time, and were not distinguishable quite up to the nucleus, might they not be regions of the circumference rendered visible under the influence of electric

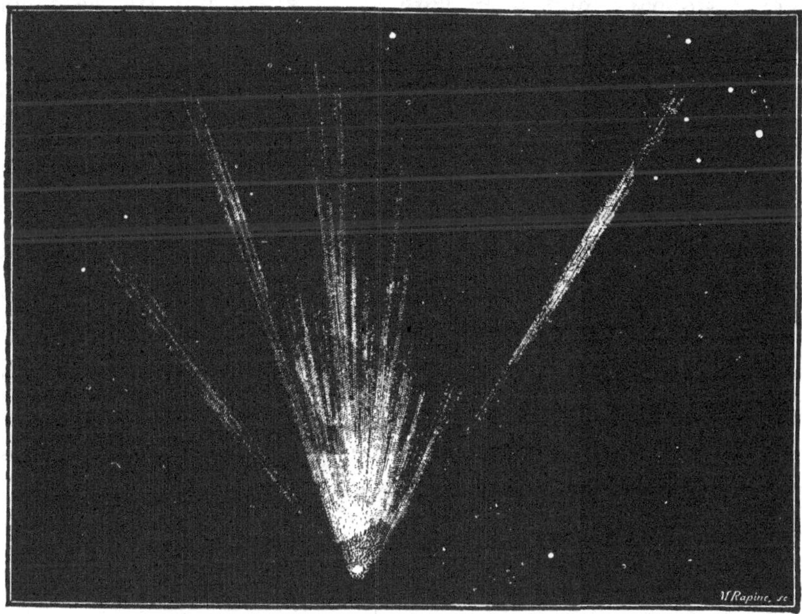

Fig. 75.—Fan-shaped tail of the great comet of 1861 on June 30.

light on leaving the direction of the earth? To gleams of electric light shining between the tenuous regions of the tail, and the limit of our atmosphere, we might with probability attribute the phosphorescence of the heavens seen on the same evening by Mr. Hind and other English observers.'

Babinet, in one of his piquant scientific notices, relates, apropos of the great comet of 1861, the following conversation: 'Monsieur, the newspapers inform us that we have a comet.'

'Yes, Madame, a very beautiful comet; the history of astronomy has never recorded one more beautiful.' 'What does it predict?' 'Nothing at all, Madame.' 'Is it a fine evening?' 'Yes, Madame, splendid, and you have only to go into the garden and you will see it.' 'Oh! if it can do one neither good nor harm it is not worth while.' The lady retires to bed. You will say to me, 'Of what use is astronomy?' 'It is of use,' I reply, 'inasmuch as we are enabled to go to bed without fear in 1861, even when a superb comet is in sight. This was not the case six hundred years ago, or even three hundred.'

Astronomy, as we have seen, has not yet produced this effect upon every one. But although the earth may have passed through the tail of a comet without its inhabitants, one or two excepted, being even conscious of it, still, were our globe to penetrate to the nucleus of one of these bodies, the event might not be so harmless. This is a distinction which M. Babinet, who clings to his theory of visible nonentities, refuses to make. If, however, the mass of a comet were so small that its action was imperceptible, it would still remain to inquire if the introduction of cometary matter into the atmosphere of the earth might not be injurious to living beings.

CHAPTER XIV.

PHYSICAL INFLUENCES OF COMETS.

SECTION I.

SUPPOSED PHYSICAL INFLUENCES OF COMETS.

The great comet of 1811; the comet wine—Prejudices and conjectures—Remarkable comets and telescopic comets—Comets are continually traversing the heavens.

In former times when a new comet was seen to project upon the sky its vaporous star and plume of light, the first question in the mouth of everyone was, What great calamity does God announce?

Even at the present day people may be heard enquiring what the comet signifies; but the greater number of enquirers are far more occupied with the physical effects likely to accrue, than with the supernatural import of the apparition. Do you think we shall have a warm and dry summer? is the question of some. Are we to anticipate foggy weather, heavy rains and inundations? ask others. It announces an abundant harvest, or a superior quality of the year's wine, is gladly remarked by those who have-not forgotten the comet and the good wine of the year 1811.

In a word, people readily believe that the passage of a comet within sight of the earth must be followed by certain consequences of a nature to influence not only the climate, temperature, and vegetation of the latter, but likewise the health of animals and man, for I have forgotten to say that the influence of comets upon the production of epidemics and other maladies was formerly an article of popular belief. To give an idea of the prejudices entertained upon this subject not more than sixty years ago, we will quote from Arago the following passage taken from the *Gentleman's Magazine* :—

'Through the influence of the comet of 1811, the winter following was very mild; the spring was wet, the summer cool, and very little appearance of the sun to ripen the produce of the earth; yet the harvest was not deficient; and some fruits not only abundant, but deliciously ripe, such as figs, melons, and wall fruit. Very few wasps appeared, and the flies became blind and disappeared early in the season. ... But what is very remarkable, in the metropolis and about it, was the number of females who produced twins, some had more, and a shoemaker's wife in Whitechapel produced four at one birth...'* This shows certainly an extravagant imagination.

In these entirely conjectural suppositions, especially in those which are advanced in the form of questions, is there any base of truth which the astronomical science of the present day might seem at all to confirm? In the majority of cases there is every reason to believe that these assumed influences amount to nothing; but then probabilities are not certainties, and a case might arise in which the apparition of a comet could be reasonably suspected of having been concerned in the production of certain terrestrial phenomena, such as, for example, meteorological phenomena.

[* This is an extract from a letter which appeared in the *Gentleman's Magazine* for November, 1813 (p. 432), and is signed J. B. It has had the distinction of being quoted by Arago and by M. Guillemin; but it should be remembered that it was merely an anonymous letter, published in an unscientific periodical. It seems to me to be characteristic of a class of letters which all who are associated with astronomy frequently receive from unscientific people, rather than representative of the prejudices prevalent at the time it was written There are always persons who write letters of this kind, and sometimes, of course, they find their way into print.—Ed.]

Let us examine the principal influences enumerated, and see if they are confirmed by facts; and, if not, whether there is reason to admit, of course with necessary reservations, a certain amount of probability.

We have already spoken of the influences which comets must exercise in virtue of their mass. These are indisputable; but, up to the present time, as we have seen, the comets of which history has made mention, and which might have been expected to produce a disturbing influence, have produced absolutely no appreciable effect. A comet passing within a short distance of the earth would in reality act upon the waters of the sea, and upon the atmosphere, for so short a time that the wave produced would be insignificant.

But may there not be comets yet unknown of masses more considerable? May there not be comets which might pass sufficiently near the earth, and remain long enough in its vicinity for their masses to occasion an appreciable disturbance? A shock or a rencontre is improbable, but possible, and, as we have seen, the consequences that would result from it are mere conjectures. As for the second question, the different velocities alone of the earth, and of any comet which might be situated for a moment in its vicinity, would, as we have already said, rapidly separate the two bodies. But this is not the kind of influence we have here to examine.

In the first place we must remember that comets are more numerous than might be supposed, that new comets make their appearance every year, that there are often several in the course of the year, and that if the influence with which they are credited belonged to them simply as comets, it would be, so to speak, continuous. This is no reason for considering it to be *nil*, but it is clear that it would be a very difficult matter to distinguish it from all other causes, regular or irregular. Those who have admitted, *à priori* so to speak, the existence of such an influence, have scarcely attempted the task of its verification.

In former times it was to comets visible to the naked eye—no others were then known—that fatal influences were alone attributed. And at the present day it is the larger and more magnificent comets, those which 'make a show' in the sky, that are supposed to exert an influence upon our globe. Nor is this surprising, since the greater or less visibility of a comet is a measure of its brilliancy and size, or, what comes nearly to the same thing, of its proximity to the earth.

SECTION II.

DO COMETS EXERCISE ANY INFLUENCE UPON THE SEASONS?

Study of the question by Arago—The calorific action of comets upon the earth appears to be inappreciable—Comparison of the meteorological statistics of various years in which comets did and did not appear—The meteorological influence of a comet is not yet proved by any authentic fact.

We have already said how general a consternation was created in 1832 by the announcement that Biela's comet would pass within a very short distance of the orbit of the earth. Arago made it the occasion of one of those brilliant and interesting notices in which he endeavoured to destroy existing prejudices, and to render the simple truths of astronomy better and more generally understood. The heading of one section of this notice was—

'*Will the future Comet modify in any appreciable degree the Course of the Seasons of the year* 1832*?*'

To this question he replies in the following terms:—

'The above title will doubtless call to mind the beautiful comet of 1811, the high temperature of that year, the abundant harvest following, and, above all, the excellent quality of the *comet wine.* I am therefore well aware that I shall have to contend with many prejudices in order to establish that neither the comet of 1811, nor any other known comet, has ever occasioned the smallest change in the seasons. This opinion is founded upon a careful examination and attentive discussion of all the elements of the problem, whilst the opposite idea, however widely spread it may be, has no foundation whatever in fact.

'It is said that comets heat our globe by their presence. Be this as it may, nothing is easier to verify. Is not the thermometer consulted many times a day in all the observatories throughout Europe? Is not an exact record kept of all the comets which appear?'

Thereupon Arago proceeds to tabulate the mean temperatures of the years between 1803 and 1831, at the same time placing by the side of them the numbers of comets observed, together with any peculiarities exhibited by them which could exercise an influence upon temperature. He has since extended this instructive table from 1735 to 1853, and proves without difficulty that no law connects the variations of mean temperature with the apparition of comets, and that years fruitful in apparitions, such as those of 1808, 1819, 1846, for example, have been marked by temperatures lower, or hardly equal to those of years in which few or no comets have been seen.

The whole of the sixty-nine comet years give a mean temperature of 51°.46 Fahr.; twenty-seven years without comets give a mean of 50°.94 Fahr. The difference of the half degree Fahrenheit, Arago explains by the fact that years without comets are most frequently cloudy; the prevalence of cloud simply concealing the comet or comets from observation. This difference becomes almost inappreciable when he compares the mean temperatures of the thirty years, in each of which only one comet appeared, and the thirty-nine years, in each of which two or several comets appeared. The difference in this case is no more than four-hundredths of a degree Fahrenheit, a quantity absolutely insignificant.

Other tables, founded upon analogous data, further establish 'that very low temperatures have frequently taken place during the apparition of comets, and very high temperatures at epochs when none of these bodies have been visible.'

Returning, then, to the comet of 1811, Arago considers how far it was possible for the brilliant train of that body to exercise an influence upon our globe? It was 102,000,000 of miles in length, it is true; but then it was not strictly directed towards the earth, and the comet at its least distance from our globe was separated by 117,000,000 of miles. Moreover, we are now assured of the extreme tenuity of these cometary appendages, and the insignificant amount of heat which they have it in their power to communicate, either at a distance by means of reflexion, or by contact. But the result might be different in the event of contact with the nucleus if, as is probably the case, the matter of which the nucleus is composed should have become heated, in the neighbourhood of the sun, to so high a temperature as to cause its partial incandescence.

Arago's demonstration did not succeed in convincing every one, for after the apparition of Halley's famous comet, the mild temperatures of the months of October and November were ascribed by many persons to the passage of the comet. 'People wish,' he observes, 'to attribute the mild temperature enjoyed by the north of France during these eight weeks to the influence of the comet! I could,' he continues, 'instance on the one hand Octobers and Novembers still milder than those of 1835, when no comets were visible, and on the other I could find instances of great cold being experienced during the same months, when brilliant comets were in sight; but to come more directly to the point, I will remark that at the end of 1835, when Paris was enjoying a very mild temperature, it was especially cold in the south, so that if the temperature were dependent upon the comet, its action would have to vary with the position of the place.'

And further, in order to judge the question fairly by this

method, that is to say, by the comparison of meteorological statistics, it is clear that we must not be content with observations relating only to one region of the earth. In order to form an impartial judgment, we must decide whether the presence or proximity of the comet corresponds to an increase of temperature over the whole of the terrestrial globe, or at least over all that portion of the globe which occupies the same relative position with regard to the comet.

The comet now in sight [July 1874] is observed by the public at the hottest time of the year, and it is probable that, without seeking further for a cause, many people attribute to the comet the high temperature from which they suffer. This present year may be in France and even throughout Europe a warm year. But is it so too for the same latitudes in America? Coggia's comet is the third of the year 1874; but we must not forget that in 1873 no less than seven comets passed their perihelia.

In conclusion, we may say that the influence of a comet upon the temperature and the seasons is generally imperceptible. It could only become sensible on the hypothesis of a collision, or a very near approach between the earth and a comet. Finally, up to the present time we have no authentic instance of such an influence. Mere opinions which are not justified by examination of the facts are but valueless hypotheses.

SECTION III.

PENETRATION OF COMETARY MATTER INTO THE TERRESTRIAL ATMOSPHERE.

Is this penetration physically possible?—Cometary influences, according to Dr. Forster—Were the dry fogs of 1783, 1831, and 1834, due to the tails of comets?—Volcanic phenomena and burning turf-beds; their probable coincidence with fogs—Probable hypothesis of Franklin—Dry fogs, atmospheric dust, and bolides.

WE perceive, then, that the influence of comets upon living beings by the action of heat is a hypothesis which, for the present, must be abandoned; in so far, at least, as the action of heat by radiation from a distance is concerned. We have throughout reserved the questions of a collision between the two bodies, and of the penetration of the earth to the heart of a mass in a state of incandescence.

Apart from the action of calorific radiation, what influence of any other kind could a comet exercise upon the meterological conditions of the earth? We know of absolutely none.

It remains, then, to consider the immediate physical or chemical influence of the cometary substance. It is not forbidden to our globe, as we have seen, to traverse the gigantic trains which form the tails of certain comets, nor to penetrate to a certain depth the vaporous atmosphere of some amongst them. Apart from these rencontres, we may suppose that cometary matter may be introduced into our atmosphere by the power of attraction. Pursuing its course in the same regions as the planets, projecting its substance far beyond its own sphere of attraction, a comet can scarcely fail to abandon fragments of its tail, which the mass of the earth, for example, may afterwards appropriate.

These fragments, it is true, by the common consent of all astronomers, are but trifling, materially speaking, and their total weight is only a very insignificant fraction of that of our atmosphere; but might not the continued introduction of these particles into the air we breathe become in the course of time a source of sickness and death to the living beings inhaling them? Might not certain kinds of epidemics be thus explained? This is a question which we scarcely have the means of answering. If the tails of comets are formed of matter so attenuated, so little coherent, it is reasonable to suppose that they may be attracted to the earth and become an integral part of it. But how are we to suppose that they descend into the depths of this envelope? At the utmost they could only float at the extreme limit of the atmosphere, and the supposed gas of which they are composed would not in any way mix with the gases of the air which human beings and animals respire.* Suppose these gaseous particles are endowed with a peculiar chemical activity, and that their contact with oxygen or nitrogen determines the formation of dense and poisonous precipitates: even then these particles of a matter so prodigiously dilated in the beginning, would contribute when condensed but an infinitesimal quantity to the air we breathe, and, unless we have faith in the homœopathic doctrine, need inspire us with no alarm.

Appeal has been made to the facts. As writers who believed in the supernatural and providential influence of comets have

[* If we suppose the tail of the comet to consist of a gas, it would mix with the other gases of the atmosphere in accordance with the known law of diffusion. If a light gas be placed upon a heavy gas, the latter will not remain floating as it were upon the former, but after a time the two will become completely mixed. —ED.]

collected all historical details attending upon each apparition which might seem to bear testimony in favour of their superstition, so have the advocates of a connexion between epidemics and comets collected a number of supposed accordances. Arago quotes from Dr. T. Forster, who had expended a large amount of erudition in forming a catalogue of so-called cometary influences.

'Mr. Forster has,' he observes, 'so extended the circle of supposed cometary actions that there is scarcely a phenomenon in nature that may not be ascribed to cometary influence. Cold and warm seasons, earthquakes, volcanic eruptions, great hail-storms, abundant snows, heavy rains, floods, droughts, famines, thick clouds of flies or locusts, the plague, dysentery, epizootic diseases, all are recorded by Mr. Forster with reference to each cometary apparition, no matter what the continent, kingdom, town, or village subjected to the ravages of the plague, famine, &c.' For example, the date of the comet of 1668 corresponds to the remark that all the cats in Westphalia were sick; that of 1746 to the earthquake in Peru which destroyed Lima and Callao ; other dates, again, correspond to the fall of an aerolite, to the passage of numerous flocks of pigeons, &c. This is truly an absurd enumeration. It recalls to mind Bayle's letter, and his parallel of the lady and the carriages in the Rue St. Honoré. But the whole is too puerile to need refutation.

Of meteorological phenomena attributed to comets because their causes have remained unknown, mention must be made of *dry fogs*, such as those of 1783, 1822, 1831, and 1834.

The appearance of this singular phenomenon, and the circumstances which, in 1783 more particularly, accompanied its long duration (it was visible more than a month), explain to a certain extent this hypothesis. Hygrometrically, this fog had not the qualities of an ordinary fog: it was not wetting.

De Saussure's hygrometer marked only 57°; the general colour of the air was that of a dull, dirty blue; distant objects were blue, or surrounded by mist, and at the distance of a league were undistinguishable. The sun, red, without brilliancy, and obscured by mist, both at his rising and setting, could be steadfastly regarded at noonday. A singular circumstance mentioned by Arago is that the dry fog of 1783 appeared to possess a certain phosphoric property, a light of its own. 'I find at least in the accounts of some observers,' he remarks, 'that it diffused, even at midnight, a light which they compare to that of the moon at its full, and which sufficed to make objects distinctly visible at a distance of more than 200 yards.'

Was the earth plunged in the tail of a comet, or had it met with the fragment of a cometary appendage abandoned in space? But why, then, was not the comet itself visible? Meteorologists (Kämtz) still continue to rank dry fogs amongst problematical phenomena; nevertheless, it was remarked that in 1783, at the two extremities of Europe, violent physical commotions took place; continued earthquakes in Calabria, and a volcanic eruption in Iceland. Could the dust and ashes projected to a distance and scattered far and wide have been the cause of the phenomenon?

Dry fogs are common in Holland, and also in the west and north of Germany. Finke tells us that they are due to the smoke produced by the combustion of the turf-beds. In 1834 the drought did in fact cause numerous fires in the forests and turf-beds of Prussia, Silesia, Sweden, and Russia.

Franklin assumed, in order to explain the dry fog of 1783, the diffusion of volcanic cinders and emanations. He likewise supposes—and this hypothesis is closely allied to that of the earth's immersion in the train of a comet—that an immense bolide might penetrate into our atmosphere, be there imperfectly consumed, and diffuse torrents of smoke or light ashes. We shall presently see that certain rains of dust can be explained in a similar manner. Many savants admit as a very probable fact that matter of extra-terrestrial origin may penetrate into the atmosphere and fall to the ground, and perhaps modify the constitution of the gaseous envelope in which we live.

SECTION IV.

CHEMICAL INFLUENCES OF COMETS.

Introduction of poisonous vapours into the terrestrial atmosphere—The end of the world and the imaginary comet of Edgar Poe; *Conversation of Eiros and Charmion* —Poetry and Science; impossibilities and contradictions.

We now come to that other cometary influence which we have already alluded to, an influence capable of changing the air we breathe by the introduction of foreign effluvia.

Nothing within the range of fact and observation, up to the present time, affords ground for belief in such an influence. But this hypothesis has had the fortune to be presented in a striking and practical form by a modern writer of powerful imagination. The American poet Edgar Poe, whose *Extraordinary Histories* are known to everyone, has placed in the mouth of a being who has suffered death, an account of the destruction of the world by the near approach of a comet. We subjoin the principal portion of this wonderful dream, in which Eiros relates to Charmion the circumstances which put an end to the world.

'The individual calamity was, as you say, entirely unanticipated, but analogous misfortunes had been long a subject of discussion with astronomers. I need scarce tell you, my friend, that, even when you left us, men had agreed to understand those passages in the most holy writings which speak of the final destruction of all things by fire as having reference to the orb of the earth alone. But in regard to the immediate agency of the ruin, speculation had been at fault from that epoch in astronomical knowledge in which the comets were divested of the terrors of flame. The very moderate density of these bodies had been well established. They had been observed to pass among the satellites of Jupiter without bringing about any sensible alteration either in the masses or in the orbits of these secondary planets.

'We had long regarded the wanderers as vapoury creations of inconceivable tenuity and as altogether incapable of doing injury to our substantial globe, even in the event of contact. But contact was not in any degree dreaded, for the elements of all the comets were accurately known. That among *them* we should look for the agency of the threatened fiery destruction had been for many years an inadmissible idea. But wonders and wild fancies had been, of late days, strangely rife among mankind; and, although it was only with a few of the ignorant that actual apprehension prevailed upon the announcement by astronomers of a *new* comet, yet this announcement was generally received with I know not what of agitation and mistrust.

'The elements of the strange orb were immediately calculated, and it was at once conceded by all observers that its path, at perihelion, would bring it into very close proximity with the earth. There were two or three astronomers of secondary note, who resolutely maintained that a contact was inevitable. I cannot very well express to you the effect of this intelligence upon the people. For a few short days they would not believe an assertion which their intellect, so long employed among worldly considerations, could not in any manner grasp. But the truth of a vitally important fact soon made its way into the understanding of even the most stolid. Finally, all men saw that astronomical knowledge lied not, and they awaited the comet. Its approach was not at first seemingly rapid ; nor was its appearance of very unusual character. It was of a dull red, and had little perceptible train. For seven or eight days we saw no material increase in its apparent diameter, and but a partial alteration in its colour. Meantime, the ordinary affairs of men were discarded, and all interest absorbed in a growing discussion, instituted by the philosophic, in respect to the cometary nature. Even the grossly ignorant aroused their sluggish capacities to such considerations. The learned *now* gave their intellect—their soul—to no such points as the allaying of fear, or to the sustenance of loved theory. They sought—they panted for right views. They groaned for perfected knowledge. *Truth* arose in the purity of her strength and exceeding majesty, and the wise bowed down and adored.

'That material injury to our globe or to its inhabitants would result from the apprehended contact, was an opinion which hourly lost ground among the wise, and the wise were now fully permitted to rule the reason and the fancy of the crowd. It was demonstrated that the density of the comet's nucleus was far less than that of our rarest gas ; and the harmless passage of a similar visitor among the satellites of Jupiter was a point strongly insisted upon, and which served greatly to allay terror. Theologists, with an earnestness fear-enkindled, dwelt upon the Biblical prophecies, and expounded them to the people with a directness and simplicity of which no previous instance had been known. That the final destruction of the earth must be brought about by the agency of fire, was urged with a spirit that enforced everywhere conviction; and that the comets were of no fiery nature (as all men now knew) was a truth which relieved all, in a great measure, from the apprehension of the great calamity foretold. It is noticeable that the popular prejudices and vulgar errors in regard to pestilence and wars—errors which were wont to prevail upon every appearance of a comet—were now altogether unknown. As if by some sudden convulsive exertion reason had at once hurled superstition from her throne. The feeblest intellect had derived vigour from excessive interest. What minor evils might arise from the contact were points of elaborate question. The learned spoke of slight geological disturbances, of probable alterations in climate, and consequently in vegetation; of possible magnetic and electric influences. Many held that no visible or perceptible effect would in any manner be produced. While such discussions were going on, the subject gradually approached, growing larger in apparent diameter, and of a more brilliant lustre. Mankind grew pale as it came. All human operations were suspended.

'There was an epoch in the course of the general sentiment when the comet had attained, at length, a size surpassing that of any previously recorded visitation. The people now, dismissing any lingering hope that the astronomers were wrong, experienced all the certainty of evil. The chimerical aspect of their terror was gone. The hearts of the stoutest of our race beat violently within their bosoms. A very few days sufficed, however, to merge even such feelings in sentiments more unendurable. We could no longer apply to the strange orb any *accustomed* thoughts. Its *historical* attributes had disappeared. It oppressed us with an hideous *novelty* of emotion. We saw it not as an astronomical phenomenon in the heavens, but as an incubus upon our hearts, and a shadow upon our brains. It had taken, with inconceivable rapidity, the character of a gigantic mantle of rare flame, extending from horizon to horizon.

'Yet a day, and men breathed with greater freedom. It was clear that we were already within the influence of the comet; yet we lived. We even felt an unusual elasticity of frame and vivacity of mind. The exceeding tenuity of the object of our dread was apparent, for all heavenly objects were plainly visible through it. Meantime, our vegetation had perceptibly altered ; and we gained faith, from this predicted circumstance, in the foresight of the wise. A wild luxuriance of foliage, utterly unknown before, burst out upon every vegetable thing.

'Yet another day—and the evil was not altogether upon us. It was now evident that its nucleus would first reach us. A wild change had come over all men ; and the first sense of *pain* was the wild signal for general lamentation and horror. This first sense of pain lay in a rigorous constriction of the breast and lungs, and an insufferable dryness of the skin. It could not be denied that our atmosphere was radically affected; the conformation of this atmosphere, and the possible modifications to which it might be subjected, were now the topics of discussion. The result of investigation sent an electric thrill of the intensest terror through the universal heart of man.

'It had been long known that the air which encircled us was a compound of oxygen and nitrogen gases, in the proportion of twenty-one measures of oxygen, and seventy-nine of nitrogen, in every one hundred of the atmosphere. Oxygen, which was the principle of combustion and the vehicle of heat, was absolutely necessary to the support of animal life, and was the most powerful and energetic agent in nature. Nitrogen, on the contrary, was incapable of supporting either animal life or flame. An unnatural excess of oxygen would result, it had been ascertained, in just such an elevation of the animal spirits as we had latterly experienced. It was the pursuit, the extension of the idea, which had engendered awe. What would be the result of a *total extraction of the nitrogen?* A combustion irresistible, all devouring, omni-prevalent, immediate ;—the entire fulfilment, in all their minute and terrible details, of the fiery and horror-inspiring denunciations of the prophecies of the Holy Book.

'Why need I paint, Charmion, the now disenchained frenzy of mankind ? That tenuity in the comet which had previously inspired us with hope, was now the source of the bitterness of despair. In its impalpable gaseous character we clearly perceived the consummation of Fate. Meantime a day again passed—bearing away with it the last shadow of Hope. We gasped in the rapid modification of the air. The red blood bounded tumultuously through its strict channels. A furious delirium possessed all men ; and with arms rigidly outstretched towards the threatening heavens, they trembled and shrieked aloud. But the nucleus of the destroyer was now upon us :—even here in Aidenn. I shudder while I speak. Let me be brief—brief as the ruin that overwhelmed. For a moment there was a wild lurid light alone, visiting and penetrating all things. Then—let us bow down, Charmion, before the excessive majesty of the great God!—then there came a shouting and pervading sound as if from the mouth itself of HIM, while the whole incumbent mass of ether in which we existed burst at once into a species of intense flame, for whose surpassing brilliancy and all-fervid heat even the angels in the high Heaven of pure knowledge have no name. Thus ended all.' *

Like most of the productions of the American poet the fragment we have quoted bears the stamp of a well marked originality. It is a curious blending of the conceptions of the poet with the philosophical descriptions and the positive, realistic analyses of the savant. This much to be desired employment of science in poetry and art is characteristic of the talent or rather, perhaps, the genius of Edgar Poe, and has the effect of producing an intense and keen emotion in the mind of the reader.

[1] 'The Conversation of Eiros and Charmion.'—*Poe's Works*, vol. ii.

Unfortunately the savant has not been equal to the poet. And we cannot read his assertions respecting cometary astronomy without smiling at the inaccuracies and even blunders into which the author has fallen. It is a great defect, since the emotion which he designs to inspire misses its effect, as soon as the reader perceives the want of accord between the fact and the dream. But on the other hand we see that Poe has designedly neglected to employ any of the ordinary catastrophes which have been supposed likely to result from the rencontre of a comet and the earth. He calls in aid neither flood nor fire, in the ordinary sense, nor the disruption of the earth. Not even poison, nor the respiration of a poisonous matter. A simple addition, in increasing proportion of oxygen gas, and all is told. It is true that he speaks, we know not why, of a total extraction of nitrogen ; we seek in vain for the scientific reason for this extraction. Nor is the final piercing sound and the explosion easier to understand. The effects are not as described, when a living creature is subjected to an increased pressure of an oxygenated atmosphere, as M. Bert's experiments have shown. But a final *coup de théâtre* was needed, and on this point Poe has made a sacrifice to the vulgar.

Other observations might be made ; but we have already explained the nature of the laws applying to cometary movements, and the reader will not fail to detect the errors of the poet, who, were he writing at the present day, would be obliged to change the form and manner of his catastrophe. The known results of spectral analysis would no longer permit him to represent a comet as an agglomeration of oxygen. Likewise nothing proves that the matter of which a comet is composed is in a gaseous state ; the nucleus on the contrary would appear to be either a solid or a liquid mass, and the atmosphere with which it is surrounded on all sides an aggregation of isolated particles.

CHAPTER XV.

SOME QUESTIONS ABOUT COMETS.

SECTION I.

ARE COMETS HABITABLE?

The inhabitants of comets as depicted in the *Pluralité des Mondes* of Fontenelle—Ideas of Lambert respecting the habitability of comets—That comets are the abode of human beings is a hypothesis incompatible with the received facts of astronomy.

AFTER NEWTON, and especially in the eighteenth century, by a not unnatural reaction of ideas from the Aristotelian doctrine of transient meteors, comets were regarded as bodies, stable and permanent as the planets ; they were obedient to the same laws of movement, and differed only as regards appearance, by their nebulosities and tails. The astronomers of that time, taken up with the verification and calculation of their positions and orbits, occupied themselves little or not at all with the study of details which were purely physical, such as are now called *cometary phenomena.* Regarding them as spheroids, solid like the planets, and similar to them in the constituents of their nuclei, to people them with inhabitants followed in the natural sequence of ideas.

Fontenelle, who, as we know, was a believer in the theory of vortices, and who, moreover, regarded the heads and tails of comets as simple optical appearances, thus expresses himself in the *Pluralité des Mondes.*

'Comets,' he observes, 'are planets which belong to a neighbouring vortex ; they move near the boundaries of it ; but this vortex, being unequally pressed upon by those that are adjacent to it, is rounder above and flatter below, and it is the part below that concerns us. Those planets which near the summit began to move in circles did not foresee that, down below, the vortex would fail them, because it is there as it were crushed. Our comet is thus forced to enter the neighbouring vortex, and this it cannot do without a shock.' Also further on, Fontenelle observes, returning to the same point : 'I have already told you of the shock which takes place when two vortices meet and repel each other. I believe that in this case the poor comet is rudely enough shaken and its inhabitants not less so. We deem ourselves very unfortunate when a comet appears in sight ; but it is the comet itself which is very unfortunate.' 'I do not think so,' said the Marquise ; 'it brings to us all its inhabitants in good health. Nothing is so delightful as thus to *change vortices.* We who never quit ours lead a life wearisome enough. If the inhabitants have sufficient knowledge to predict the time of their entrance into our world, those who have already made the voyage announce beforehand to others what they will see.' 'You will soon discover a planet which has a great ring about it, they will say perhaps,' speaking of Saturn. 'You will see another which will be followed by four little ones. Perhaps even there are people appointed to look out for new worlds as they appear in sight, and who cry immediately, *A new sun! a new sun!* as sailors cry, *Land! land!* Believe me, we have no need to pity the inhabitants of a comet.'

Lambert in his *Lettres Cosmologiques* (1765) devotes a chapter to the question, *Are comets habitable* ? Guided by considerations foreign to science, and dominated by a preconceived idea that all globes must be inhabited, he seeks to discover reasons which may permit us to believe that comets, more numerous than the planets in the solar system, are habitable celestial bodies.

A first difficulty arises from the extremes of temperature to which comets are subjected at their aphelia and perihelia. 'How are we to conceive,' he observes, 'that beings can exist in an abode which is subjected to the utmost extremes of heat and cold ? The comet which appeared in 1759 (that of Halley) and which returns the quickest of all those whose periods are known undergoes a winter 70 years long. But there is even a greater extreme of heat.' Although Lambert objects to Newton's calculation as to the heat to which the comet of 1680 must have been subjected during its perihelion passage, still he is obliged to admit that on the 8th December, 1680, 'the comet being one hundred and sixty times nearer to the sun than we are ourselves, must have been subjected to a degree of heat twenty-five thousand six hundred times as great as we are. Whether this comet was of a more compact substance than our globe, or was protected in some other way, it made its perihelion passage in safety, and we may suppose all its inhabitants also passed safely. No doubt they would have to be of a more vigorous temperament and of a constitution very different from our own. But why should all living beings necessarily be constituted like ourselves ? Is it not infinitely more probable that amongst the different globes of the universe a variety of organizations exist, adapted to the wants of the people who inhabit them, and fitting them for the places in which they dwell, and the temperatures to which they will be subjected? Have we not in like manner abandoned the prejudice which for a length of time caused the torrid and frigid zones to be regarded as uninhabitable? Is man the only inhabitant of the earth itself ? And if we had never seen either bird or fish, should we not believe that the air and water were uninhabitable? Are we sure that fire has not its invisible inhabitants, whose bodies, made of asbestos, are impenetrable to flame? Let us admit that the nature of the beings who inhabit comets is unknown to us ; but let us not deny their existence, and still less the possibility of it.'

Thus regarded as a matter of pure hypothesis, it is plain that the question of the habitability of comets may always be answered in the affirmative. But we must not forget that at the time when Lambert wrote, comets were regarded as solid bodies enveloped by a considerable atmosphere, and the tendency to assimilate them to the planets was general ; add to this a few vague ideas upon the subject of final causes—such as Lambert held—and it was natural to people all the stars of heaven, and even the sun himself with inhabitants.

Andrew Oliver published nearly about the same date (1772) an *Essay upon Comets*, wherein he seeks to explain the formation of tails by a mutual repulsion of electric origin, between the atmospheres of the sun and the comet: he devotes the second part of his curious work to showing that the tails of comets are probably intended to render their bodies habitable worlds. The enormous variations of temperature to which a comet is subjected in passing from one extremity of its orbit to the other, are exactly or at least suitably compensated by variations in the density of its atmosphere. This, together with the movements due to the action of the sun and the supposed velocity of rotation, prevents the extremes of heat and cold from becoming intolerable. At the aphelion both its atmosphere and tail are condensed about the comet, and the air is in a state of perfect calm. In proportion as it approaches perihelion, the atmosphere becomes rarified, the equilibrium is constantly broken, and currents of fresh air temper the extreme ardour of the solar rays.

These, as we see, are but physical romances composed by the partisans of a preconceived idea of the habitability of these bodies. Neither Fontenelle, nor Lambert, nor Andrew Oliver would probably write at the present day as they did a hundred or a hundred and fifty years ago. And for this two reasons may be assigned, the one philosophical, the other scientific. In the first place the *à priori* has by common consent been banished from science, which leaves to metaphysics the task of supporting theses by arguments based upon ideas such as that of final causes. We no longer ask for example how comets must be constituted to permit the existence of living beings, which Providence could not have withheld from bodies so numerous and important. But we seek by the study of observed facts and by the discussion of the probable physical consequences which must follow from these facts to form an approximate idea of the conditions—physical, luminous, calorific, and chemical—of known comets. And should he then enter upon the question of the habitability of these bodies, we do not consider it in the absolute and unconditional manner in which it was entertained by Lambert. We merely compare the probable conditions as determined by observation with those which seem to be compatible on the surface of the terrestrial globe with the existence of organized living beings. In short, there has been a total change of method.

A second reason which would have brought about a change in the opinions of the eminent savants whose theories we have just quoted, is that within the last hundred years—as we have seen in detail—the physical and even chemical constitution of comets has been carefully studied. We no longer assimilate them to the planets except as regards their movement of translation. Everything leads us to believe that the agglomerations of which they are composed are in a rudimentary state analogous to the *rudis indigestaque moles* of chaos. The incessant transformations which take place in their nuclei, their atmospheres and their tails indicate an equilibrium eminently unstable, and which would be very difficult to reconcile with the known conditions of life.

After this, let all who please picture to themselves the comet which has lately paid us so brief a visit [July, 1874], peopled with astronomers such as those of whom Lambert* speaks. We will not cavil with them; we do not fight with shadows.

* 'I like to picture to myself,' he says, 'these globes, voyaging in space, and peopled with astronomers, who are there on purpose to contemplate nature on a grand scale, as we contemplate it on a small scale. From their moving observatory, as it is wafted from sun to sun, they see all things pass successively before their view, and can determine the positions and motions of all the stars, measure the orbits of the comets and the planets which glide by them, see how the particular laws develop into general laws, and know, in a word "the details of the universe." In truth, I picture to myself that astronomy must be for the inhabitants of such a comet a terribly complicated science. But doubtless their intelligence is proportional to the difficulties.'

SECTION II.

WHAT WOULD BECOME OF THE EARTH IF A COMET WERE TO MAKE IT ITS SATELLITE?

Conditions of temperature to which the earth would be subjected if it were compelled by a comet to describe the same orbit as the latter—The comets of Halley, and of 1680, examined from this point of view—Extremes of heat and cold: opinion of Arago: impossibility of living beings resisting such changes.

Arago has examined, in an indirect manner, the question of the habitability of comets; that is to say, he has considered how far the enormous distances through which a body passes in describing a very eccentric orbit around the sun, such as that of a cometary orbit, are compatible with the existence of inhabitants similar to man. Could the earth, he enquires, ever become the satellite of a comet, and, if so, what would be the fate of its inhabitants?

Arago, basing his reasoning upon the comparative smallness of the masses of comets, regards the transformation of the earth into the satellite of a comet, as an event 'within the bounds of possibility, but which is very improbable,' an opinion no one at the present day will be inclined to dispute. He next supposes our earth successively made tributary to the comet of Halley and to that of 1680, and proceeds to consider the conditions of temperature to which our globe would be subjected whilst travelling in company with each.

With the comet of Halley our year would be sixty-five times longer than at present. 'In this period of seventy-five years, which the new year of the earth would include, five would be expended in passing over that portion of the curve comprised within the orbit of Saturn. Let us regard these five years as equivalent to the summer and the temperate seasons; there would still remain seventy (the number given by Lambert), which would belong entirely to winter. At the time of the comet's perihelion passage, the earth, the satellite of the comet, would receive from the sun a quantity of rays three times greater than that which it now receives. At its aphelion, thirty-eight years after, the quantity of rays received would be twelve hundred times less than now.'

With the great comet of 1680, the year would be equal to 575 of our years, if we assume Whiston's period. The distance of the earth from the sun would vary, during this long period, from the $\frac{6}{1000}$th part of its actual mean distance, to more than 138 times this distance. Our least distance from the sun would thus be to our greatest distance in the proportion of 6 to 138,296, or of 1 to 23,050. The intensity of the heat received at the perihelion would be 28,000 times as great as the actual mean heat. What would be the effect of such a condition of things? We can no longer admit with Newton that the heat acquired by the comet was 2,000 times that of red-hot iron, when on December 17, 1680, it passed within so small a distance of the sun. 'This last result,' says Arago, 'was founded upon incorrect data. The problem was much more complicated than was supposed by Newton, or could have been believed at the time when the *Principia* was published. It is now known that in order to find the temperature which a determinate quantity of heat could communicate to a planetary body it is necessary to know the state of the surface of that body and of its atmosphere.' On these points nothing is known as regards the comet of 1680, but the case is different with the earth. Arago, however, merely confines himself to the following remarks:

'There is no doubt that at first the solid envelope of the earth would experience a degree of heat 28,000 times greater than that of summer; but soon the seas would turn into vapour, and the thick beds of clouds arising therefrom would protect it from the conflagration, which at first would seem to be inevitable. It is therefore certain that the vicinity of the sun would cause a great increase of temperature, the numerical value of which, from the nature of things, we should be unable to assign.'

At the aphelion, the distance being 138 times the actual distance, the heat received from the sun by the earth would be about 19,000 times less than the mean heat at present. 'Concentrated in the focus of the largest lens,' Arago justly observes, 'it would certainly produce no sensible effect even upon an air thermometer. The temperature of our globe would then depend only upon the heat which might remain undissipated of that which it had received during its perihelion passage, and upon the intrinsic heat of the regions of space in the neighbourhood of the aphelion.'

To take the worst possible case, he assumes the loss to be complete, and the heat of the perihelion to be entirely dissipated. The extreme degree of cold would then be hardly more than — 58° Fahrenheit. Reasoning upon the fact that voyagers to the Polar regions, Franklin for example, in 1820, have endured a cold of — 57°·5 * Fahrenheit, and likewise upon

* [In the Polar Expedition that has just returned the *Alert* experienced a temperature of −73°·7′ Fahr., and the mean temperature for thirteen days was −58°·9′, and for five days and nine hours −66°·29′. Although I mention these facts, I need scarcely say that I do not share Arago's opinion that it would be possible for human beings to support the extreme temperatures that would result if the earth were to become a satellite of the comet of 1680.—Ed.]

experiments, which have shown that man can, under special conditions, support a heat of 266° Fahrenheit, he arrives at the unexpected conclusion 'that there is nothing to prove that if the earth had become a satellite of the comet of 1680, the human race would have disappeared through the effects of temperature.'

Let us in the first place consider only the extreme values—viz. the quantity of heat received by our globe at its perihelion and aphelion, independently of the physiological effects. At the perihelion, the amount of heat would be equal to 28,000 times the actual heat that we now experience; at the aphelion it would be 19,000 times less. Thus in the first case it would then be 532 millions of times greater than in the second.

What human organization could bear these inconceivable extremes of heat and cold? How could Arago believe, taking into account only the immediate action of a temperature so high, and of a cold so intense, that our constitution would not be infallibly destroyed between them? But if we ask ourselves what on such an hypothesis would become of our globe itself, its land and sea, its climate, vegetation, &c., could the answer be for a moment doubtful? As regards vegetation alone, do we not see that a few degrees, more or less, a little dryness or humidity in excess or defect, causes death to it, or renders it unproductive? Corn does not yield grain in the tropics, and man, although by his industry and by artificial means he is enabled to endure climates to which he is unaccustomed, with all the precautions in the world does not become acclimatized when he passes from one zone to the other.

Now all that lives on the surface of the earth is constituted to exist under certain conditions of day and night, and alternations of the seasons, in an atmosphere whose chemical composition, density and hygrometrical conditions remain constant or vary only within very restricted limits. What subversions, what revolutions in the habits and conditions most indispensable to existence must there be, if the earth were compelled to follow the movements of a comet, such as that of 1680 ! To a certainty, one revolution of the comet would see the annihilation of the human race, together with the greater part of the fauna and flora of the world. Naturalists of the present day have every reason to believe that the transformations revealed by palæontological study have been produced in different ages by corresponding modifications, whether slow or sudden it matters little, in the physical state of the atmosphere and earth. Nevertheless, to explain these changes, there is no need to attribute to the modifications in question an extent at all comparable to those which would result from the transformation of the earth into the satellite of a comet experiencing a range of temperature varying from 28,000 times the actual heat we habitually experience to a quantity 19,000 times less.

Happily for our globe, as Arago has himself admitted, the probability of an event of this kind is so extremely small, that we may dismiss it from our minds. It is one of those problems of pure curiosity, the examination of which merely furnishes the mind with terms of comparison between what is and what might be, between what is around us in the world that we inhabit, and what may be in worlds different to ours.

SECTION III.

IS THE MOON AN ANCIENT COMET ?

Hypothesis of Maupertuis: the planetary satellites originally comets, which have been retained by the attractions of the planets—The Arcadians and the moon—Refutation of this hypothesis by Dionys du Séjour.

In the same spirit of speculative enquiry, it has likewise been asked if the moon is not an ancient comet which the earth has diverted from its orbit about the sun, and forced to gravitate about itself. 'Not only,' says Maupertuis, 'might a comet carry away our moon, but it might itself become our satellite, and be condemned to perform its revolutions about our earth and illuminate our nights. Our moon might have been originally a small comet which, in consequence of having too nearly approached the earth, has been made captive by it. Jupiter and Saturn, bodies much larger than the earth, and whose power extends to a greater distance, and over larger comets, would be more liable than the earth to make such acquisitions; consequently Jupiter has four moons revolving about him, and Saturn five.'

Upon what foundation, upon what serious reasoning has Maupertuis erected this ingenious hypothesis? He does not tell us. Pingré, who records it, observes that the partisans of this opinion based it upon an ancient tradition mentioned by Ovid and Lucian. The Arcadians were persuaded that their ancestors inhabited Arcadia before the moon existed. But such a belief furnishes a very poor argument. The arguments, or rather the calculations, by which Dionys du Séjour has annihilated the hypothesis of Maupertuis are more difficult to refute. I will briefly state them as given by Pingré : 'Subjected to the test of analytical reasoning,' he observes, 'the whole theory falls to the ground. Dionys du Séjour has proved :— 1st. That it is absolutely impossible for a comet moving in a trajectory either parabolic or hyperbolic to become a satellite of the earth; 2nd. That for a comet whose orbit is elliptic to become a satellite of the earth it would be necessary that when it entered into the sphere of the earth's attraction, its relative motion, that is to say, the difference between its velocity and that of the earth, should be only 2,176 feet per second. But is it possible that a comet whose orbit, although elliptic, approaches, nevertheless, very nearly to a parabola should have a velocity relative to the earth of only 2,176 feet per second, whilst it is demonstrated that the relative velocity of a parabolic comet placed at the same distance, that is to say, at the distance of the earth from the sun, must, under the most unfavourable circumstances, be 39,000 feet per second? Besides,' adds Pingré, 'even if this were so, the comet, when transformed into our moon, would pass in each of its revolutions to the extremity of the sphere of the earth's attraction, and the least force would suffice to detach it from us. . . It would then recommence its orbit about the sun.'

These reasons are derived from the laws of cometary and planetary motion, and from the principle of gravitation, of which the laws are the expression; but it is also evident that in physical constitution nothing can be more unlike a comet than the moon. Everything leads us to believe that our satellite is entirely, at its surface at least, reduced to a solid condition. If it has an atmosphere it is the least vaporous possible and is of extremely slight density. Now, all known comets, those at least which have been subjected to telescopic scrutiny, appear to have been characterised by a predominance of nebulous atmosphere about the nucleus. The savants of the eighteenth century, who regarded comets as planetary globes, nevertheless recognised the entire absence of analogy between the physical constitution of the moon and that of a comet. Maupertuis, in order to explain how our satellite, an ancient comet disguised, might have lost its coma and tail, had only to invent a new hypothesis, to the effect that some other comet might in its passage have swept away the atmosphere of the moon. If I remember rightly, it is not Maupertuis who makes this new supposition, but some other author whose name has escaped me.

TABLE I.

ELLIPTIC ELEMENTS OF THE RECOGNISED PERIODICAL COMETS OF THE SOLAR SYSTEM.

No.	Name of Comet	Sidereal Revolution	Semi-Major Axis	Perihelion Distance	Aphelion Distance	Day of Perihelion Passage
		Years				
1	Encke . . .	3·285	2·209701	0·332875	4·086528	1871 Dec. 29
2	Brorsen . .	5·483	3·109618	0·596762	5·622475	1868 Apr. 17
3	Winnecke .	5·591	3·149900	0·781538	5·518260	1869 June 30
4	Tempel . .	5·963	3·291415	1·770548	4·812282	1873 May 9
5	D'Arrest. .	6·567	3·506698	1·280280	5·733117	1870 Sept. 23
6	Biela North	6·587	3·513740	0·860161	6·167319	1852 Sept. 24
	Biela South	6·629	3·528733	0·860592	6·196874	1852 Sept. 23
7	Faye. . . .	7·413	3·801849	1·682173	5·921525	1866 Feb. 14
8	Tuttle . . .	13·811	5·75652	1·03011	10·48294	1871 Nov. 30
9	Halley. . .	76·37	18·00008	0·58895	35·41121	1835 Nov. 15

No.	Name of Comet	Longitude of Perihelion	Longitude of Ascending Node	Inclination	Eccentricity	Direction of Movement
		° ′ ″	° ′ ″	° ′ ″		
1	Encke . .	158 12 24	334 34 9	13 7 35	0·8493573	D
2	Brorsen .	116 2 3	101 14 6	29 22 39	0·8080916	D
3	Winnecke	275 56 1	113 33 21	10 48 19	0·7518847	D
4	Tempel .	238 1 6	78 43 19	9 45 49	0·4620711	D
5	D'Arrest.	318 40 5 0	146 25 23	15 39 12	0·6349044	D
6	Biela. . .	109 20 24	246 5 16	12 33 25	0·7552007	D
	Biela. . .	109 13 21	246 9 11	12 33 47	0·7561187	D
7	Faye . . .	50 0 27	209 45 28	11 22 6	0·5575383	D
8	Tuttle . .	116 4 36	269 17 12	54 17 0	0·8210540	D
9	Halley . .	304 58 41	55 38 3	17 44 45	0·9672807	R

TABLE II.

GENERAL CATALOGUE OF THE ELEMENTS OF THE ORBITS OF COMETS.*

Numbers	Year of Appearance	Perihelion Passage (Greenwich Mean Time)	Longitude of Perihelion	Longitude of Ascending Node	Inclination	Eccentricity	Perihelion Distance	Direction of Motion	Computer	Name of Comet
	A.D.	DATE H. M. S.	° ′	° ′	° ′					
1	66	Jan. 14 4 48 0	325 0	32 40	40 30	. .	0·4446	R	Hind . .	Halley's
2	141	Mar. 29 2 24 0	251 55	12 50	17 0	. .	0·7200	R	,, . .	Halley's
3	240	Nov. 9 23 51 0	271 0	189 0	44 0	. .	0·3715	D	Burckhardt .	
4	539	Oct. 20 14 51 0	313 30	58 or 238	10 0	. .	0·3412	D	,, . .	
5	565	July 8 23 51 0	88 0	158 0	62 0	. .	0·7192	R	,, . .	
6	568	Aug. 28 6 28 49	316 47	294 36	4 2	. .	0·8894	D	Hind . .	
7	574	April 7 6 43 14	143 39	128 17	46 31	. .	0·9630	D	,, . .	
8	770	June 6 14 6 1	357 7	90 59	61 49	. .	0·6422	R	Laugier . .	
9	837	Feb. 28 23 51 0	289 3	206 33	10 or 12	. .	0·5800	R	Pingré . .	
10	961	Dec. 30 3 50 25	268 3	350 35	79 33	. .	0·5518	R	Hind . .	
11	989	Sept. 11 23 51 0	264 0	84 0	17 0	. .	0·5683	R	Burckhardt .	Halley's
12	1066	April 1 0 0 0	264 55	25 50	17 0	. .	0·7200	R	Hind . .	Halley's
13	1092	Feb. 15 0 0 0	156 20	125 40	28 55	. .	0·9281	D	,, . .	
14	1097	Sept. 21 21 26 39	332 30	207 30	73 30	. .	0·7384	D	Burckhardt .	
15	1231	Jan. 30 7 12 40	134 48	13 30	6 5	. .	0·9478	D	Pingré . .	
16	1264	July 12 13 31 0	241 38	157 40	35 5	. .	0·3117	D	Hoek . .	
17	1299	Mar. 31 7 29 0	3 20	107 8	68 57	. .	0·3179	R	Pingré . .	
18	1337	June 15 1 46 0	2 20	93 1	40 28	. .	0·8282	R	Laugier . .	
19	1366	Oct. 13 0 0 0	66 0	212 0	6 0	. .	0·9581	D	Peirce . .	
20	1378	Nov. 8 18 19 27	299 31	47 17	17 56	. .	0·5835	R	Laugier . .	Halley's

21	1385	Oct.	16	6	14	25	101	47	268	31	52	15	. .	0·7737	R	Hind . .	
22	1433	Nov.	4	10	9	51	281	2	133	49	79	1	. .	0·3395	R	Laugier . .	
23	1456	June	8	22	0	40	301	0	48	30	17	56	. .	0·5855	R	Pingré . .	Halley's
24	1468	Oct.	7	9	49	40	356	3	61	15	44	19	. .	0·8533	R	Laugier . .	
25	1472	Feb.	28	5	13	13	48	3	207	32	1	55	. .	0·5646	R	,, . .	
26	1490	Dec.	24	11	16	50	58	40	288	45	51	37	. .	0·7376	D	Hind . .	
27	1491	Jan.	4	21	35	0	113	0	268	0	75	0	. .	0·7550	R	Peirce . .	
28	1506	Sept.	3	15	52	34	250	37	132	50	45	1	. .	0·3860	R	Laugier . .	
29	1531	Aug.	25	19	0	40	301	12	45	30	17	0	0·967391	0·5799	R	Halley . .	Halley's
30	1532	Oct.	19	14	53	0	135	44	119	8	42	27	. .	0·6125	D	Méchain . .	
31	1533	June	14	21	11	25	217	40	299	19	28	14	. .	0·3269	D	Olbers . .	
32	1556	April	22	0	25	0	274	15	175	26	30	12	. .	0·5049	D	Hind . .	
33	1558	Aug.	10	12	24	45	329	49	332	36	73	29	. .	0·5773	R	Olbers . .	
34	1577	Oct.	26	22	44	36	129	42	25	20	75	10	. .	0·1775	R	Woldstedt . .	
35	1580	Nov.	28	13	6	39	108	29	19	7	64	34	0·998631	0·6025	D	Schjellerup .	
36	1582	May	6	9	51	22	256	15	229	18	60	47	. .	0·1683	R	D'Arrest . .	
37	1585	Oct.	8	0	38	44	9	8	37	41	6	6	. .	1·0948	D	Peters & Sawitsch	
38	1590	Feb.	8	0	39	4	217	57	165	37	29	30	. .	0·5677	R	Hind . .	
39	1593	July	18	13	39	0	176	19	164	15	87	58	. .	0·0891	D	La Caille . .	
40	1596	July	25	5	8	38	270	55	330	21	51	58	. .	0·5672	R	Hind . .	
41	1607	Oct.	26	17	10	58	301	38	48	40	17	12	0·967089	0·5880	R	Bessel . .	Halley's
42	1618 I.	Aug.	17	3	2	0	318	20	293	25	21	28	. .	0·5130	D	Pingré . .	
43	1618 II.	Nov.	8	8	25	1	3	5	75	44	37	12	. .	0·3895	D	Bessel . .	
44	1652	Nov.	12	15	41	0	28	19	88	10	79	28	. .	0·8475	D	Halley . .	
45	1661	Jan.	26	21	9	0	115	16	81	54	33	1	. .	0·4427	D	Méchain . .	
46	1664	Dec.	4	11	36	5	130	33	81	16	21	18	. .	1·0256	R	Lindelôf . .	
47	1665	April	24	5	16	0	71	55	228	2	76	5	. .	0·1065	R	Halley . .	
48	1668	Feb.	24	18	46	6	40	9	193	26	27	7	. .	0·2511	D	Henderson . .	
49	1672	Mar.	1	8	38	0	46	59	297	30	83	22	. .	0·6974	D	Halley . .	
50	1677	May	6	0	38	0	137	37	236	49	79	3	. .	0·2806	R	,, . .	

* The dates of the Comets up to 1582 are given according to the Old Style, and, after 1582, according to the New Style.

GENERAL CATALOGUE OF THE ELEMENTS OF THE ORBITS OF COMETS—*continued.*

Numbers	Year of Appearance	Perihelion Passage (Greenwich Mean Time)					Longitude of Perihelion		Longitude of Ascending Node		Inclination		Eccentricity	Perihelion Distance	Direction of Motion	Computer	Name of Comet
	A.D.	DATE.		H.	M.	S.	°	′	°	′	°	′					
51	1678	Aug.	18	7	34	0	322	48	163	20	2	52	0·626970	1·1453	D	Le Verrier .	
52	1680	Dec.	17	23	46	9	262	49	272	9	60	40	0·999985	0·00622	D	Encke . .	
53	1682	Sept.	14	19	4	53	301	56	51	11	17	45	0·967920	0·5829	R	Rosenberger .	Halley's
54	1683	July	12	17	25	15	86	31	173	18	83	48	0·983247	0·5535	R	Clausen . .	
55	1684	June	8	10	17	0	238	52	268	15	65	49	. .	0·9602	D	Halley . .	
56	1686	Sept.	16	14	34	0	77	0	350	35	31	22	. .	0·3250	D	” . .	
57	1689	Nov.	29	4	48	1	269	41	90	25	59	5	. .	0·0189	R	Vogel . .	
58	1695	Nov.	9	17	0	0	60	0	216	0	22	0	. .	0·8435	D	Burckhardt .	
59	1698	Oct.	18	16	58	0	270	51	267	44	11	46	. .	0·6913	R	Halley . .	
60	1699	Jan.	13	8	23	0	212	31	321	46	69	20	. .	0·7440	R	La Caille . .	
61	1701	Oct.	17	9	51	0	133	41	298	41	41	39	. .	0·5926	R	Burckhardt .	
62	1702	Mar.	13	14	33	22	138	47	188	59	4	25	. .	0·6468	D	” . .	
63	1706	Jan.	30	4	57	0	72	36	13	11	55	14	. .	0·4269	D	Struyck . .	
64	1707	Dec.	11	23	30	0	79	55	52	47	88	36	. .	0·8597	D	La Caille . .	
65	1718	Jan.	14	21	44	16	121	40	127	55	31	8	. .	1·0254	R	Argelander .	
66	1723	Sept.	27	15	4	9	42	53	14	14	50	0	. .	0·9988	R	Spoerer . .	
67	1729	June	13	6	19	27	320	31	310	38	77	5	1·0050334	4·0435	D	Burckhardt .	
68	1737 I.	Jan.	30	8	21	0	325	55	226	22	18	21	. .	0·2228	D	Bradley . .	
69	1737 II.	June	8	7	39	0	262	37	123	54	39	14	. .	0·8670	D	Daussy . .	
70	1739	June	17	10	0	0	102	39	207	25	55	43	. .	0·6736	R	La Caille . .	

71	1742	Feb. 8 4 21 14	217 34	185 35	67 4	. .	0·7655	R	Struyck . .	
72	1743 I.	Jan. 10 20 20 16	92 58	67 32	2 16	. .	0·8382	D	Olbers . .	
73	1743 II.	Sept. 20 14 11 13	247 16	6 15	45 38	. .	0·5236	R	D'Arrest . .	
74	1744	Mar. 1 7 55 39	197 14	45 48	47 8	. .	0·2223	D	Wolfers . .	
75	1747	Mar. 3 9 57 19	277 2	147 19	79 7	. .	2·1985	R	Maraldi . .	
76	1748 I.	April 28 19 25 24	215 1	232 52	85 27	. .	0·8406	R	" . .	
77	1748 II.	June 18 21 18 1	278 47	33 8	67 3	. .	0·6254	D	Bessel . .	
78	1757	Oct. 21 7 54 39	122 58	214 13	12 50	. .	0·3375	D	Bradley . .	
79	1758	June 11 3 17 39	267 38	230 50	68 19	. .	0·2154	D	Pingré . .	
80	1759 I.	Mar. 12 13 14 34	303 10	53 50	17 37	0·9676844	0·5845	R	Rosenberger .	Halley's
81	1759 II.	Nov. 27 0 33 58	53 38	139 40	79 3	. .	0·8021	D	Chappe . .	
82	1759 III.	Dec. 16 12 48 51	139 4	79 20	4 42	. .	0·9618	R	" . .	
83	1762	May 28 8 1 42	104 2	348 33	85 38	. .	1·0090	D	Burckhardt .	
84	1763	Nov. 1 20 54 58	84 57	356 18	72 34	0·9954268	0·4983	D	Lexell . .	
85	1764	Feb. 12 13 42 15	15 15	120 5	52 54	. .	0·5552	R	Pingré . .	
86	1766 I.	Feb. 17 8 41 0	143 15	244 11	40 50	. .	0·5053	R	" . .	
87	1766 II.	April 26 23 43 55	251 13	74 11	8 2	0·864000	0·3990	D	Burckhardt .	
88	1769	Oct. 7 14 53 22	144 11	175 4	40 46	0·9992490	0·1227	D	Bessel . .	
89	1770 I.	Aug. 14 0 38 36	356 16	132 0	1 35	0·786839	0·6743	D	Le Verrier .	
90	1770 II.	Nov. 22 5 39 0	208 23	108 42	31 26	. .	0·5282	R	Pingré . .	
91	1771	April 19 5 6 19	104 3	27 52	11 15	1·0093698	0·9035	D	Encke . .	
92	1772	Feb. 16 15 43 40	110 9	257 16	17 3	0·724510	0·9860	D	Hubbard . .	Biela's
93	1773	Sept. 5 14 1 50	75 17	121 8	61 15	1·002490	1·1283	D	Lexell . .	
94	1774	Aug. 15 19 55 21	317 28	180 45	83 20	1·028295	1·4328	D	Burckhardt .	
95	1779	Jan. 4 2 4 20	87 14	25 4	32 31	. .	0·7132	D	Zach . . .	
96	1780 I.	Sept. 30 22 13 53	246 36	123 41	54 23	0·999946	0·0963	R	Clüver . .	
97	1780 II.	Nov. 28 20 21 0	246 52	141 1	72 3	. .	0·5153	R	Olbers . .	
98	1781 I.	July 7 4 31 59	239 11	83 1	81 43	. .	0·7759	D	Méchain . .	
99	1781 II.	Nov. 29 12 33 25	16 3	77 23	27 12	. .	0·9610	R	Legendre . .	
100	1783	Nov. 19 11 50 50	50 3	55 45	44 53	0·5395345	1·4544	D	Burckhardt .	

GENERAL CATALOGUE OF THE ELEMENTS OF THE ORBITS OF COMETS—*continued.*

Numbers	Year of Appearance	Perihelion Passage (Greenwich Mean Time)	Longitude of Perihelion	Longitude of Ascending Node	Inclination	Eccentricity	Perihelion Distance	Direction of Motion	Computer	Name of Comet
	A.D.	DATE. H. M. S.	° ′	° ′	° ′					
101	1784	Jan. 21 4 47 26	80 44	56 49	51 9	. .	0·7079	R	Méchain . .	
102	1785 I.	Jan. 27 7 48 43	109 52	264 12	70 14	. .	1·1434	D	„ . .	
103	1785 II.	April 8 8 58 51	297 30	64 34	87 32	. .	0·4273	R	„ . .	
104	1786 I.	Jan. 30 20 57 51	156 38	334 8	13 36	0·84836	0·3348	D	Encke . .	Encke's
105	1786 II.	July 8 13 37 10	158 38	195 24	50 59	. .	0·3942	D	Reggio . .	
106	1787	May 10 19 48 39	7 44	106 52	48 16	. .	0·3489	R	Saron . .	
107	1788 I.	Nov. 10 7 25 26	99 8	156 57	12 28	. .	1·0630	R	Méchain . .	
108	1788 II.	Nov. 20 7 15 39	22 50	352 24	64 30	. .	0·7573	D	„ . .	
109	1790 I.	Jan. 16 18 58 9	58 25	172 50	29 44	. .	0·7473	R	Saron . .	
110	1790 II.	Jan. 28 7 36 9	111 45	267 9	56 58	. .	1·0633	D	Méchain . .	Tuttle's
111	1790 III.	May 21 5 46 54	273 43	33 11	63 52	. .	0·7980	R	„ . .	
112	1792 I.	Jan. 13 12 50 15	36 21	190 42	39 46	. .	1·2926	R	Zach . .	
113	1792 II.	Dec. 27 7 47 9	135 53	283 15	49 7	. .	0·9668	R	Piazzi . .	
114	1793 I.	Nov. 4 20 12 0	228 42	108 29	60 21	. .	0·4034	R	Saron . .	
115	1793 II.	Nov. 28 5 6 21	71 54	2 0	51 31	0·9734211	1·4951	D	D'Arrest . .	
116	1795	Dec. 21 10 35 1	156 41	334 39	13 42	0·8488828	0·3344	D	Encke . .	Encke's
117	1796	April 2 19 47 42	192 44	17 2	64 55	. .	1·5781	R	Olbers . .	
118	1797	July 9 2 31 10	49 27	329 16	50 41	. .	0·5266	R	„ . .	
119	1798 I.	April 4 11 58 16	105 7	122 12	43 45	. .	0·4846	D	„ . .	
120	1798 II.	Dec. 31 13 17 3	34 27	249 30	42 26	. .	0·7795	R	Burckhardt .	

121	1799 I.	Sept. 7 5 50 36	3 38	99 23	51 2	. .	0·8403	R	Wahl . .	
122	1799 II.	Dec. 25 18 3 46	190 23	326 30	77 5	. .	0·6244	R	,, . .	
123	1801	Aug. 8 13 23 0	183 49	44 28	21 20	. .	0·2617	R	Burckhardt .	
124	1802	Sept. 9 21 23 5	332 9	310 16	57 1	. .	1·0942	R	Olbers . .	
125	1804	Feb. 13 14 6 55	148 45	176 48	56 29	. .	1·0711	D	Gauss . .	
126	1805	Nov. 21 11 59 50	156 47	334 20	13 33	0·8461753	0·3404	D	Encke . .	Encke's
127	1806 I.	Jan. 1 22 1 10	109 28	251 16	13 37	0·7457068	0·9070	D	Hubbard . .	Biela's
128	1806 II.	Dec. 28 22 9 2	97 3	322 23	35 3	1·010182	1·0819	R	Hensel . .	
129	1807	Sept. 18 17 43 59	270 55	266 47	63 10	0·9954878	0·6461	D	Bessel . .	
130	1808 I.	May 12 22 52 4	69 13	322 59	45 43	. .	0·3899	R	Encke . .	
131	1808 II.	July 12 4 0 58	252 39	24 11	39 19	. .	0·6080	R	Bessel . .	
132	1810	Sept. 29 2 23 31	52 45	310 21	61 11	. .	0·9758	D	Iriesnecker .	
133	1811 I.	Sept. 12 6 10 32	75 1	140 25	73 2	0·9950933	1·0355	R	Argelander .	
134	1811 II.	Nov. 10 23 46 17	47 27	93 2	31 17	0·9827109	1·5821	D	Nicolai . .	
135	1812	Sept. 15 7 31 31	92 19	253 1	73 57	0·9545412	0·7771	D	Encke . .	
136	1813 I.	Mar. 4 12 38 10	69 56	60 48	21 14	. .	0·6991	R	Nicollet . .	
137	1813 II.	May 19 10 1 7	197 43	42 40	81 2	. .	1·2160	R	Gerling . .	
138	1815	April 25 23 48 42	149 2	83 29	44 30	0·9312197	1·2128	D	Bessel . .	Olbers's
139	1816	Mar. 1 8 18 0	267 36	323 15	43 5	. .	0·0485	D	Burckhardt .	
140	1818 I.	Feb. 7 10 55 0	95 7	250 4	20 2	. .	0·7333	D	Pogson . .	
141	1818 II.	Feb. 25 23 0 49	182 45	70 26	89 44	. .	1·1977	D	Encke . .	
142	1818 III.	Dec. 4 2 10 2	103 7	90 7	62 41	. .	0·8479	R	Bessel . .	
143	1819 I.	Jan. 27 6 8 53	156 59	334 33	13 37	0·8485841	0·3353	D	Encke . .	Encke's
144	1819 II.	June 27 16 52 9	287 5	273 44	80 46	. .	0·3411	D	Brinkley . .	
145	1819 III.	July 18 21 36 18	274 41	113 11	10 43	0·7551904	0·7736	D	Encke . .	Winnecke's
146	1819 IV.	Nov. 20 5 53 34	67 19	77 14	9 1	0·6867458	0·8926	D	,, . .	
147	1821	Mar. 21 12 52 39	239 29	48 41	73 33	. .	0·0918	R	Rosenberger .	
148	1822 I.	May 5 13 34 52	192 48	177 25	53 36	. .	0·5042	R	Gambart . .	
149	1822 II.	May 23 23 6 40	157 12	334 25	13 20	0·8444643	0·3460	D	Encke . .	Encke's
150	1822 III.	July 16 0 35 2	219 54	97 51	37 43	. .	0·8461	R	Heiligenstein .	

GENERAL CATALOGUE OF THE ELEMENTS OF THE ORBITS OF COMETS—*continued.*

Numbers	Year of Appearance	Perihelion Passage (Greenwich Mean Time)	Longitude of Perihelion	Longitude of Ascending Node	Inclination	Eccentricity	Perihelion Distance	Direction of Motion	Computer	Name of Comet
	A.D.	DATE. H. M. S.	° ′	° ′	° ′					
151	1822 IV.	Oct. 23 18 28 29	271 40	92 45	52 39	0·9963021	1·1451	R	Encke . .	
152	1823	Dec. 9 10 39 29	274 34	303 3	76 12	. .	0·2265	R	,, . .	
153	1824 I.	July 11 12 18 40	260 17	234 19	54 34	. .	0·5913	R	Rümker . .	
154	1824 II.	Sept. 29 1 23 58	4 31	279 16	54 37	1·0017345	1·0501	D	Encke . .	
155	1825 I.	May 30 13 6 39	273 55	20 6	56 41	. .	0·8891	R	Clausen . .	
156	1825 II.	Aug. 18 17 3 55	10 14	192 56	89 42	. .	0·8835	D	,, . .	
157	1825 III.	Sept. 16 6 33 18	157 15	334 27	13 21	0·8448885	0·3449	D	Encke . .	Encke's
158	1825 IV.	Dec. 10 16 7 28	318 47	215 43	33 33	0·9954285	1·2408	R	Hubbard . .	
159	1826 I.	Mar. 18 10 43 9	109 49	251 27	13 34	0·7466012	0·9024	D	,, . .	Biela's
160	1826 II.	April 21 23 27 46	117 11	197 30	39 57	1·0089597	2·0029	D	Nicolai . .	
161	1826 III.	April 29 0 56 13	35 48	40 29	5 17	. .	0·1882	R	Clüver . .	
162	1826 IV.	Oct. 8 22 51 14	57 48	44 6	25 57	. .	0·8528	D	Argelander .	
163	1826 V.	Nov. 18 9 47 55	315 30	235 6	89 22	. .	0·0269	R	Gambart . .	
164	1827 I.	Feb. 4 22 7 4	33 30	184 28	77 36	. .	0·5065	R	Heiligenstein .	
165	1827 II.	June 7 20 11 15	297 32	318 10	43 39	. .	0·8082	R	,, ,,	
166	1827 III.	Sept. 11 16 37 44	250 57	149 39	54 5	0·9992730	0·1378	R	Clüver . .	
167	1829	Jan. 9 17 54 7	157 18	334 30	13 21	0·8446245	0·3456	D	Encke . .	Encke's
168	1830 I.	April 9 6 43 30	212 12	206 22	21 16	. .	0·9214	D	Carlini . .	
169	1830 II.	Dec. 27 15 50 58	310 59	337 53	44 45	. .	0·1259	R	Wolfers . .	
170	1832 I.	May 3 23 24 45	157 21	334 32	13 22	0·8454141	0·3435	D	Encke . .	Encke's

171	1832 II.	Sept.	25	12	38	58	227	55	72	27	43	19	. .	1·1835	R	Peters . .	
172	1832 III.	Nov.	26	9	36	44	109	56	248	12	13	12	0·7513780	0·8793	D	Santini . .	Biela's
173	1833	Sept.	10	9	29	30	224	21	323	28	7	18	. .	0·4643	D	Hartwig . .	
174	1834	April	2	15	55	11	276	34	226	49	5	57	. .	0·5150	D	Petersen . .	
175	1835 I.	Mar.	30	16	29	51	206	9	58	56	9	3	. .	2·0514	R	Rümker . .	
176	1835 II.	Aug.	26	8	39	32	157	23	334	35	13	21	0·8450356	0·3444	D	Encke . .	Encke's
177	1835 III.	Nov.	15	22	32	1	304	32	55	10	17	45	0·9673909	0·5866	R	Westphalen .	Halley's
178	1838	Dec.	19	0	17	38	157	27	334	37	13	21	0·8451775	0·3440	D	Encke . .	Encke's
179	1840 I.	Jan.	4	10	13	42	192	12	119	58	53	6	1·0002050	0·6185	D	Peters & O. Strüve	
180	1840 II.	Mar.	12	23	46	32	80	18	236	49	59	13	0·9978836	1·2214	R	Plantamour .	
181	1840 III.	April	2	11	53	27	324	12	186	3	79	52	. .	0·7483	D	Rümker . .	
182	1840 IV.	Nov.	13	15	27	55	22	32	248	56	57	57	0·9698526	1·4808	D	Götze . .	
183	1842 I.	April	12	0	26	9	157	29	334	39	13	20	0·8447904	0·3450	D	Encke . .	Encke's
184	1842 II.	Dec.	15	22	57	39	327	16	207	49	73	34	. .	0·5043	R	Laugier . .	
185	1843 I.	Feb.	27	9	51	9	278	40	1	15	35	41	0·9999157	0·0055	R	Hubbard . .	
186	1843 II.	May	6	1	20	33	281	30	157	15	52	45	1·0001798	1·6163	D	Götze . .	
187	1843 III.	Oct.	17	3	33	46	49	34	209	30	11	23	0·5558997	1·6922	D	Möller . .	Faye's
188	1844 I.	Sept.	2	11	22	36	342	31	63	49	2	55	0·6176539	1·1864	D	Brünnow . .	
189	1844 II.	Oct.	17	8	15	15	179	36	31	39	48	36	0·9996083	0·8554	R	Plantamour .	
190	1844 III.	Dec.	13	16	11	42	296	2	118	19	45	39	1·0003530	0·2517	D	Bond . .	
191	1845 I.	Jan.	8	3	58	19	91	20	336	44	46	51	. .	0·9053	D	Hind . .	
192	1845 II.	April	21	0	44	37	192	33	347	7	56	24	. .	1·2546	D	Faye . . .	
193	1845 III.	June	5	16	9	44	262	3	337	49	48	42	0·9898742	0·4016	R	D'Arrest . .	
194	1845 IV.	Aug.	9	15	1	50	157	44	334	20	13	8	0·8474362	0·3382	D	Encke . .	Encke's
195	1846 I.	Jan.	22	2	15	11	89	6	111	8	47	26	0·9924026	1·4807	D	Jelinek . .	
196	1846 II.	Feb.	10	22	10	22	109	3	245	54	12	35	0·7566060	0·8564	D	Hubbard . .	Biela's
197	1846 III.	Feb.	25	8	58	39	116	28	102	41	30	56	0·7933880	0·6501	D	Brünnow . .	Brorsen's
198	1846 IV.	Mar.	5	13	5	18	90	27	77	34	85	6	0·9620891	0·6637	D	Van Deinse .	
199	1846 V.	May	27	19	44	55	82	39	161	18	57	36	. .	1·3747	R	Graham . .	
200	1846 VI.	June	1	5	5	53	240	8	260	29	30	24	0·7213385	1·5286	D	Peters . .	

GENERAL CATALOGUE OF THE ELEMENTS OF THE ORBITS OF COMETS—*continued.*

Numbers	Year of Appearance	Perihelion Passage (Greenwich Mean Time)				Longitude of Perihelion	Longitude of Ascending Node	Inclination	Eccentricity	Perihelion Distance	Direction of Motion	Computer	Name of Comet
	A.D.	DATE.	H.	M.	S.	° ′	° ′	° ′					
201	1846 VII.	June 5	11	30	5	162 6	261 53	29 19	0·9899389	0·6337	R	Oudemans . .	
202	1846 VIII.	Oct. 29	21	59	57	98 47	4 38	49 39	0·9933127	0·8294	D	Quirling . .	
203	1847 I.	Mar. 30	6	49	29	276 2	21 42	48 40	0 9999129	0·0426	D	Hornstein . .	
204	1847 II.	June 12	9	1	39	137 42	173 26	80 17	. .	2.1172	R	D'Arrest . .	
205	1847 III.	Aug. 9	8	50	44	246 45	338 17	83 26	0·9985879	1·7661	R	Mauvais . .	
206	1847 IV.	Aug. 9	6	13	10	21 21	76 42	32 38	0·9974348	1·4843	R	Schweizer . .	
207	1847 V.	Sept. 9	13	1	31	79 12	309 49	19 8	0·972560	0·4879	D	D'Arrest . .	
208	1847 VI.	Nov. 14	9	36	39	274 13	190 50	71 51	1·0001326	0·3290	R	Rümker . .	
209	1848 I.	Sept. 8	1	20	40	310 35	211 35	84 28	. .	0·3198	R	" . .	
210	1848 II.	Nov. 26	2	44	10	157 47	334 22	13 9	0·8478280	0·3370	D	Encke . .	Encke's
211	1849 I.	Jan. 19	8	31	37	63 15	215 13	85 2	. .	0·9596	D	Hensel . .	
212	1849 II.	May 26	12	37	26	235 45	202 33	67 8	0·9978863	1·1590	D	Weyer . .	
213	1849 III.	June 8	4	53	15	267 6	30 32	66 55	0·997830	0·8943	D	D'Arrest . .	
214	1850 I.	July 23	12	40	16	273 25	92 53	68 11	0·9988519	1·0814	D	Carrington .	
215	1850 II.	Oct. 19	8	14	20	89 16	206 0	40 9	. .	0·5653	D	Mauvais . .	
216	1851 I.	April 1	22	25	17	49 42	209 31	11 22	0·5549601	1·6999	D	Möller . .	Faye's
217	1851 II.	July 9	2	39	15	322 56	148 25	13 55	0·6592674	1·1733	D	Schulze . .	D'Arrest's
218	1851 III.	Aug. 26	5	37	52	310 59	223 41	38 9	0·9968586	0·9843	D	Brorsen . .	
219	1851 IV.	Sept. 30	19	8	58	338 46	44 21	73 59	. .	0·1420	D	Klinkerfues .	
220	1852 I.	Mar. 14	19	6	25	157 51	334 23	13 8	0·8476726	0·3375	D	Encke . .	Encke's

221	1852 II.	April	19	15	15	6	278	42	317	29	49	11	1·0525041	0·9129	R	Hartwig . .	
222	1852 III.	Sept.	22	22	38	25	109	8	245	51	12	33	0·7558650	0·8606	D	Hubbard . .	Biela's
223	1852 IV.	Oct.	12	18	0	57	43	14	346	10	40	55	0·9189170	1·2500	D	Westphal . .	
224	1853 I.	Feb.	23	23	55	49	153	44	69	34	20	13	0·990412	1·0919	R	Hartwig . .	
225	1853 II.	May	9	19	39	59	201	45	40	58	57	49	0·9893194	0·9087	R	Rümker . .	
226	1853 III.	Sept.	1	16	54	26	310	57	140	31	61	30	0·7294246	0·3070	D	Stockwell . .	
227	1853 IV.	Oct.	16	14	31	44	302	15	220	6	61	0	1·0012289	0·1727	R	D'Arrest . .	
228	1854 I.	Jan.	2	17	19	36	56	39	227	1	66	1	. .	2·0456	R	Klinkerfues .	
229	1854 II.	Mar.	24	0	20	41	213	49	315	27	82	33	. .	0·2771	R	Mathieu . .	
230	1854 III.	June	22	2	1	43	272	58	347	49	71	8	. .	0·6473	R	Bruhns . .	
231	1854 IV.	Oct.	27	12	13	4	94	24	324	29	40	55	0·9933246	0·7987	D	Lesser . .	
232	1854 V.	Dec.	15	17	11	27	165	9	238	8	14	9	0·9864041	1·3575	D	Adam . .	
233	1855 I.	Feb.	5	1	8	11	226	38	189	44	51	24	0·9651850	2·1935	R	Tiele . . .	
234	1855 II.	May	29	10	58	0	239	29	260	11	23	10	0·9039970	0·5649	R	Schulze . .	
235	1855 III.	July	1	4	40	0	157	53	334	26	13	8	0·8477869	0·3371	D	Encke . .	Encke's
236	1855 IV.	Nov.	25	9	8	58	86	2	51	35	10	11	0·997255	1·2323	R	Hoek . .	
237	1857 I.	Mar.	21	8	43	38	74	44	313	10	87	56	0·9992144	0·7725	D	Schulze . .	
238	1857 II.	Mar.	28	16	4	19	115	46	101	45	29	49	0·8022946	0·6206	D	Bruhns . .	Brorsen's
239	1857 III.	July	17	23	33	10	249	36	23	41	58	58	0·9989984	0·3675	R	Villarceau . .	
240	1857 IV.	Aug.	23	23	54	59	21	47	200	49	32	46	0·9803714	0·7468	D	Möller . .	
241	1857 V.	Sept.	30	21	7	5	250	8	14	58	56	3	0·9969135	0·5629	R	Linsser . .	
242	1857 VI.	Nov.	19	1	42	31	44	13	139	19	37	49	0·9969918	1·0090	R	Auwers . .	
243	1857 VII.	Nov.	28	19	36	14	323	3	148	27	13	56	0·6598094	1 1704	D	Schulze . .	D'Arrest's
244	1858 I.	Feb.	23	12	34	20	115	52	269	3	54	24	0·820903	1·0255	D	Bruhns . .	Tuttle's
245	1858 II.	May	2	1	24	32	275	40	113	31	10	48	0·7541036	0·7688	D	Hänsel . .	Winnecke's
246	1858 III.	May	2	7	42	37	195	59	170	43	23	0	. .	1·2097	D	Watson . .	
247	1858 IV.	June	5	7	5	39	226	6	324	58	80	3	. .	0·5443	R	Auwers . .	
248	1858 V.	Sept.	29	23	8	51	36	13	165	19	63	2	0·9962933	0·5785	R	Hill . . .	
249	1858 VI.	Oct.	18	8	41	33	157	57	334	29	13	4	0·8463915	0·3407	D	Encke . .	Encke's
250	1858 VII.	Sept.	13	21	26	37	49	52	209	40	11	22	0·5577360	1·6948	D	Möller . .	Faye's

GENERAL CATALOGUE OF THE ELEMENTS OF THE ORBITS OF COMETS—*continued.*

Numbers	Year of Appearance	Perihelion Passage (Greenwich Mean Time)	Longitude of Perihelion	Longitude of Ascending Node	Inclination	Eccentricity	Perihelion Distance	Direction of Motion	Computer	Name of Comet
	A.D.	DATE. H. M. S.	° ′	° ′	° ′					
251	1858 VIII.	Oct. 12 19 26 46	4 13	159 45	21 17	. .	1·4270	R	Weiss . .	
252	1859	May 29 5 25 38	75 21	357 21	83 32	. .	0.2010	R	Hertzsprung .	
253	1860 I.	Feb. 16 16 9 30	173 45	324 3	79 36	. .	1·1973	D	Liais . . .	
254	1860 II.	March 5 17 12 25	50 16	8 56	48 13	. .	1·3083	D	Seeling . .	
255	1860 III.	June 16 0 20 56	161 31	84 43	79 18	0·997240	0·2921	D	Liais . . .	
256	1860 IV.	Sept. 28 6 49 0	111 59	104 14	28 14	. .	0·9539	R	Valz . . .	
257	1861 I.	June 3 9 21 30	243 22	29 56	79 46	0·9834631	0·9207	D	Oppolzer . .	
258	1861 II.	June 11 12 17 7	249 4	278 58	85 26	0·9853832	0·8224	D	Sawitsch . .	
259	1861 III.	Dec. 7 4 17 18	173 31	145 7	41 57	. .	0·8391	R	Pape . .	
260	1862 I.	Feb. 6 4 7 49	158 0	334 31	13 5	0·8467094	0·3399	D	Encke . .	Encke's
261	1862 II.	June 22 0 43 59	229 20	326 33	7 54	. .	0·9813	R	Seeling . .	
262	1862 III.	Aug. 22 21 53 32	344 41	137 27	66 26	0·9612708	0·9626	R	Oppolzer . .	
263	1862 IV.	Dec. 28 8 33 28	125 10	355 45	42 23	. .	0·8026	R	Engelmann .	
264	1863 I.	Feb. 3 11 47 16	191 23	116 56	85 22	0·9999470	0·7948	D	,, . .	
265	1863 II.	April 4 21 42 13	247 15	251 16	67 22	. .	1·0681	R	Frischauf . .	
266	1863 III.	April 20 20 39 7	305 31	249 59	85 29	. .	0·6284	D	Karlinski . .	
267	1863 IV.	Nov. 9 11 35 16	94 43	97 30	78 5	. .	0·7066	D	Stampfer . .	
268	1863 V.	Dec. 27 18 19 44	60 24	304 43	64 29	. .	0·7715	D	Weiss . .	
269	1863 VI.	Dec. 29 4 0 45	183 7	105 1	83 19	1·000650	1·3131	D	Rosén . .	
270	1864 I.	July 27 19 50 29	185 32	174 51	65 1	. .	0·6640	R	Valentiner .	

271	1864 II.	Aug. 15	13 46 54	304 12	95 14	1 52	0·9967771	0·9093	R	Kowalczyk .	
272	1864 III.	Oct. 11	9 41 54	159 18	31 45	70 18	0·9999532	0·9312	R	Von Asten .	
273	1864 IV.	Dec. 22	11 7 31	321 42	203 13	48 52	. .	0·7709	D	Teitjen . .	
274	1864 V.	Dec. 27	17 16 20	162 24	160 54	17 7	. .	1·1146	R	Valentiner .	
275	1865 I.	Jan. 14	8 10 23	141 16	253 3	87 32	. .	0·0260	R	Tebbutt . .	
276	1865 II.	May 27	23 1 55	158 8	334 37	13 4	0.8463016	0·3410	D	Von Asten .	Encke's
277	1866 I.	Jan. 11	3 12 47	60 29	231 26	17 18	0·9054198	0·9765	R	Oppolzer . .	Faye's
278	1866 II.	Feb. 14	0 29 48	50 0	209 45	11 22	0·5575383	1·6822	D	Möller . .	Faye's
279	1867 I.	Jan. 19	20 39 15	75 52	78 36	18 13	0·8490551	1·5725	D	Searle . .	
280	1867 II.	May 23	22 5 25	236 9	101 10	6 25	0·5097048	1·5636	D	Sandberg .	Tempel's
281	1867 III.	Nov. 6	23 5 22	213 36	64 59	96 34	. .	0·3304	R	Oppolzer . .	
282	1868 I.	April 17	10 28 58	116 3	101 14	29 22	0·8079556	0·5965	D	Schulze . .	Brorsen's
283	1868 II.	June 25	22 40 56	287 8	53 40	48 12	. .	0·5824	R	Plummer . .	
284	1868 III.	Sept. 14	16 34 11	158 13	334 33	13 7	0·8491305	0·3336	D	Von Asten .	Encke's
285	1869 I.	June 29	22 35 31	275 55	113 33	10 48	0·7518847	0·7815	D	Oppolzer . .	Winnecke's
286	1869 II.	Oct. 9	20 23 29	139 43	311 30	111 40	. .	1·2307	R	Doberck . .	
287	1869 III.	Nov. 20	19 36 33	41 17	292 40	6 55	. .	1·1026		Bruhns . .	
288	1870 I.	July 14	18 37 43	303 32	141 45	58 12	. .	1·0087	R	Dreyer . .	
289	1870 II.	Sept. 2	4 43 17	7 53	12 56	99 21	. .	1·8167	D	Gerst . .	
290	1870 III.	Sept. 22	13 35 10	318 41	146 26	15 39	0·6350040	1·2780	D	Leveau . .	D'Arrest's
291	1870 IV.	Dec. 19	21 1 34	185 21	94 45	147 16	. .	0·3893	R	Schulhof . .	
292	1871 I.	June 10	14 23 47	141 50	279 19	87 36	0·997814	0·6543	D	Holetschek .	
293	1871 II.	July 27	13 33 14	308 14	211 55	101 59	. .	1·0759	R	Schulhof . .	
294	1871 III.	Dec. 1	19 8 16	116 5	269 17	54 17	0·8210550	1·0301	D	Tischler . .	Tuttle's
295	1871 IV.	Dec. 20	8 4 33	29 33	147 2	98 24	. .	0·6945	R	Schulhof . .	
296	1871 V.	Dec. 28	20 17 31	158 13	334 33	13 7	0·8493481	0·3330	D	Von Asten .	Encke's
297	1873 I.	May 9	1 12 51	237 39	78 45	9 44	0·4619980	1·7695	D	Sandberg . .	Tempel's
298	1873 II.	June 25	9 3 59	306 10	120 54	12 43	0·5441551	1·3445	D	Plummer . .	
299	1873 III.	July 18	11 40 48	50 3	209 39	11 22	0·5573828	1·6826		Möller . .	Faye's
300	1873 IV.	Sept. 10	19 3 0	64 26	230 38	96 0	. .	0·7944	R	Weiss . .	

GENERAL CATALOGUE OF THE ELEMENTS OF THE ORBITS OF COMETS—*continued.*

Numbers	Year of Appearance	Perihelion Passage (Greenwich Mean Time)	Longitude of Perihelion	Longitude of Ascending Node	Inclination	Eccentricity	Perihelion Distance	Direction of Motion	Computer	Name of Comet
	A.D.	DATE. H. M. S.	° ′	° ′	° ′					
301	1873 V.	Oct. 1 18 18 44	50 28	176 43	121 29	. .	0·3849	R	Weiss . .	
302	1873 VI.	Oct. 10 11 16 56	116 6	101 16	29 23	0·8089037	0·5935	D	Plummer . .	Brorsen's
303	1873 VII.	Dec. 1 5 15 19	85 43	250 20	30 1	. .	0·7345	D	Weiss . .	
304	1874 I.	Mar. 9 21 59 20	300 36	31 31	58 17	. .	0·4394		Schulhof . .	
305	1874 II.	Mar. 13 22 56 56	245 53	274 7	148 25	. .	0·8860	R	Weiss . .	
306	1874 III.	July 5 3 5 53	347 20	213 12	. .	. .	1·4398		Schulhof . .	
307	1874 IV.	July 8 20 34 38	271 5	118 44	66 23	. .	0·6757		Schulhof . .	
308	1874 V.	Aug. 26 20 23 59	344 9	251 29	41 51	0·9992305	0.9827	D	Grützmacher .	
309	1874 VI.	Oct. 18 16 50 43	298 47	281 38	99 26	. .	0·5197		Holetschek .	
310	1875 I.	April 13 2 31 38	158 22	334 41	13 7	0·8494248	0·3329	D	Von Asten .	Encke's

NOTE

ON THE DESIGNATION OF COMETS, AND ON THE CATALOGUE OF COMETS.

By the Editor.

The numbers of the comets mentioned in the text of this work (as *ex. gr.* in the tables on pp. 143 and 145) do not always agree with the numbers given in the preceding catalogue. Thus, Comet II., 1852, appears in the catalogue as Comet IV., 1852. This arises from the fact that the comets in the catalogue are numbered in the order of date of their perihelion passages, while the numbers in the text often refer to the order of discovery. There is a good deal of uncertainty with regard to the numbering of comets, in consequence of this double system, which, however, seems to have been unavoidable. In the 78th volume of the leading astronomical journal, the *Astronomische Nachrichten*, No. 1,868 (1872), the editor (Dr. Peters), then director of the Observatory at Altona, and now of that at Kiel, refers to the designation of comets, and states that the system of denoting them by the year and the Roman number was first introduced in the general index to the first twenty volumes of the *Nachrichten*, and that the numbers referred to the order of discovery. In the index to the 26th volume the comets in the year were first numbered according to the order of date of their perihelion passages, and *generally* this arrangement has been followed since; but, Dr. Peters observes, this method is inconvenient, as the designation of comets whose perihelion passages are near together are liable to be changed upon each recalculation of the elements, and as comets whose elements have not been calculated thus receive no numbers; he, therefore, announced that in future he should number the comets in each year according to their dates of discovery. But in Nos. 1,871 and 1,872 Dr. Peters prints two letters of Dr. Oppolzer and Dr. Littrow upon the subject; and chiefly in consequence of the German Astronomical Society having in 1867 decided that the best numbering of comets was according to the dates of their perihelion passages, he modified his previous announcement in so far that he designates the comets in each year *a*, *b*, *c*, . . . in order of discovery, leaving I., II., III. . . . for the order of the perihelion passages, and this arrangement has been followed in the *Astronomische Nachrichten* since 1872. This will serve to explain the existing confusion, which is even greater than appears, as the numbering in the titles is frequently different to the numbering in the indexes; and the same want of uniformity happens, of course, also with regard to the designation of comets in other works.

Thus, the comet discovered by Westphal on July 24, 1852, was for some time known as Comet II., 1852, and would have to be sought for under this designation in the volumes of the *Astronomische Nachrichten* about that date; but now it would be quoted as Comet IV., 1852. It might have been, perhaps, better if M. Guillemin had altered all the designations of comets, so as to bring them into uniformity with the catalogue at the end; but, practically, the inconvenience is but trifling, as it is generally very easy to identify the comet alluded to, and the difficulty such as it is must arise in its full force whenever there was occasion to refer to the original calculations in the *Astronomische Nachrichten*, the *Monthly Notices* of the Royal Astronomical Society, &c.

It will sometimes be found that the elements used in the text are slightly different from those in the catalogue; this is due to the fact that there are often, indeed generally, several calculations of a comet's orbit. To take as an example Westphal's Comet II., 1852: In vol. 35 of the *Astronomische Nachrichten* there are three sets of parabolic elements, two by Sonntag and one by Rümker; and there are also three sets of elliptic elements, one by Sonntag and two by Marth; while in vol. 50 there is a complete discussion of all the observations, with the elements deduced therefrom, by Westphal. It may happen, therefore, that the values used are not always identical; but these slight discrepancies are not of any consequence, and I have not thought it necessary to remedy them; but in any case where I have found that an error of importance has crept in I have corrected it (as *ex. gr.* in the elements of Westphal's comet, which by an accident were incorrect in the original work).

It must be borne in mind that the catalogue contains only those comets whose elements have been calculated with some approach to accuracy. Thus, the comet of 1746, mentioned in the table on p. 145, will not be found in the catalogue, because only rough elements were computed for this comet by Mr. Hind; but these were sufficient to lead him to consider it to be identical with the comet of 1231. Also, in some cases of periodical comets, the eccentricity is not given in the catalogue. This, of course, happens when only the parabolic elements have been obtained, and the periodicity has been determined by their accordance with those of some other comet.

The above remarks refer chiefly to the relation of the catalogue to the designations of the comets in the body of the work, and it now remains to speak of the catalogue itself.

The table of the periodical comets (p. 531) is copied from M. Guillemin's table, except that I have corrected the elements of Tempel's comet, which by an obvious accident were erroneous, and have also corrected one or two slight accidental errors. As for the general catalogue (Table II.) M. Guillemin states (ch. V., sec. 5) that it is extracted from Mr. James C. Watson's *Theoretical Astronomy* (Philadelphia, 1868), and I have reprinted it without alteration, except that I have added from Mr. Watson's catalogue the name of the calculator and the hours and minutes of the perihelion passages that were omitted by M. Guillemin. The latter addition is of slight consequence, but the former is of some importance, as, in case it should be desired to make further investigations in regard to any cometary orbit, the name of the calculator would be of great assistance. I should state, also, that I have carefully compared Mr. Watson's catalogue with M. Guillemin's reprint, and with Mr. Hind's catalogue, and have so been able to correct several misprints. Thus (except for the introduction of one comet mentioned below), the general catalogue up to the end of 1866 is due to Mr. Watson. M. Guillemin has continued it up to nearly the end of 1874; but as, owing to the short time that had elapsed since the apparitions of some of the comets, several of the orbits were only provisional, while several had not been calculated at all, I have thought it better to complete Mr. Watson's catalogue myself *de novo*, so that for the portion subsequent to 1866 I am solely responsible.

This portion which I have formed is the result of a careful study of the *Astronomische Nachrichten* and the *Monthly Notices* of the Royal Astronomical Society from 1866 to the end of 1875, and is, I believe, as accurate as it can be rendered by means of these data. When an orbit has been calculated independently by several computers it must, of course, remain a matter of opinion which set of elements should be preferred, and in my choice I have been influenced by a variety of circumstances, such as the value of the observations employed, the length of arc over which they extend, &c. I may observe that after a comet has been discovered a rough parabolic orbit is frequently obtained at once from the first three available observations, merely for the purpose of ascertaining the general path of the comet, &c. This orbit is soon superseded as the observations are multiplied, and it may appear ultimately that no parabolic orbit will satisfactorily represent the comet's motion, in which case elliptic or hyperbolic elements will have to be calculated. When the comet has left us some time, and all the observations of it have been published, the work of calculating the definitive elements of the orbit (in which all the available observations are taken into account) is usually undertaken by some astronomer, and these, of course, are to be preferred when they exist. Thus, between the different sets of elements there may be very wide discrepancies, and, if the definitive elements have not been calculated, it is sometimes difficult to decide which has the greatest probability of accuracy. Also, in regard to the periodical comets, the orbits calculated are of two kinds, viz. those obtained from some previous apparition by calculating the perturbations to which the comet has been subjected in the interval, and those obtained from observations of the comet during the apparition in question. In my portion of the catalogue (and, I presume, in the portion previous to 1867 also) the elements are sometimes of one class and sometimes of the other; in fact, I have merely chosen the orbit which seemed to me to be most likely to be the nearest to the truth. It is for this reason that I have thought it proper not to alter the elements given by M. Guillemin for the periodical comets in Table I., although it will be seen that I have sometimes preferred slightly different elements. The elements for the four apparitions of Encke's Comet in 1865, 1868, 1871, and 1875 were deduced from the elements given in the able discussion of the motion of this comet by Prof. Von Asten in No. 2,038 of the *Astronomische Nachrichten* (1875). The insertion of Encke's Comet, 1865, is the only change that has been made in Mr. Watson's catalogue. At its apparition in 1865 the comet was only observed in the southern hemisphere.

The great comet of 1874, discovered by Coggia on April 17 of that year, and which has been so often referred to in this book as Comet III., 1874, appears in the catalogue as Comet IV.; this is because another comet, discovered on August 19, also by Coggia, passed its perihelion on July 5—three days earlier than the great comet—according to Schulhof's elements (*Ast. Nach.*, vol. 84, p. 262), which have been adopted in the catalogue. But the comet discovered on August 19 was very faint, and but few observations were made of it, so that there is much uncertainty with regard to its orbit. According to the elements of Holetschek (*Ast. Nach.*, vol. 84, p. 269), the perihelion passage took place on July 19, eleven days after that of the great comet. It is, therefore, doubtful which comet first passed its perihelion.

What has been said will show the nature of the uncertainties attending cometary orbits; and I need scarcely add that for purposes of exact astronomical research no catalogue, however excellent, can supersede the necessity of referring to the original calculations and observations. For example, it may happen that neither an elliptic, parabolic, nor hyperbolic orbit satisfies the whole of the observations satisfactorily, and in this case the selection of any of the sets of elements may be all but arbitrary.

In conclusion, I will repeat that the catalogue only contains comets whose orbits have been calculated; so that, for example, the comet seen by Mr. Pogson and referred to in the note on Biela's comet (p. 265) is not included, as, since only two observations were made of it, no orbit could be calculated. A list of comets whose orbits have not been calculated will be found in Mr. G. F. Chambers's *Handbook of Descriptive and Practical Astronomy* (1861.)

The reader will notice the extraordinary dearth of comets in the last two years. No other comet except Encke's was seen in 1875, and none have been seen during the present year. This complete absence of comets, following years so rich in comets as were 1873 and 1874, is very remarkable.

Printed in the United States
By Bookmasters